PRECISION AGRICULTURE '97

Volume I : Spatial Variability in Soil and Crop

PRECISION AGRICULTURE '97
Volume I : Spatial Variability in Soil and Crop

Edited by
John V. Stafford, Silsoe Research Institute, UK

Papers presented at the First European Conference on Precision Agriculture
Warwick University Conference Centre, UK
7-10 September 1997

Organised by
The SCI Agriculture and Environment Group

Co-sponsored by
The Fertiliser Society
The European Society of Agriculture Engineers
The American Society of Agronomy
The Soil Science Society of America

First published 1997
Reprinted 1998 (twice)

A CIP Catalogue for this book is available from the British Library.

Papers presented at the first European Conference on Precision Agriculture, Warwick, 8-10 September 1997, organised by the SCI Agriculture and Environment Group.

ISBN 1 85996 136 3 vol. I
ISBN 1 85996 231 9 vol. II
ISBN 1 85996 236 X set of two vols

BIOS Scientific Publishers Ltd
9 Newtec Place, Magdalen Road, Oxford OX4 1RE, UK
Tel: +44 (0)1865 726286 Fax: +44 (0)1865 246823
World Wide Web home page: http://www.Bookshop.co.uk/BIOS/

DISTRIBUTORS

Australia and New Zealand
    Blackwell Science Asia
    54 University Street
    Carlton, South Victoria 3053

India
    Viva Books Private Limited
    4325/3 Ansari Road,
    Daryaganj
    New Dehli 110002

Singapore and South East Asia
    Toppan Company (S) PTE Ltd
    38 Liu Fang Road, Jurong
    Singapore 2262

USA and Canada
    BIOS Scientific Publishers
    PO Box 605, Herndon
    VA 22070

Printed by Print in Black, Midsomer Norton, Bath, UK

Contents

3 3 8. 16

VOLUME I

VOLUME II

TECHNOLOGY 449

System Modelling

Economics

Practical Experience and Management

Author Index

Conference Organisation

Scientific Co-ordinator

Dr. J. V. Stafford, Silsoe Research Institute, UK

Scientific and Technical Program Committee

Prof. J. de Baerdemaeker, Katholiek University, Leuven, Belgium
Dr. S. Christensen, Danish Institute of Agricultural Science, Flakkebjerg, Denmark
Prof. W. Day, Silsoe Research Institute, UK
Prof. R. Godwin, Silsoe College, UK
Dr. D. Goense, IMAG/WAU, Wageningen, Netherlands
Dr. M. Griffin, ADAS, Cambridge, UK
Dr. J. Hummell, USDA-ARS, Urbana, USA
Dr. I. Richards, Levington Agriculture, Ipswich, UK
Dr. P. Robert, University of Minnesota, Minneapolis, USA
Prof. E. Schnug, Institute of Plant Nutrition and Soil Science, Braunschweig, Germany
Dr. J. Schueller, University of Florida, Gainesville, USA
Prof. F. Sevila, ENSAM, Montpellier, France
Dr. J. Stafford, Silsoe Research Institute, UK

Organising Committee

Dr. J. V. Stafford, Silsoe Research Institute
Mrs. E. Holden, Silsoe Research Institute
Mr. S. Cox, Hitchen
Mr. S. Blackmore, Silsoe College
Mr. C. Dawson, Dawson Associates, Strensall

Introduction

The lack of spatial uniformity in the factors that influence the growth of field crops and hence their productivity has been known and appreciated since early times. The parable of the sower in the New Testament of the Bible refers to, 'seed falling on good ground and producing a crop of a hundred, sixty or thirty times what was sown.'! Until recent times, agricultural practice has only been able to cope to a very limited extent with spatial variability in factors such as soil type, field aspect or weed competition.

With the advent of accurate, reliable and affordable positioning systems, particularly the Global Positioning System (GPS), the possibility of taking into account within-field spatial variability was born, with the promise of both economic and environmental benefit. Parallel advances in sensing systems, in precise application mechanisms and control systems and in the information processing power of computers have led to the adoption of the concept of precision agriculture. Otherwise known by phrases such as site-specific farming, farming-by-the-metre or spatially variable application, precision agriculture may be defined as the targeting of inputs to field crops according to locally determined requirements at the metre - or even centimetre - level.

The concept has been technology-led, with sufficient technology in place so that spatially variable application of inputs could be implemented in production agriculture. However, interpretation of spatial variability and the necessary understanding of soil science, crop science and agronomy on a site-specific basis are very much less well developed. It is widely recognised that much research is necessary in these areas in order to give a sound foundation to the whole concept of precision agriculture and its implementation in production agriculture. It was with this in mind that the 1st European Conference on Precision Agriculture was conceived with the intention of bringing soil scientists, crop scientists, agronomists, environmental scientists, technologists and agricultural economists together to present and discuss current research.

The Conference follows on from three very successful biennial conferences on precision agriculture organised in Minneapolis, USA by Professor Pierre Robert, an early pioneer of the 'site-specific' approach who has contributed significantly to the progress and current high profile of the topic. This 'high profile' has extended both within the agricultural industry and the research community, in Europe, the Americas and Australia with increasing interest from other areas of the world. It is reflected in the interest shown in this first European Conference which has attracted submission of over 180 paper abstracts, resulting in more than 110 full papers.

The two volumes presented here collect together these papers. The papers, each of which was assessed/refereed by a Scientific Committee, report on research and development covering all the range of disciplines that constitute precision agriculture. Volume I covers one of the three main topics of the Conference - 'Spatial Variability in Soil and Crop'. The other two main topics - 'Technology' and 'IT and Management' - are covered in Volume II.

John V. Stafford
Silsoe, May 1997

Acknowledgements

The administration, editing, assessment, collating and production of these two volumes has entailed considerable work and the input of the Scientific Committee, the Publications Department of SCI (particularly, Anne Borcherds) and secretarial assistance at Silsoe Research Institute (Niki Robinson) is very gratefully acknowledged. Extra editorial assistance by Sidney W.R. Cox enabled deadlines to be met.

The smooth and efficient planning and organisation of the conference owes everything to SCI Conference Department led by Anne Potter. The input of the Organising Committee has been invaluable. The inspiration of Rosemary Briars who designed the conference logo and proceedings cover is appreciated.

KEYNOTE PAPERS

SPATIAL VARIABILITY IN SOIL – IMPLICATIONS FOR PRECISION AGRICULTURE

A.B. McBRATNEY, M.J. PRINGLE

Australian Centre for Precision Agriculture, Ross St Building A03, The University of Sydney, NSW 2006, Australia

ABSTRACT

For precision agriculture to be successfully implemented, efficient and accurate methods for assessing the spatial variability of soil, and relating it to yield quantity and quality components, must be developed and utilised. This paper discusses methods of describing the spatial variability of soil; past research on its nature and magnitude of the variability present, sampling, interpolation and proximal sensing techniques for the measurement of variability and, finally, work to establish the links between the spatial variability of soil and the spatial variability of crop yield components is discussed. Gathering the requisite soil information to adequately describe soil spatial variability may be a major impediment to the successful implementation of precision agriculture. Some concepts are presented that may be of considerable use in the future. These include methods of data analysis such as wavelets or Kalman filtering, or techniques that improve the quality of information by finding optimal combinations of sampling and proximal sensing techniques or discovering the geographic regions in which pedotransfer functions and yield response curves are effective.

INTRODUCTION

The aim of precision agriculture in relation to cropping is to match "resource application and agronomic practices with soil attributes and crop requirements as they vary across a site" (McBratney and Whelan, 1995). As soil serves as a growth medium for crops, its condition is a major factor affecting potential and actual yield. Traditional cropping has regarded the field as the smallest convenient unit of management. Considered relatively homogeneous, the field was generally managed by practices such as uniformity in sowing, fertiliser and pesticide applications which ignored the spatial variability of the soil, and hence the site-specific crop requirements. This has been exacerbated by the development of modern agriculture increasing field sizes and thereby encompassing increased soil spatial variability, particularly in Europe where fields were originally quite small. Burrough (1993), in addressing the question of why information on soil spatial variability is of such importance, stated: "The driving force behind wanting to understand soil variability is a need for improving control over the world's physical environment". A precision agriculture system matches resource availability to crop capability and then knowledge of soil spatial variability is essential.

This paper aims to discuss spatial variability of soil with reference to the implications for precision agriculture. Methods for describing soil spatial variability are given. The magnitude of variability for some common soil properties is reviewed and collated. Some techniques for mapping soil spatial variability are given with a brief discussion of work that has tried to establish the relationship between soil spatial variability and crop spatial variability.

DESCRIBING SOIL SPATIAL VARIABILITY

In precision agriculture, the spatial scales of interest are in the range 1 m to say 1 km. Traditional methods of describing spatial variability of soil have been through chloropleth soil maps that divide soil into seemingly homogeneous units similar to the assumptions used in traditional agriculture. The variability within these mapping units is often quite high and knowledge about the nature of the variability vague. Additionally, the minimum area represented on conventional soil maps is in the order of tens to hundreds of square metres. The development of geostatistics (Webster and Oliver, 1990) has been a prime force in allowing soil spatial variability to be better quantified, especially at finer scales. Precision agriculture demands an accurate method of quantifying the spatial variability of soil properties. These applications include the variogram and its relatives, spectral analysis, wavelets, and the state-space approach.

Semivariance and the variogram

Variograms measure the spatial dependence of soil properties using semivariance. The average variance between any pair of sampling points (*i.e.* the semivariance) for soil property S at any vector h apart can be given by the formula (Webster and Oliver, 1990):

$$\gamma(h) = \frac{1}{2m} \sum_{i=1}^{m} \{s(\mathbf{x}_i) - s(\mathbf{x}_i + \mathbf{h})\}^2 \tag{1}$$

where $\gamma(h)$ is the average semivariance of the soil property, m is the number of pairs of sampling points, s is the value of the property S, x is the coordinate of the point, and h is the lag (separation distance of the pairs). Ideally, pairs of sampling points closer together should show smaller semivariance (*i.e.* they are more alike for the property), whereas pairs of points farther away from each other should display larger semivariance.

A variogram is generated by plotting 'average semivariance with lag', $\gamma(h)$, against the lag. Various models can be fitted to the data to describe the variogram; two of the more common ones are the spherical model and the exponential model. It is important to choose the correct model to fit the data (McBratney and Webster, 1986).

The spherical model is given by the formula (Webster and Oliver, 1990):

$$\begin{cases} \gamma(h) = C_0 + C\left\{\left(\frac{3h}{2a}\right) - \frac{1}{2}\left(\frac{h}{a}\right)^3\right\} & \text{for } h \leq a \\ \gamma(h) = C_0 + C & \text{for } h > a \end{cases} \tag{2}$$

where $C_0 + C$ is the sill (the maximum variance of the property), h is the scalar of h in Equation 1 and describes the isotropic form of the variogram, and a is the range (the lag at which the sill is obtained). In reaching a sill the spherical model implies that the variance of the property only reaches a certain level. The significance of the range parameter is that lags lower than a infer spatial dependence for the property whilst values above a have no spatial dependence.

The exponential model is given by the formula (Webster and Oliver, 1990):

$$\gamma(h) = C_0 + C\{1 - \exp(-h / r)\} \tag{3}$$

The essential difference between the spherical model and the exponential model is that the range is approached asymptotically. In the exponential model, the term r is similar to a in describing the range and, although in theory the spatial dependence of the soil for the property should be infinite, the 'effective' range is taken as $3r$.

An important feature of any variogram is the nugget effect. In theory, samples taken at zero lag (or the closest practical distance) should show zero variance. However this is not always seen to be the case, with some properties showing large variation at very small lags. This is known as the nugget effect and the value of the nugget effect is the 'nugget variance'. The nugget effect describes the inherent random variation in the soil. In some cases, the nugget variance may constitute almost all of the sill variance in which case it can be thought that the soil shows very little spatial dependence. The amount of nugget variance can be a function of a too-large sampling interval and can sometimes be decreased by sampling more densely.

When the variation is more complex, *e.g.* strong global and local trends or periodicities, more sophisticated approaches for describing spatial variation are necessary. Trends are accounted for by trend surfaces or intrinsic random functions (Matheron, 1973). Strong periodicities (which may be natural but more likely caused by management effects) are best determined and described in the frequency domain by spectral analysis (Webster, 1977). Recently, a new technique called 'wavelet analysis' has been developed to deal with fields that are both non-stationary and periodic (Foufoula-Georgiou and Kumar, 1994; McBratney, 1997). Thus far, this technique has not been applied in precision agriculture.

The state-space approach (Kalman Filter)

An alternative method for describing spatial variation is the state-space approach. In terms of precision agriculture, the state-space approach is useful for its essentially one-dimensional nature, *i.e.* it works well characterising variability of transects or lines. This implies that it could be successfully applied to 'on-the-go' operations that sense and map soil variability in real-time as farm machinery operates over a field.

The state-space approach was developed as a means of linear estimation. It uses regression coefficients to filter noisy data in order to extract the signal and predict ahead of the last observation (Morkoc et al., 1985). The general form of the state-space model is given by (Shumway, 1985; Wendroth et al., 1992):

$$\mathbf{y}_i = \mathbf{M}_i \mathbf{x}_i + \mathbf{v}_i \quad \text{for } i = 1, 2, ..., n \text{ observation points} \tag{4}$$

where $\mathbf{y}_i$ is a vector of observed values $y_{i1}, ..., y_{iq}$ at point i. $\mathbf{M}_i$ is a $q{\times}p$ matrix of measurements (where p is the number of soil attributes of interest and q is the number of soil attributes observed), and v_i is a vector of 'noise' in the observations. The observed noise is assumed to have a mean of zero, and be uncorrelated and normally distributed. The vector x_i is given by the relationship:

$$x_i = \Phi x_{i-1} + w_i \tag{5}$$

which describes how the state of the system, x_i, at location i is related to the state of the system at x_{i-1} at location i-1. The parameter Φ is the state-space coefficient and the parameter w_i is another noise vector, but acts independently of v_i.

Also known as 'Kalman Filtering', the state-space approach has applications in studies of soil spatial variability. Morkoc et al. (1985) were among the first to apply the state-space approach to spatial variability of soil temperature and water content. They found it a useful means of describing the spatial nature of these characteristics. Wendroth et al. (1992) later used the state-space approach to model spatial variability in crop yield. They observed that there were not enough parameters tested in the experiment to accurately determine the factors affecting the spatial variability of yield.

Shumway (1985) noted that the state-space approach has advantages over other spatial prediction techniques such as kriging in that data does not have to show stationarity, and missing data can be estimated. A further advantage is that no matrices have to be inverted, an operation that is essential for kriging and one that requires considerable computational power (Webster and Oliver, 1990).

The state-space approach, especially Kalman Filtering, may make a significant contribution to precision agriculture although more research is required on its applications.

THE MAGNITUDE OF SOIL SPATIAL VARIABILITY

Spatial variability can occur on a variety of scales, between regions, between fields, or within fields (Bouma and Finke, 1993) – variation in soil components can sometimes be discerned on a sub-millimetre scale (Burrough, 1993). Soil variability naturally arises through complex interactions between time, parent material, topography, climate, and organisms (Jenny, 1941). Anthropogenic actions can also lead to changes in soil variability; actions such as tillage and fertilising can affect soil structure and patterns of nutrient status (Bouma and Finke, 1993; Cattle et al., 1994). Bouma and Finke (1993) state that variability in soil properties can be seen as either static (e.g. texture, organic matter) or dynamic (e.g. moisture status, temperature), although those of most importance depend on the relevance to the soil user. Soil chemical and physical properties are of interest to users of precision agriculture as their spatial variability can lead to variability in yield.

Chemical properties of soil that vary spatially include nutrients, pH, salinity, and organic matter. Soil nutrients are essential for crop growth, and the advent of fertiliser technology makes them easily manipulated. Nitrogen, phosphorus and potassium are three important macro-nutrients required by crops, and effects on yield are readily seen if deficiencies occur. Spatial variation in nutrients can occur due to uptake by plants, leaching or a change to the system such as a liming operation altering the chemical balance of the soil. Bouma and Finke (1993) note that fertiliser applications can increase further the spatial variability of soil nutrients through uneven spreading patterns. Plants have optimum ranges for pH and salinity at which they can physiologically function and outside these ranges effects on growth and yield can be observed. Both pH and salinity can vary spatially due to parent material, climate, and management effects. Organic matter is important in soil water retention, storage and release of plant nutrients, and also adsorption of pesticides. Organic matter can vary according to the spatial variation of the factors that affect its accumulation and decay: temperature, pH, moisture content, oxygen availability, soil texture, and management (Jenkinson, 1988).

Physical properties of soil that vary spatially are structure and texture. Soil structure and texture are important as they have an indirect influence on other soil physical properties such as bulk density, water holding capacity, and soil strength. Soil structure is the arrangement of particles and voids in any given volume of soil. Structure affects physical properties such as gas diffusion, water movement and root anchorage while texture influences yield by affecting nutrient availability and water and gas movement (McBratney and Whelan, 1995). Spatial variation in soil texture is largely determined by pedogenic processes and is not easily manipulated by anthropogenic actions, while structure can be affected by operations such as cultivation (Cattle et al., 1994).

The magnitude of spatial variability as described by published variograms

There has been much research quantifying the spatial nature of soil, especially since the advent of geostatistics in soil science. Figures 1a to 1e show variograms of soil properties of interest to precision agriculture which appear in the literature. The variograms are plotted on the same scale in order to make comparisons about the spatial dependence of soil properties found in different studies.

The following rigorous method was applied to the generation of each variogram from the published information:

- Only those data are shown where there have been several studies conducted on the same soil property. Thus there are 14 entries for pH, 9 for clay content (topsoil + subsoil), 6 for carbon content, 5 for potassium, and 7 for phosphorus.
- All data for the same property were converted to the same units (*e.g.* mg/kg for nutrients, dag/kg for clay). This entailed making corresponding changes to the estimates of nugget and sill variance.
- If an exponential model (Equation 3) was used to describe the variogram in the original research, an 'equivalent' spherical model was used here to approximate the exponential model where:

$$C_{0sph.} = C_{0exp.}$$
$$C_{sph.} = 0.95 C_{exp.}$$
$$a_{sph.} = 3r$$

- Linear models and double spherical models were also approximated by the spherical model for purposes of comparison. In converting the linear model to the spherical model, the largest lag and semivariance given were used to estimate the sill variance and range. In converting the double spherical model to the spherical model, the two C parameters were summed to estimate the sill variance.

- Using the parameter estimates of C_0, C, and a, Equation 2 was used to estimate semivariance and produce graphs up to a lag of 500 m.
- To derive an effective sampled 'area' from soil sampled in transects, it was assumed that the transect was 0.5 m wide. This was essential for obtaining Figures 2b and 2c.

We thought it would be useful to construct average variograms of the properties for future reference. These were constructed by averaging the semivariance for the curves present at each lag. Both spherical and exponential models were fitted to the 'average semivariance at each lag'. The model with lowest root mean square error was deemed to be the most suitable. Average variograms are shown on the appropriate figures and the respective parameter values are given in Table 1.

pH. Data concerning the spatial nature of soil pH seem to be most numerous in the literature and Figure 1a shows fourteen variograms derived by various authors since 1978.

One of the earliest studies illustrating the variogram was by Campbell (1978) who looked at the spatial nature of pH over two soil types ('Pawnee' and 'Ladysmith') in the USA. Both showed nugget variance was 100% of the total variance present which suggests that the sample area was too small. Variogram data generated by Webster and McBratney (1987) at Broom's Barn, (UK) for pH had quite a large sill variance compared to some of the other studies. Uehara *et al.* (1985) found spatial dependence of

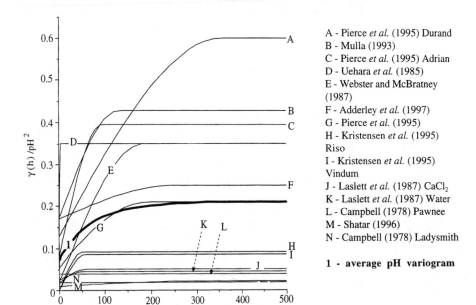

A - Pierce *et al.* (1995) Durand
B - Mulla (1993)
C - Pierce *et al.* (1995) Adrian
D - Uehara *et al.* (1985)
E - Webster and McBratney (1987)
F - Adderley *et al.* (1997)
G - Pierce *et al.* (1995)
H - Kristensen *et al.* (1995) Riso
I - Kristensen *et al.* (1995) Vindum
J - Laslett *et al.* (1987) CaCl$_2$
K - Laslett *et al.* (1987) Water
L - Campbell (1978) Pawnee
M - Shatar (1996)
N - Campbell (1978) Ladysmith

1 - average pH variogram

FIGURE 1a. pH variograms. Lettering scheme is in order of decreasing sill variance.

pH had a very short range at a site in Indonesia. Laslett *et al.* (1987), in trying to find what methods of spatial prediction are most suitable for given situations, used pH in water and then in calcium chloride to derive variograms. The variograms obtained show that the difference in these two methods is not great for a soil that has relatively small nugget and sill variance. Mulla (1993) used a soil with a very large nugget variance and a relatively high sill variance for a study on differential management. Pierce *et al.* (1995) conducted studies at three different sites that displayed vast differences in the spatial nature of pH, with 'Plainwell' being quite close to the 'average' for this collection of data, and 'Adrian' and 'Durand' showing much higher variance, with the Durand site showing the highest sill variance of all. Kristensen *et al.* (1995) found two sites in Denmark, Vindum and Riso, were very similar in terms of their spatial nature for pH. Shatar (1996) in a site-specific management study at Moree in Australia found the soil had small total variance, but a large proportion of nugget variance. The pH variogram of Adderley *et al.* (1997) showed high nugget variance and an above average sill variance for a soil in west Africa. The 'average' variogram for soil pH is an exponential model.

Clay content. Figure 1b shows variograms for some clay contents of soil found in the literature. Studies done on clay content were divided into topsoil and subsoil and average variograms were plotted for each because the topsoil was found to be much more homogeneous in nature.

Kristensen *et al.* (1995) found the soil at Vindum displayed small total variance with a range that went off the scale of the graph to 1350 m. The site at Riso also displayed a

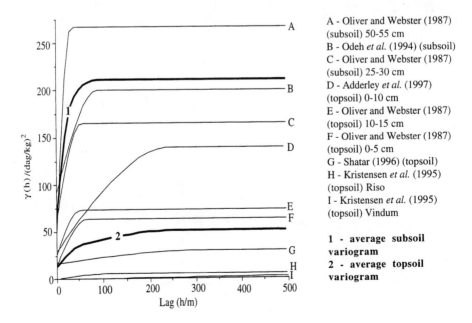

A - Oliver and Webster (1987) (subsoil) 50-55 cm
B - Odeh *et al.* (1994) (subsoil)
C - Oliver and Webster (1987) (subsoil) 25-30 cm
D - Adderley *et al.* (1997) (topsoil) 0-10 cm
E - Oliver and Webster (1987) (topsoil) 10-15 cm
F - Oliver and Webster (1987) (topsoil) 0-5 cm
G - Shatar (1996) (topsoil)
H - Kristensen *et al.* (1995) (topsoil) Riso
I - Kristensen *et al.* (1995) (topsoil) Vindum

1 - average subsoil variogram
2 - average topsoil variogram

FIGURE 1b. Topsoil and subsoil clay content (dag/kg) variograms.

small total variance although it had a much shorter spatial dependence with a range of about 100 m. Oliver and Webster (1987) found that the variance of clay content increased down a profile with the subsoil at 50-55 cm having approximately four times the total variance of the top 0-5 cm. The average variograms for both the topsoil and subsoil were best approximated by exponential models.

Carbon. There is not a lot of published data on spatial variability of soil carbon. The data in Figure 1c represents the 0-30 cm layer of soil. Although it is known that carbon content decreases rapidly with depth in the profile and that the top few centimetres tend to be very variable, the models are shown here for the purposes of some comparison. The variability of carbon content in the soil profile is reduced by cultivation (Cattle *et al.*, 1994) and some of the variation in the variograms displayed in Figure 1c may be explained by cultivation history. The site with the most variable carbon content was Riso, whilst that with the least variability was studied by Goovaerts and Chiang (1993). The average variogram for carbon content of the soil was approximated by an exponential model.

Potassium (K). Figure 1d shows variograms for soil potassium content. Pierce *et al.* (1995), although not specifying what form of analysis was used, found 'Plainwell' and 'Durand' displayed similar total variance, but there was quite a difference in the amounts of nugget variance. The site at Adrian had a very high total variance. Kristensen *et al.* (1995) again found that there were relatively small amounts of total variance for available K on their study sites although, again, the ranges differed. The average variogram for potassium was approximated by a spherical model.

Phosphorus (P). Figure 1e shows the variograms for phosphorus content. As with potassium, Pierce *et al.* (1995) again did not specify the form of chemical analysis used. The site Adrian displayed a small total variance, although quite a large proportion of

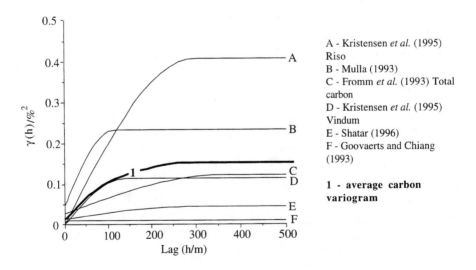

FIGURE 1c. Carbon (%) variograms. Organic carbon unless otherwise specified.

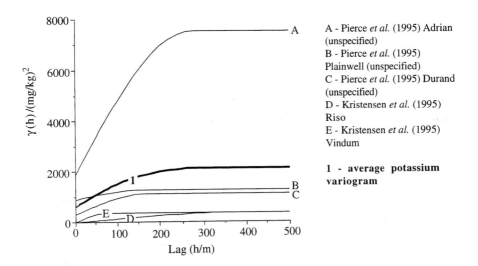

FIGURE 1d. Potassium (mg/kg) variograms. Available K unless otherwise specified.

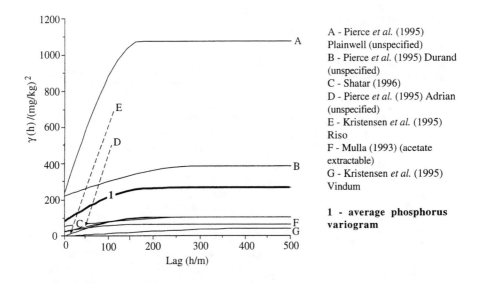

FIGURE 1e. Phosphorus (mg/kg) variograms. Available P unless otherwise specified.

nugget variance. The site at Durand showed a higher total variance, while 'Plainwell' was higher again. Mulla (1993) used acetate-extractable P and the site subsequently shows a low amount of variability; this may be due to the incompatibility of acetate-extractable P with available P. The site with the lowest variability in P was Vindum. The average variogram for phosphorus content of soil was approximated by a spherical model.

Average variogram parameters and within-block variance. Table 1 shows some general spatial relationships, especially for topsoil. For example, the nugget variance, C_0, is about half of the spatially-correlated variance, C, and the range, a or $3r$, is around 200 m. Of the topsoil characteristics, carbon content is an exception to these generalisations; its nugget variance is about a tenth of the spatially-correlated variance. Most variograms suggest that differential management at scales between 1 m and 100 m would be efficacious. Soil nutrients such as potassium and phosphorus seem to be better described by spherical models, whereas other properties such as pH, clay and carbon tend to be better described by an exponential model. The within-block variance columns show that the variance within a 1 ha area of a field is approximately double that within a 0.01 ha area.

TABLE 1. The average variogram behaviour parameters. Both exponential and spherical models were fitted. The model shown in the table is the one with the lower root mean square error.

Property	Model	C_0 (unit^2)	C (unit^2)	r or a (m)	Within-block variance (10×10 m)	Within-block variance (100×100 m)
pH	Exp.	0.07	0.14	69.79	0.08	0.14
Clay (tops.) (dag/kg)	Exp.	14.74	37.82	65.32	17.61	34.23
Clay (subs.) (dag/kg)	Exp.	53.92	157.5	19.97	89.10	188.34
Carbon (%)	Exp.	0.012	0.15	97.34	0.02	0.07
Potassium (mg/kg)	Sph.	631.6	1502.9	255.2	677.6	1080.60
Phosphorus (mg/kg)	Sph.	81.10	188.4	193.0	88.65	153.36

Some further remarks about variograms. There is some suggestion in the literature that observed variograms and their parameters are a function of sampling interval and the area of study. The numerous pH variograms which we presented in Figure 1a allow us to now test these ideas. In Figure 2a, the nugget variance is plotted against the minimum sampling distance. A moderately increasing linear relationship is found, suggesting that the nugget variance depends on the minimum sample spacing. The smaller the minimum sampling distance, the smaller the nugget variance, with the points showing increasing scatter as the distance increases beyond about 15 m. Likewise, an increasing linear relationship is evident in Figure 2b which shows the sill variance plotted against the square root of the sampling area (which gives an estimate of an equivalent transect length). In other words, longer transects or larger areas are likely to show larger sill variances. In Figure 2c, the range is plotted against the square root of the sample area divided by the sample interval. Dividing the square root of the sample area by the sample interval represents, effectively, the number of samples taken. Only those studies which were based on grid sampling, and therefore have a fixed minimum sample interval, are shown on Figure 2c. There seems to be an increase in range with the square root of the sample area divided by the sample interval. This implies that the more samples taken, the longer the range parameter will be in the final variogram. A tentative caveat is that the observed variogram is a function of the sampling scheme. This arises probably because the underlying variogram increases continuously with distance in a fractal model, similar to the deWijsian variograms employed by McBratney (1992) in describing the spatial variation of exchangeable magnesium.

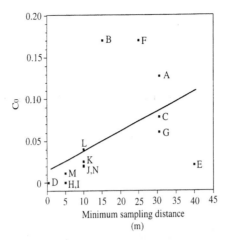

FIGURE 2a. Nugget variance plotted against minimum sampling distance. Lettering scheme is same as that used for Figure 1a and follows in Figures 2b and 2c.

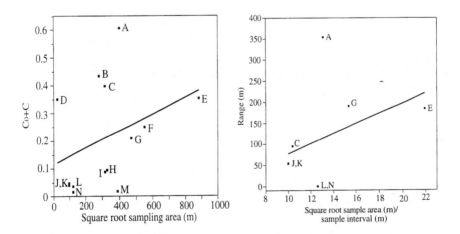

FIGURE 2b. Sill variance plotted against the square root of the sampling area.

FIGURE 2c. Range plotted against the square root of the sampling interval.

TECHNIQUES FOR MAPPING SOIL SPATIAL VARIABILITY

Given that soil variation will affect yield, it is important to have soil information at the same level of resolution as crop yield information. There are basically two techniques for obtaining soil information: sampling techniques and scanning techniques. Sampling techniques involve destructively taking specimens and analysing them in the laboratory. Because of the costs of this technique, sample spacings are usually large and interpolation is needed to make fine-scale maps. Scanning techniques, on the other

13

hand, offer the possibility of observations at very small intervals, the disadvantage being that the quality of the estimate at observation points will probably be lower than conventional techniques. This section reviews briefly both sampling and scanning techniques. It seems that some combination of both will be required ultimately.

<u>Sampling techniques</u>

<u>Interpolation - Kriging.</u> Kriging is a form of moving weighted average that estimates the amount of a property at an unsampled point depending on its distance from a sampled point. The variogram is essential for kriging in that semivariance is a determinant of the weighting system, and the more accurate the variogram, the more accurate the kriging. Kriging can be used to construct isarithmic maps (Burgess and Webster, 1980). A kriged map has advantages over traditional chloropleth maps made from soil survey as the estimates are made with minimum bias and variance, and continuity in the data can be expressed over the sampled landscape.

The kriging equation can be represented by the equation (after Webster and Oliver, 1990):

$$\mathbf{A}\begin{bmatrix}\mathbf{w}\\\psi\end{bmatrix}=\mathbf{b} \qquad (6)$$

where $\mathbf{A}$ is a matrix of semivariances of a property at each sampled point, and $\mathbf{b}$ is a vector of average semivariances between the sampled area and the sampled points. $\mathbf{w}$ is a vector of weights between 0 and 1 assigned to each point according to the Lagrange multiplier, ψ, which minimises the sum of squares.

Kriging also produces estimation variances given by (after Webster and Oliver, 1990):

$$\hat{\sigma}^2(A)=\mathbf{b}\begin{bmatrix}\mathbf{w}\\\psi\end{bmatrix}-\bar{\gamma}(A,A) \qquad (7)$$

where A is the area of the block over which the estimate is made (for point kriging this is simply a point). The expression $\bar{\gamma}(A,A)$ is the variance within a block of a given shape (usually a square) and of area A. It is therefore possible to map the expected variance in the estimates.

Maximum kriging variance is a criterion for the quality of sampling and interpolation (Webster and Oliver, 1990). This has been calculated for the average variograms given in Table 1. The results are shown in Table 2a which compares sampling on a 20 m and a 100 m grid for point kriging and kriging onto 10 m × 10 m blocks. The point kriging variances, in most cases, are shown to almost double when grid spacing increases from 20 m to 100 m. The kriging variances for the 10 m blocks inside the grids is shown to decrease in comparison to their respective point kriging estimates, although the kriging variances for the 100 m blocks are over three times larger than the 20 m blocks. This implies that the best option for soil sampling is to sample at small intervals and then perform block kriging so as to obtain the most accurate estimate.

14

TABLE 2a. Variances for point kriging grids of 20 m and 100 m for each property, respectively. Variances are also shown for block kriging (10 m x 10 m blocks within both of the grids).

Property	Kriging variances (20 m grid)	Kriging variance (100 m grid)	Kriging variance (10 m blocks in 20 m grid)	Kriging variance (10 m blocks in 100 m grid)
pH	0.11	0.18	0.03	0.10
Clay (tops.) (dag/kg)	24.75	44.80	7.14	27.19
Clay (subs.) (dag/kg)	146.97	247.70	57.87	158.60
Carbon (%)	0.03	0.09	0.01	0.07
Potassium (mg/kg)	883.90	1305.00	206.29	627.39
Phosphorus (mg/kg)	117.10	188.80	28.45	100.15

This suggested approach is probably better appreciated by perusal of Table 2b in which the kriging variances have been transformed to approximate 95% confidence intervals (CI) by taking the square root and multiplying by two. We see, for example, for the average variogram from sampling on a 100 m grid (the *de facto* standard in North American precision agriculture) that the estimates are within ± 0.85 pH units. Decreasing the sample grid spacing to only 20 m marginally improves the estimates to ± 0.67 pH units. Making estimates over 10 m × 10 m blocks improves matters considerably, however, with the 95% CI for the 10 m block in the 100 m grid being less than that for point kriging at 20 m. The 10 m block in the 20 m grid is even better.

TABLE 2b. 95% confidence intervals of predicted values shown for both point and block kriging.

Property	95% CI of predictions (20 m grid)	95% CI of predictions (100 m grid)	95% CI of predictions (10 m blocks in 20 m grid)	95% CI of predictions (10 m blocks in 100 m grid)
pH	0.67	0.85	0.34	0.63
Clay (tops.) (dag/kg)	9.95	13.39	5.34	10.43
Clay (subs.) (dag/kg)	24.25	31.48	15.21	25.19
Carbon (%)	0.36	0.60	0.22	0.53
Potassium (mg/kg)	59.46	72.25	28.73	50.10
Phosphorus (mg/kg)	21.64	27.48	10.67	20.02

More sophisticated forms of kriging are available including universal kriging, disjunctive and indicator kriging, but the same principles apply.

Interpolation - Other methods for spatial prediction. Kriging is only one form of spatial prediction. There are many other methods available including global means/medians, moving averages, inverse squared distance interpolation, Akima's interpolation, natural neighbour interpolation, quadratic trend surfaces, and Laplacian smoothing splines. These, and the relative virtues of each, are detailed in Laslett *et al.* (1987).

McBratney *et al.* (1996) proposed criteria for deciding which method of interpolation is most suitable for precision agriculture, given a certain sample size and intensity. The results are summarised in Table 3.

In Table 3, the term *NA* means interpolation of any form is not applicable to that situation. This occurs when sampling is sparse or the sample size is small. The term *?IS* refers to an inverse square or some informal prediction method. These can be used at moderate sample sizes and moderate to high sampling intensities. However there may be problems with the accuracy of estimates. *GS* refers to using a geostatistical method such as ordinary kriging or universal kriging with a global variogram, or Laplacian smoothing splines. These are useful in situations where sample number is high at between 101-500 samples and the intensity is moderate to high. For sparse, low intensity sampling with very high numbers of samples (>501) *GS* may be the only applicable method. *LGS* is a local neighbour method of kriging or Laplacian smoothing splines, applicable to situations where there are very high numbers of samples and a moderate to high intensity. *NR** implies that sample size is so large and intensity is so great that the soil will be adequately and accurately described by the observations and so spatial prediction will only be necessary if the sampling is uneven.

TABLE 3. Summary table from McBratney *et al.* (1996). Provisional methods of spatial prediction for precision agriculture in relation to sample size and intensity.

| | Sample Intensity (No. per minimum area of interest) | | |
| | Sparse | Moderate | Intense |
Sample size	0.0001 – 0.01	0.01 – 1	>1
<10	NA	NA	NA
10 – 100	NA	?IS	?IS / NR*
101 – 500	NA	GS	GS / NR*
>501	NA / GS	LGS	LGS / NR*

Interpolation with covariates. One of the criticisms of sampling and interpolation techniques for precision agriculture is that, with the sheer number of samples and the subsequent laboratory analysis involved, it is simply too expensive. A possible approach is to find related, or surrogate, and much cheaper measures of the soil properties of interest.

The most formal approach is to use a technique like co-kriging. This will usually be expensive except when it can be used in conjunction with fine-resolution data such as digital elevation models and remotely-sensed imagery (McBratney and Webster, 1983). Another problem with co-kriging is the assumption that the underlying relationship between variables is linear. This is often not the case. An alternative approach is to find relationships between the soil property of interest and more-easily, or cheaply, measured soil properties and environmental variables which are known at a great many locations, *e.g.* altitude. This can be done by various 'regression' techniques. The residuals from the regression can then be interpolated using a variety of techniques, including kriging, to assess the accuracy of the estimates. This combination has been called 'regression kriging' (Odeh *et al.*, 1994; 1995).

The 'regression' techniques that can be used include:

- Generalised Linear Models (GLMs) are used in situations where classical linear regression models cannot be applied. Unlike classical models, GLMs can accommodate non-normal distributions, transformations to linearity (Venables and Ripley, 1994), and non-constant variance (Hastie and Pregibon, 1992). GLMs can also accommodate data that is either continuous or only able to exist in a certain state (Webster and Oliver, 1990), *e.g.* multi-state variables such as the type of parent material.
- Generalised Additive Models (GAMs) have many features in common with linear regression models and GLMs but with the advantage of more flexibility in that they can easily identify non-linearities and are easier to interpret (Hastie, 1992). Functions are primarily fitted in GAMs non-parametrically by scatter-plot smoothers, which are a kind of regression tool. Some examples of scatter-plot smoothers are smoothing splines, locally-weighted polynomials, and near-neighbour smoothers (Hastie, 1992).
- Classification and Regression Trees are a means of hierarchical classification on data whereby predicted values are given based on probability of outcome. Similar to the classification systems used in botany, the regression tree bifurcates from a starting point, with each subsequent split in the data forming a 'node'. The data will keep on being partitioned into increasingly homogeneous units until it not feasible to continue. Clark and Pregibon (1992) state that it can be easier to interpret information from tree-based models than from linear regression models. An example of a Classification and Regression Tree applied to soil science is given by Baker *et al.* (1993).
- Neural networks are a form of regression capable of modelling non-linear, multivariate relationships (Goh, 1995) assuming that output data from each layer can be generated from its inputs (Tamari *et al.*, 1996). Figure 3 shows a neural network in its simplest form, consisting of three layers to perform a calculation: an input layer, a 'hidden' layer, and an output layer (Venables and Ripley, 1994). Each circle in Figure 3 is a processing element known as a neuron and computes output from input data. The input neurons carry the observed data, while the hidden and the output layers process the inputs, multiplying each by a weighting factor and summing the product. In practice there may be many 'hidden' layers that independently calculate output data and pass it onto the next layer for evaluation. The output layer makes the final calculation and determines the output value (Tamari *et al.*, 1996). The sums can then be further processed by a non-linear transfer function (Goh, 1995).

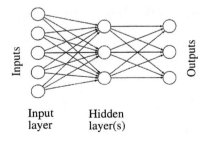

FIGURE 3. A simple neural network (after Venables and Ripley, 1994).

As stated above, it is advantageous to find relationships between the soil property of interest and more-easily, or cheaply, measured properties and variables. Terrain analysis is a tool that has been developed to describe landforms in this fashion. As the nature of soil at any point can be seen as a function of its place in the landscape (Walker *et al.*, 1968), terrain analysis takes advantage of this relationship using the Digital Elevation Model (DEM), a digital representation of continuous variation in relief over space (Burrough, 1986). DEMs serve as a starting point for terrain analysis as algorithms can be used to calculate terrain properties such as slope gradient, plan and profile curvature, aspect, and upslope area (Bell *et al.*, 1995) which can then be applied to soil properties. Moore *et al.* (1993) used terrain analysis to quantify and predict the spatial nature of soil attributes such as phosphorus, organic matter and pH according to the DEM and a wetness index for a site in Colorado, USA. As terrain analysis and DEMs becomes more widely used it is foreseen that they will have a significant application in precision agriculture, and so the provision of farm DEMs (and GIS) will become a major activity.

Considerable work has been done by soil scientists to develop relationships between easily-measured soil attributes that can be used in simulation models and decision-support systems. These relationships are known in the soil science literature as pedotransfer functions. Pedotransfer functions for the prediction of difficult-to-measure properties such as hydraulic conductivity, moisture retentivity and moisture content of soil have been developed for many data sets with varying degrees of success (*e.g.* Vereecken *et al.* 1989; Aina and Periaswamy, 1985; Dijkerman, 1988). Wösten and van Genuchten (1988) derived regression models for prediction of unsaturated hydraulic conductivity functions from soil texture, organic matter, and bulk density data, whilst Bell and van Keulen (1995) developed pedotransfer functions using soil pH, texture data, and organic matter content to predict cation exchange capacity (CEC). They accounted for 96% of the variability. The disadvantage of using pedotransfer functions is that one cannot describe the properties of all the soil in all the world; they are largely site-specific. Pidgeon (1972) found that equations derived in England fitted less well when applied to data from Ugandan soil, and Gupta and Larson (1979) found that when their pedotransfer function was applied to data collected in a different study there was a loss in accuracy of prediction. There is an intensive labour component involved, however, in that the relationships have to be found before they can be applied. Also, the area that the model is valid for has to be found due to their site-specificity. There is a trade-off between the advantages and disadvantages of using pedotransfer functions. However, the implication for precision agriculture is that once they are found they can be used indefinitely.

Proximal sensing or scanning techniques (complete enumeration)

Proximal sensing or scanning techniques are an alternative to sampling and interpolative techniques in assessing the spatial variability of soil. Although maybe not as accurate, proximal sensing or scanning is less time-consuming and expensive and is therefore advantageous. Proximal sensing or scanning can be described as 'complete enumeration' because there is so little physical distance between sampling points that properties of a field can be determined almost continuously.

Proximal sensing or scanning techniques can be either invasive or non-invasive, where 'invasive' refers to the sensor being contained within the soil volume to take a measurement, while 'non-invasive' refers to the sensor staying outside, usually above, the soil. Table 4 summarises some of the soil sensors that have been developed. Some of the methods are described in more detail in the following section.

In order to successfully apply proximal sensing and scanning to precision agriculture, mathematical methods of profile reconstruction will need to be developed to give real-time soil profile information. Two possible methods that might be used here are Tikhonov regularisation (Borchers *et al.*, 1997) and computed tomography (Rosenfeld and Kak, 1982), both of which are mathematical techniques that could be applied to data generated by variable-geometry sensor systems on field machinery.

TABLE 4. Some examples of the types of soil sensors and scanners developed for soil properties that may be of interest to precision agriculture (after Viscarra Rossel and McBratney, 1997).

Soil property	Sensor/scanner	Type
Moisture	Near Infra-Red (NIR)	Invasive
	Capacitance	Invasive
	Microwave	Invasive
	Ground-penetrating Radar (GPR)	Non-invasive
Nitrogen	NIR	Invasive
	ISFET	Invasive
Clay content	NIR	Invasive
Organic matter	Optical (660 nm)	Invasive
Cation Exchange Capacity (CEC)	NIR	Invasive
Salinity	Electromagnetic induction (EM)	Non-invasive
Strength	Penetrometer	Invasive
Potassium	Gamma Radiometry	Non-invasive

Invasive sensing or scanning. Near infra red (NIR) seems to be the most important invasive method for sensing or scanning soil. NIR techniques use wavelengths in the range 700-2500 nm to measure the diffuse reflectance of a material's chemistry. NIR techniques are particularly applicable as soil minerals and organic matter display characteristic behaviour in the NIR spectral region which can correspond to amounts of carboxyl, hydroxyl, sulphate, carbonate, and amine functional groups present in the structure (Ben-Dor and Banin, 1995). Ben-Dor and Banin (1995) successfully measured soil clay content, specific surface area, cation exchange capacity, organic matter content and calcium carbonate content using NIR. NIR can also be used to measure soil moisture content at wavelengths of 1450, 1950 and 2950 nm (Whalley and Stafford, 1992). Viscarra Rossel and McBratney (1997) used NIR to simultaneously measure clay and moisture content at four wavelengths with success.

Non-invasive sensing or scanning. Ground penetrating radar (GPR) is a non-invasive method for estimating soil moisture content which uses EM radiation in the microwave range (Whalley and Stafford, 1992). GPR works on a principle similar to time-domain reflectometry (TDR). With TDR, the delay between transmission and reflection of a pulse along a transmission line is determined. It is then possible to calculate the dielectric constant of the soil and hence the volumetric water content. There are two essential differences between TDR and GPR. Firstly, in TDR the length of the waveguides is known and hence calculating the dielectric constant is easy, whereas in GPR this distance is not known (Whalley and Stafford, 1992). Secondly, methods such as TDR and neutron probes are restricted by having to be inserted into the ground, whilst GPR can be attached to a field machine, and therefore used for continuous mapping of soil moisture on a large scale. Chanzy et al. (1996) compared the accuracy of GPR for ground measurements and airborne measurements at 15 m above the ground. It was found that measurements correlated well with soil moisture and conductivity effects were small, as opposed to the ground mode of operation. The airborne method is also relatively unaffected by vegetation and microtopography.

Gamma-radiometry refers to the measurement of naturally-occurring isotopes of potassium, uranium and thorium in the top 30-45 cm of soil by radioactive decay (Bierwirth, 1996). Measurements can be taken either at the ground surface or by an aeroplane flying at low altitude (~60 m – Cook et al., 1996). The intensity of emission of gamma radiation is proportional to the abundance of the source. Measurements of gamma radiation are made by a thallium-activated sodium iodide crystal detector. A characteristic frequency of the isotope can be derived from Planck's equation (Cook et al., 1996):

$$E = nh = hc / l \tag{8}$$

where E is the energy, n is the characteristic frequency of the isotope, h is Planck's constant, c is the speed of light, and l is the wavelength. The frequencies of interest for soil are 1.46 MeV for ^{40}K, 1.76 MeV for ^{238}U, and 2.615 MeV for ^{232}Th. For precision agriculture, the most relevant element is potassium. Other uses of gamma radiometry in precision agriculture might include establishing the identity of underlying parent material which may give an indication of soil mineralogy. While gamma radiometry can be used to quickly and effectively map potassium concentrations in the soil, there are disadvantages in that water content of the soil suppresses the readings, and only total K is measured, as opposed to that which is plant-available (Cook et al., 1996).

Electromagnetic induction (EM) offers a means to estimate soil salinity by non-invasive proximal measurements (Rhoades and Corwin, 1981). An electromagnetic induction unit consists of an inducer coil that produces an electromagnetic field which induces a current in the conductor (i.e. the soil). This current is proportional to the electrical conductivity of the soil (Corwin and Rhoades, 1990). A receiver coil measures the strength of the secondary field and gives a value of the bulk EC_a of the soil (Corwin and Rhoades, 1990). The major advantages of using EM sensors are that, unlike electrodes and salinity probes, measurements are proximal and thus avoid problems associated with soil/electrode contact, e.g. electrode souring and insertion into dry, compacted or rocky soil. A further benefit of EM measurements is the increased speed at which salinity surveys may be conducted (Corwin and Rhoades, 1984). The disadvantage of using EM sensors is that they require soil to be relatively uniform with regard to moisture and structure (Corwin and Rhoades, 1990). There are various types of electromagnetic induction sensors available for use (e.g. EM-31, EM-34), but for precision agriculture the most relevant is the EM-38 which measures the conductivity of the root zone (Corwin and Rhoades, 1990). This also gives an indication of not only salinity but also water content and nutrient status. The EM-38 could be used to map areas of conductivity quickly and efficiently through being connected to a tractor with a Global Positioning System (GPS) and a data logger. Instead of making static measurements and interpolating from sampled points, the EM sensor can make measurements continuously and produce maps accordingly.

ESTABLISHING RELATIONSHIPS BETWEEN SOIL SPATIAL VARIABILITY AND CROP SPATIAL VARIABILITY

Spatial variability of crop yield is often correlated with spatial variability in the underlying soil, although the relationship can be a complex one. This section discusses some ways of modelling crop yield response to soil factors and the methods that can be employed. An example from a sorghum field in Australia is given.

Yield response curves

When inputs are manipulated in the hope of matching outputs, as is the case in a precision agriculture system, knowledge of crop response to soil factors becomes vital. Models can be derived which attempt to relate the response of yield to soil factors. These are known as 'yield response curves' and can be fitted to the mean of the data or to the maximum yield. Both types of curve can be fitted to relate single or multiple soil variables to yield using the modern 'regression' techniques outlined above.

Most yield response curves relate the mean yield over a range of the soil factor of interest, be it pH, a textural property, or a nutrient. One of the earliest examples of a mean yield response to soil factors was the Mitscherlich equation (Gregory, 1988). Yield response can also be measured according to maximum yield. This is known as the 'boundary line'. The boundary line relationship is based on the principle that all "biological material has an upper limit of development or response in a given situation" (Webb, 1972). Especially useful for determining yield relationships to soil nutrients, the upper limit can be seen as the line of best performance when 'amount of nutrient' is plotted as abscissa and 'relative yield (%)' is plotted as ordinate. Data points in a plot of nutrient level against yield show much scatter due to variation and interactions in crop growth factors and errors in measurement (Webb, 1972), but the boundary line is drawn over the highest observed yields for a range of nutrient levels, thereby making discernible the range of nutrient concentrations where yield will be greatest. Yield observations closest to the line show that the nutrient is non-limiting to crop growth, whilst those that are farther away have some other (unknown) factor affecting growth (Schnug et al., 1995). Webb (1972), in developing the boundary line method, conceded that it does not explain *why* yield variation occurs, only that it *does* occur. Another early crop response model was Liebig's 'Law of the Minimum' (Gregory, 1988) which is effectively a boundary line in that the curve is at a maximum when the factor affecting yield is not limiting to growth and *vice versa*.

Similar to pedotransfer functions, the issue that has to be considered when constructing yield response curves of any type is the area in which they are effective. The spatial variability of soil often means that one yield response curve cannot be accurate for a field, let alone a whole farm or region.

An example of the relationship between the spatial variability of clay content and crop spatial variability – Creek Field, Moree, Australia

Shatar (1996) conducted a study on the site-specific soil factors that controlled sorghum yield in a field near Moree in north-west NSW. Using the data obtained for clay content of the soil, an example is given here of its effect on yield. Clay content showed a reasonable correlation to yield ($r = 0.51$) relative to other soil properties and is of importance because it is an intrinsic soil property that is very hard to change and affects many other soil properties such as cation exchange capacity (CEC) and water-holding capacity.

Figure 4 is a map of the sampling sites in relation to the field. These sampling sites are located at 101 centiles of the yield distribution, plus 13 extra sites which fill in spatial gaps. We have divided up the area into four arbitrary areas that might represent 'differential' management. Mean and boundary line response curves have been derived according to clay content for the whole field (Figure 5) and for the four areas of management (Figures 6a and 6b). Results for the actual yield and the potential yield (as given by the boundary line) were mapped using kernel smoothing. These are shown in Figures 7a and 7b. A map of difference in yield was also derived (Figure 7c).

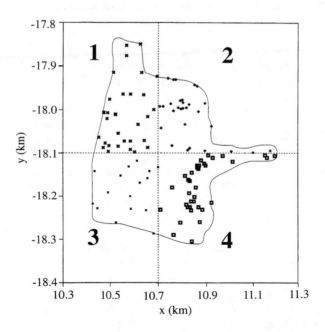

FIGURE 4. Creek field, Moree, Australia, for the sorghum crop, 1996. The field has been divided into four arbitrary areas of management by the hatched line and the 114 soil sampling sites are shown with different markers according to the area of belonging.

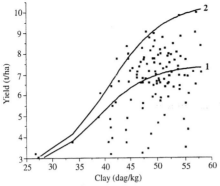

FIGURE 5. Yield plotted against clay content for Creek field, 1996. A mean response curve and a boundary line has been fitted to the data.

Figure 5 shows the mean yield response curve and the boundary line for the data. Both curves in Figure 5 were fitted using cubic smoothing splines (a special form of generalised additive model). The mean yield of sorghum in the field increases with clay content to approximately 7.3 t/ha. The boundary line fitted to the maximum yields for each clay content shows that, potentially, sorghum can yield up to 10 t/ha when clay content is approximately 55%. When yields are below the boundary line it can be thought that there is some other factor affecting yield, as stated by Schnug *et al.* (1995). At low clay contents it would not be worth putting extra effort into increasing yield; the lower water-holding capacity of the soil is limiting growth. As clay content is one of the factors Bouma and Finke (1993) describe as 'static', *i.e.* not easily changed by management, there is not a lot that can be done to improve yield in these parts of field.

The curves shown in Figure 5, while technically correct, may be on too broad a scale for the resolution required by precision agriculture. When the field is broken down into four areas (Figure 4) the following yield response curves are obtained.

Figures 6a and 6b show that when a mean yield curve and a boundary line are fitted to the observations from each area, there are many different functions to be considered. The mean yield of the field, as related to its clay content, cannot be taken as representative, *e.g.* Area 2 has a vastly different yield response function to clay than Area 3. The same can be seen for the boundary lines for the different areas. There is much variation between the curves, especially in Area 2 compared to the other three areas. Areas 3 and 4 show quite a bit of similarity in their boundary lines, suggesting that these two areas have a similar yield potential in terms of clay content; when one area is yielding differently to another it may be assumed that there is some other unknown factor affecting yield.

Figures 7a to 7c show the maps derived from the data for actual yield, potential yield (as derived from the boundary line) and the difference between the two for the whole field. The actual yield map for the field shows that there is quite a bit of variation in yield, with a range between about 3.5 t/ha and 9.5 t/ha. The areas of the field that have lighter texture yield least and so it can be concluded that clay content has an effect on the yield of sorghum.

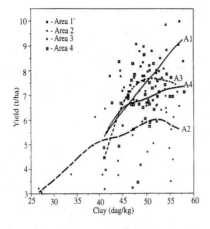

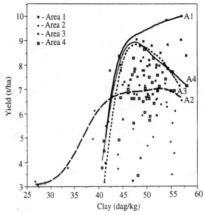

FIGURE 6a. Mean yield response to clay content fitted to data from the four arbitrary areas of management. The response lines for different areas are denoted by 'A' followed by the area's number.

FIGURE 6b. Boundary line responses of yield for the four areas as determined by clay content. Labelling is the same as for Figure 6a.

Figure 7b shows the potential yield for the field. An equation was fitted to the boundary line shown in Figure 5 which enabled predictions to be made of yield with clay content. The range of yield is not so wide for this map, with the maximum being about 9.5 t/ha and the minimum being around 4.0 t/ha. Figure 7c shows the spatial differences in Figure 7a and 7b. The difference between the potential yield and the actual yield is the need, or opportunity, for improved management of the system. The biggest difference in the yields is about 4.5 t/ha. This represents a significant 'hidden' economic loss from the field and differential management is obviously required. The method we have employed is empirical, using the data to model a decision-support system. Another option available would be to use a crop simulation model, *e.g.* CERES-Sorghum, a crop simulation model for sorghum growth (Alargarswamy and Ritchie, 1991), however this is beyond the scope of this paper.

CONCLUSIONS

The topic of spatial variability of soil and its implications for precision agriculture is a large one and only a few of the more important topics are dealt with in this paper. Indeed, spatial variability is only half of the issue the application of precision agriculture cannot be successful from a soil management point of view until temporal soil variability is recognised, described and manipulated. The major point of the discussion in this paper has been how the gathering of precise soil information could be an impediment to decision making in precision agriculture. Some possible areas where more research is needed may include wavelet analysis, Kalman filtering, optimal

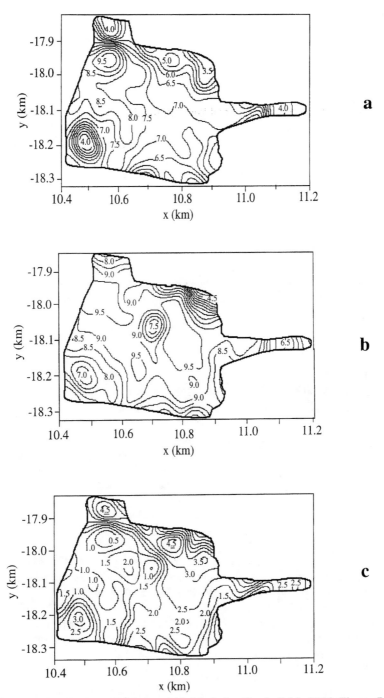

FIGURES 7a - 7c. Maps of sorghum yield (t/ha) for Creek field, 1996. Figure 7a is actual yield, 7b is potential yield as determined by the boundary line in Figure 5, and 7c is the difference between 7a and 7b.

combinations of sampling and proximal sensing, soil profile reconstruction techniques, farm DEMs for terrain analysis and the eternal questions about the relationships between crop yield and soil variability. Established methods for describing the spatial variability of soil such as variograms are powerful tools at precision agriculture's disposal and they indicate there is considerable variation present in fields in the spatial range of interest. This gives scope for differential management to proceed.

There is a need to develop a data base cataloguing the spatial variation of soil properties for all management regimes in different regions. This will give an idea of the variability present that has to be dealt with and will allow comparisons to be made. Applying precision agriculture is a double-edged sword; we want to be site-specific but want minimal data acquisition and analysis to describe the yield. This is the paradox we must overcome!

ACKNOWLEDGMENTS

We thank the Australian Grains Research and Development Corporation (GRDC) for financial support to M. Pringle (through a research grant to CSIRO Land & Water) and Tamara Shatar for the use of her data relating soil properties to sorghum yield.

REFERENCES

Adderley, W.P., Jenkins, D.A., Sinclair, F.L., Stevens, P.A., Verinumbe, I. (1997) The influence of soil variability on tree establishment at an experimental agroforestry site in north east Nigeria. *Soil Use and Management*, **13**, 1-8.

Aina, P.O., Periaswamy, S.P. (1985) Estimating available water-holding capacity of western Nigerian soils from soil texture and bulk density, using core and sieved samples. *Soil Science*, **140**, 55-58.

Alargarswamy, G., Ritchie, J.T. (1991) Phasic development in CERES-Sorghum model. *Predicting Crop Phenology*, T. Hodges, (Ed.), Boca Raton, Florida, CRC Press, pp 143-152.

Baker, F.A., Verblya, D.L., Hodges, C.S., Jr., Ross, E.W. (1993) Classification and regression tree analysis for assessing hazard of pine mortality caused by *Heterobasidion annosum*. *Plant Disease*, **77**, 136-139.

Bell, J.C., Butler, C.A., Thompson, J.A. (1995) Soil-terrain modelling for site-specific agricultural management. *Site-Specific Management For Agricultural Systems*, P.C. Robert, R.H. Rust, W.E. Larson, (Eds), Madison, Wisconsin, ASA, CSSA, SSSA, pp 209-227.

Bell, M.A., van Keulen, H. (1995) Soil pedotransfer functions for four Mexican soils. *Soil Science Society of America Journal*, **59**, 865-871.

Ben-Dor, E., Banin, A. (1995) Near-infrared analysis as a rapid method to simultaneously evaluate several soil properties. *Soil Science Society of America Journal*, **59**, 364-372.

Bierwirth, P.N. (1996) Gamma-radiometrics, a remote sensing tool for understanding soils. *Australian Collaborative Land Evaluation Program Newsletter*, **5**, 12-14.

Borchers, B., Uram, T., Hendrickx, J.M.H. (1997) Tikhonov regularisation of electrical conductivity depth profiles in field soils. *Soil Science Society of America Journal* (in press).

Bouma, J., Finke, P.A. (1993) Origin and nature of soil resource variability. *Soil Specific Crop Management*, P.C. Robert, R.H. Rust, W.E. Larson, (Eds), Madison, Wisconsin, ASA, CSSA, SSSA, pp 3-14.

Burgess, T.M., Webster, R. (1980) Optimal interpolation and isarithmic mapping of soil properties. I. The semivariogram and punctual kriging. *Journal of Soil Science*, **31**, 315-331.

Burrough, P.A. (1986) *Principles of geographic information systems for land resources assessment*, New York, Oxford University Press, Chapter 3.

Burrough, P.A. (1993) Soil variability: a late 20th century view. *Soils and Fertilizers*, **56**, 531-562.

Campbell, J.B. (1978) Spatial variation of sand content and pH within a single contiguous delineation of two soil mapping units. *Soil Science Society of America Journal*, **42**, 460-464.

Cattle, S.R., Koppi, A.J., McBratney, A.B. (1994) The effect of cultivation on the properties of a rhodoxeralf from the wheat/sheep belt of New South Wales. *Geoderma*, **63**, 215-225.

Chanzy, A., Tarussov, A., Judge, A., Bonn, F. (1996) Soil water content determination using a digital ground-penetrating radar. *Soil Science Society of America Journal*, **60**, 1318-1326.

Clark, L.A., Pregibon, D. (1992) Tree-based models. *Statistical Models in S*, J.M. Chambers, T.J. Hastie, (Eds), AT & T Laboratories, pp 377-419.

Cook, S.E., Corner, R.J., Groves, P.R., Grealish, G.J. (1996) Use of airborne gamma radiometric data for soil mapping. *Australian Journal of Soil Research*, **34**, 183-194.

Corwin, D.L., Rhoades, J.D. (1984) Measurement of inverted electrical conductivity profiles using electromagnetic induction. *Soil Science Society of America Journal*, **48**, 288-291.

Corwin, D.L., Rhoades, J.D. (1990) Establishing soil EC–depth relations from electromagnetic induction measurements. *Communications in Soil Science and Plant Analysis*, **21**, 861-901.

Dijkerman, J.C. (1988) An Ustult-Aquult-Tropept catena in Sierra Leone, West Africa, II. Land qualities and land evaluation. *Geoderma*, **42**, 29-49.

Foufoula-Georgiou, E., Kumar, P. (1994) *Wavelets in Geophysics*, San Diego, Academic Press.

Fromm, H., Winter, K., Filser, J., Hantschel, R., Beese, F. (1993) The influence of soil type and cultivation system on the spatial distributions of the soil fauna and microorganisms and their interactions. *Geoderma*, **60**, 109-118.

Goh, A.T.C. (1995) Modelling soil correlations using neural networks. *Journal of Computing in Civil Engineering*, **9**, 275-278.

Goovaerts, P., Chiang, C.N. (1993) Temporal persistence of spatial patterns for mineralizable nitrogen and selected soil properties. *Soil Science Society of America Journal*, **57**, 372-381.

Gregory, P.J. (1988) Crop growth and development. *Russell's Soil Conditions and Plant Growth*, A. Wild, (Ed.), Longman Group UK Limited, 11th edition, pp 31-68.

Gupta, S.C., Larson, W.E. (1979) Estimation of soil water retention characteristics from particle size distribution, organic matter percent and bulk density. *Water Resources Research*, **15**, 1633-1635.

Hastie, T.J. (1992) Generalized additive models. *Statistical Models in S*, J.M. Chambers and T.J. Hastie, (Eds), AT & T Laboratories, pp 249-307.

Hastie, T.J., Pregibon, D. (1992) Generalised linear models. *Statistical Models in S,* J.M. Chambers, T.J. Hastie, (Eds), AT & T Laboratories, pp 195-247.

Kristensen, K., Simmelsgaard, S.E., Djurhuus, J., Olesen, S.E. (1995) Spatial variability of soil physical and chemical parameters. *Proceedings of the Seminar on Site Specific Farming*, S.E. Olesen, (Ed.), Danish Institute of Plant and Soil Science, SP-report No. 26, pp 39-55.

Jenkinson, D. (1988) Soil organic matter and its dynamics. In: *Russell's Soil Conditions and Plant Growth*, A. Wild, (Ed.), Longman Group UK Limited, 11th edition, pp 564-607.

Jenny, H. (1941) *Factors of Soil Formation*, McGraw-Hill Book Company, p 15.

Laslett, G.M., McBratney, A.B., Pahl, P.J., Hutchinson, M.F. (1987) Comparison of several spatial prediction methods for soil pH. *Journal of Soil Science*, **38**, 325-341.

Matheron, G. (1973) The intrinsic random functions and their applications. *Advances in Applied Probability*, **5**, 439-468.

McBratney, A.B. (1992) On variation, uncertainty and informatics in environmental soil management. *Australian Journal of Soil Research*, **30**, 913-935.

McBratney, A.B. (1997) Some considerations on methods for spatially aggregating and disaggregating soil information. *Nutrient Cycling in Agroecosytems* (in press).

McBratney, A.B., Webster, R. (1983) Optimal interpolation and isarithmic mapping of soil properties. V: Co-regionalization and multiple sampling strategy. *Journal of Soil Science*, **34**, 137-162.

McBratney, A.B., Webster, R. (1986) Choosing functions for semivariograms of soil properties and fitting them to sampling estimates. *Journal of Soil Science*, **37**, 617-639.

McBratney, A.B., Whelan, B.M. (1995) *The potential for site-specific management of cotton farming systems*. Discussion Paper No.1, Co-operative Research Centre for Sustainable Cotton Production, Australia.

McBratney, A.B., Whelan, B.M., Viscarra Rossel, R.A. (1996) Spatial prediction for Precision Agriculture. *Precision Agriculture: Proceedings of the 3rd International Conference*, P.C. Robert, R.H. Rust, W.E. Larson, (Eds), Madison, Wisconsin, ASA, CSSSA, SSSA, pp 331-342.

Moore, I.D., Gessler, P.E., Nielsen, G.A., Peterson, G.A. (1993) Soil attribute prediction using terrain analysis. *Soil Science Society of America Journal*, **57**, 443-452.

Morkoc, F., Biggar, J.W., Nielsen, D.R., Rolston, D.E. (1985) Analysis of soil water content and temperature using state-space approach. *Soil Science Society of America Journal*, **49**, 798-803.

Mulla, D.J. (1993) Mapping and managing spatial patterns in soil fertility and crop yield. *Soil Specific Crop Management*, P.C. Robert, R.H. Rust, W.E. Larson, (Eds), Madison, Wisconsin, ASA, CSSA, SSSA, pp 15-26.

Odeh, I.O.A., McBratney, A.B., Chittleborough, D.J. (1994) Spatial prediction of soil properties from landform attributes derived from a digital elevation model. *Geoderma*, **63**, 197-214.

Odeh, I.O.A., McBratney, A.B., Chittleborough, D.J. (1995) Further results on prediction of soil properties from terrain attributes: heterotopic cokriging and regression-kriging. *Geoderma*, **67**, 215-226.

Oliver, M.A., Webster, R. (1987) The elucidation of soil pattern in the Wyre Forest of the west Midlands, England. II Spatial distribution. *Journal of Soil Science*, **38**, 293-307.

Pidgeon, J.D. (1972) The measurement and prediction of available water capacity of ferrallitic soils in Uganda. *Journal of Soil Science*, **23**, 431-441.

Pierce, F.J., Warnke, D.D., Everett, M.W. (1995) Yield and nutrient variability in glacial soils of Michigan. *Site-Specific Management For Agricultural Systems*, P.C. Robert, R.H. Rust, W.E. Larson, (Eds), Madison, Wisconsin, ASA, CSSA, SSSA, pp 133-152.

Rhoades, J.D., Corwin, D.L. (1981) Determining soil electrical conductivity–depth relations using an inductive electromagnetic soil conductivity meter. *Soil Science Society of America Journal*, **40**, 225-260.

Rosenfeld, A., Kak, A.C. (1982) *Digital picture processing*, New York, Academic Press Inc., 2nd edition, Volume 1, Chapter 8.

Schnug E., Heym, J., Murphy, D.P. (1995) Boundary line determination technique (BOLIDES). *Site-Specific Management for Agricultural Systems*, P.C. Robert, R.H. Rust, W.E. Larson, (Eds), Madison, Wisconsin, ASA, CSSA, SSSA, pp 899-908.

Shatar, T.M. (1996) *Site-Specific Crop Management – Relationships Between Edaphic Factors and Sorghum Yield Within a Paddock*. Bachelor of Science in Agriculture Thesis (Hons.), Department of Agricultural Chemistry and Soil Science, The University of Sydney, Australia.

Shumway, R.H. (1985) Time series in the soil sciences: Is there life after kriging? *Soil Spatial Variability*, D.R. Nielsen, J. Bouma, (Eds), Netherlands, PUDOC, pp 35-56.

Tamari, S., Wösten, J.H.M., Ruiz-Suarez, J.C. (1996) Testing an artificial neural network for predicting soil hydraulic conductivity. *Soil Science Society of America Journal*, **60**, 1732-1741.

Uehara, G., Trangmar, B.B., Yost, R.S. (1985) Spatial variability of soil properties. *Soil Spatial Variability*, D.R. Nielsen, J. Bouma, (Eds), Netherlands, PUDOC, pp 61-92.

Viscarra Rossel, R.A., McBratney, A.B. (1997) Laboratory evaluation of a proximal sensing technique for simultaneous measurement of clay and water content. *Geoderma*, (in press).

Venables, W.N., Ripley, B.D. (1994) *Modern Applied Statistics with S-Plus*. New York, Springer-Verlag.

Vereecken, H., Maes, J., Darius, P. (1989) Estimating the soil moisture retention characteristic from texture, bulk density, and carbon content. *Soil Science*, **148**, 389-403.

Walker, P.H., Hall, G.F., Protz, R. (1968) Relationship between landform parameters and soil properties. *Soil Science Society of America Proceedings*, **32**, 101-104.

Webb, R.A. (1972) Use of the boundary line in the analysis of biological data. *Journal of Horticultural Science*, **47**, 309-319.

Webster, R. (1977) Spectral analysis of gilgai soil. *Australian Journal of Soil Research*, **15**, 191-204.

Webster, R., McBratney, A.B. (1987) Mapping soil fertility at Broom's Barn by simple kriging. *Journal of the Science of Food and Agriculture*, **38**, 97-115.

Webster, R., Oliver, M.A. (1990) *Statistical Methods in Soil and Land Resource Survey*, New York, Oxford University Press.

Wendroth, O., Al-Omran, A.M., Kirda, C., Reichardt, K., Nielsen, D.R. (1992) State-space approach to spatial variability of crop yield. *Soil Science Society of America Journal*, **56**, 801-807.

Whalley, W.R., Stafford, J.V. (1992) Real-time sensing of soil water content from mobile machinery: options for sensor design. *Computers and Electronics in Agriculture*, **7**, 269-284.

Wösten, J.H.M, van Genuchten, M.T.H. (1988) Using texture and other soil properties to predict the unsaturated soil hydraulic functions. *Soil Science Society of America Journal*, **54**, 186-192.

TECHNOLOGY FOR PRECISION AGRICULTURE

J.K. SCHUELLER

University of Florida, Department of Mechanical Engineering, P.O. Box 116300, Gainesville, FL 32611-6300, USA

ABSTRACT

Machinery systems for spatially-variable crop production are described along with some of their historical development and the contemporary utilisation of such systems. Equipment to perform field mapping, especially of crop yields, is reviewed. Equipment to vary crop inputs according to predetermined maps is also discussed. Some of the needed technical characteristics and the contemporary situation are presented, along with areas of likely further development.

INTRODUCTION

It has long been recognised that crops and soils (e.g., Nielsen and Bouma, 1984; Carr *et al.*, 1991; Cassman and Plant, 1992[1]) are not uniform within a given field. Astute farmers have always responded to such variability by taking appropriate actions. But the large areas and high level of mechanisation of modern crop production make such actions less frequent than they should be.

For over a decade now, there have been some technical methods to utilise modern electronics to respond to field variability. These methods go by many names, often a two word combination. The first may be such terms as spatially-variable, GPS-based, prescription, site-specific, or precision. The second might be crop production, agriculture, or farming. It seems that the correct term should probably be spatially-variable crop production as that is more accurate and descriptive than the term precision farming.

Whatever the technology area is termed, the concept is similar. Variations occur in crop or soil properties within a field. These variations are noted, and often mapped. There then may or may not be some management action taken as a consequence of the spatial variability within the field. A description and taxonomy of some of the types of responses to the variability are included in Schueller (1992). Five types are identified as:

1. Homogeneous -- in which the entire field is treated in a uniform manner.
2. Automatic -- in which a piece of equipment measures some variable quantity and immediately adjusts accordingly.
3. Temporally Separate -- in which some variable quantity is measured and recorded and later some appropriate action occurs.
4. Multivariate -- in which multiple measurements and actions are taken.
5. Historical -- in which measurements are taken from multiple crop years or the like.

---

[1]Precision farming has spawned a diverse and voluminous literature. Due to space considerations, only representative, primarily recent, citations are presented here.

Homogeneous is not responding to the spatial variability and treating the field as a uniform entity. It is the default system of most contemporary crop production.

The automation of field equipment to respond to spatial variation of crops and soils has a long history. The self-levelling hillside combine responded to topographic variations. The famous Ferguson system varied tractor three-point hitch height to maintain a relatively constant load on the tractor in varying soil draft conditions. Notice the philosophy of such equipment's design. The automation is machinery-centred in that it counters difficulties the machinery is having in separating the grain under varying yields or providing enough power for tough spots. However, when advocates of spatially-variable crop production refer to Automatic control, they are referring to something that is agronomy-centred. For example, herbicide application may be controlled with input from an organic matter sensor. Or anhydrous ammonia side-dressing may be guided by real-time sensing of nitrate levels. Another example would be a planter which varies planting depth in response to soil moisture. The machine is responding to the agronomic needs in these situations.

The Automatic Control response to spatial variability is being widely researched and marketed. In fact, a real-time, simple, reliable, accurate nitrate sensor is the Holy Grail of agricultural sensor research for maize (corn) production. Such a sensor reduces crop production costs and nitrate pollution by allowing nitrogen side-dressing to be variably applied according to localised, current need.

However, most discussions of spatially-variable crop production or precision farming deal with Temporally-Separate control. Some quantity such as crop yield or soil nutrient levels is measured and mapped. At a later time, some cropping operation is controlled based upon the maps or their derivatives. The operations in Temporally-Separate spatially-variable crop production will therefore be the principal topic of this paper.

The Multivariate and Historical types of responses are enhancements of the Temporally Separate and a discussion of them is beyond the scope of this paper.

To a large extent, spatially-variable crop production has been technology-driven. Advances in electronics and computers, such as the global positioning system (GPS) and geographic information systems (GIS), have led to new tools which clever individuals have applied to agriculture. As technology advanced, a gathering momentum of activity and development can be seen through the literature, such as Schueller (1988), Schueller (1991), Schueller (1992), Robert et al. (1993), Auernhammer (1994), Robert et al. (1995), Robert et al. (1996), and Stafford (1996). This conference continues that trend. As this short paper cannot repeat that which fills books, those desiring to be knowledgeable of spatially-variable crop production technologies should be familiar with such literature.

MAPPING PROGRAMS

The generation of maps of crop or soil properties is the first and most important step in spatially-variable crop production. These maps allow the spatial variability to be appreciated and provide the basis for spatially-variable control of the crop production. The mapping operations can be classified into:

1. Remote Sensing
2. Field Operations
3. Manual

based upon how the information used to generate the maps is gathered.

Remote Sensing measures the visible or non-visible optical properties of a field or groups of fields. The most-known procedure is to take images from satellites such as LANDSAT or SPOT. These images, if properly ground-truthed, may allow mapping of crop, pest, or soil properties within a field. Remote sensing images may be gathered from a wide variety of platforms, including satellites, aeroplanes, remotely piloted vehicles (RPVs), or even bucket trucks. They may be gathered by different devices including sensors, film cameras, digital cameras, and video recorders. The gathered images must then be manipulated to correct for errors, such as geometric or chromatic distortion. Finally a useful map is generated, assuming that point measurements have been taken in the field to ground-truth (verify) the accuracy and calibration of the measurement. Images of USA agricultural fields are available from various commercial firms. However, the commercial extent of agricultural remote sensing has not broadened as quickly as many expected. Perhaps recent improvements in resolution, timeliness, and cost will facilitate wider commercial utilisation.

Measurements may also be taken during normal field operations. The most common is the measurement of yield during harvesting. Measuring soil properties during tillage or planting is another example. Since a field operation is already being performed, the costs involved in the measurement operations are primarily marginal. Unfortunately, this then requires that the measurement for mapping be completed under the constraints of the other operation. Usually this implies that the operations are performed at specific times and that the mapping measurement activities cannot affect the completion of a time-critical field operation.

Manual measurements may be taken by farmers or farm workers specifically to gather information. Currently, the most common measurement of this type is the taking of a soil sample. Manual mapping measurements can be taken at the particular time and location to provide the most accurate and useful information. Unfortunately, the time and financial cost of such activities is often problematic. So the number or density of measurements which can be economically justified is limited. Another type of manual measurement is the mapping of pest infestations or other crop problems (Stafford et al., 1996).

Grain yield mapping

One of the areas of greatest interest in machinery-centred automatic controls for agricultural equipment is the automatic control of grain harvester travel speed. Schueller et al. (1983) is an example. The speed of the harvester is varied so that a harvester encounters a constant flow rate of material despite the spatial variation of yield within the field being harvested. The constant flow rate is desirable due to the highly non-linear relationship between flow rate and threshing and separation losses within the harvester. The control of the harvester speed is rather straightforward. The greater difficulty is to accurately measure the flow rate. So flow rate sensors (e.g., Schueller et al., 1985) have been researched.

35

These same flow rate sensors[2] can be used to generate yield maps. If the flow rate is divided by the harvester travel speed multiplied by the effective harvester harvesting width, the yield per unit of area is determined. Since harvesting width is usually an approximately fixed number, and width sensors and harvester path post-processing are under development, the easy measurement of harvester travel speed makes mapping easy once the grain flow rate is accurately measured.

The most popular method of measuring flow rate in the USA is to direct the stream of clean grain on its way from the separating mechanism to the grain holding tank (bin) against some sort of target or plate. The force of the grain hitting the plate at a fixed velocity is then proportional to the mass of the grain. This is the basis of the Case-IH and John Deere sensors, as well as some popular after-market sensors, such as Ag-Leader and Micro-Trak.

The flow rate may also be measured on a volumetric basis. One such method is to measure the amount of time a light beam is intercepted by grain on the clean grain elevator. Another measurement method is to measure the speed of a paddle wheel of fixed displacement in the grain stream.

Tests (and theory) indicate that an accurate method of flow rate sensing is to pass the grain between a gamma radiation source and a radiation detector. The attenuation of the radiation is proportional to the mass of grain.

Yield mapping on grain harvesters continues to be widely researched (Stafford *et al.*, 1996; Auernhammer *et al.*, 1994). There have been some comparison tests (Birrell *et al.*, 1996).

Other crop yield maps

Yield mapping for other crops has occurred later than for grains and soybeans. The greatest effort in the USA appears to be in cotton. The impact force method has been less successful in this crop. Other methods include light interruption and measuring changes in the weight of the basket which holds the harvested cotton.

Potato yields have been measured with load cells on the harvester (Schneider *et al.*, 1996). Sugar beet (Walter *et al.*, 1996) and sugar cane (Cox, 1997) yield have similarly been measured. Peanut combine yield mapping has also been demonstrated. In all these systems, either the mass flow rate of the separated harvested material is measured on some conveyor chain or the change rate of the weight of the storage tank or bin is measured. The difficulty with the latter approach is that a small change in a large value must be accurately measured.

Crops which are harvested in discrete, rather than continuous, quantities can also be mapped. This is done by recording the locations of such items as hay bales or full citrus bins.

---

[2]Actually there is a difference in that the material flowrate most affecting harvesting efficiency is typically the material-other-than-grain (MOG) flowrate while yield mapping should use just the grain flowrate.

Pest and crop maps

The locations and relative levels of pests can also be mapped by sensors or by recording human indications. In fallow fields or young crops, the presence of weeds can be easily detected by the presence of green vegetation. In more mature crops, they can only be detected by more sophisticated means, such as vision or sensor systems which discriminate on the basis of plant physical characteristics or colour shade.

Other pests, as well as nutritional or water problems, can generally only be detected by observing the health or growth of the crop plants. This observation may occur by the operation of sensors during other field operations, by remote sensing measurement of light reflectance characteristics, or by manual samples. Since many potential causes exist for health or growth problems, significant management intervention would seem to be necessary for calibration, cause determination, and corrective action selection.

Soil maps

Soil maps are available for agricultural land in the United States and other countries. However, these were generally made for purposes other than spatially-variable crop production. Accordingly, the scale, accuracy, and usefulness may be limited. Maps of such quantities as nutrient levels, moisture, topography, soil type, and texture have also been made by some researchers.

A common simple type of mapping uses a GPS-equipped all-terrain vehicle (ATV) to circumnavigate the boundary and thereby generate a map of the boundary of the field. But the most common spatially-variable soil map is the map of soil characteristics (nutrient, pH, cation exchange capacity, organic matter, etc.) based upon manual individual soil tests. Due to the costs for taking the samples and processing them, interpolation, including geostatistical methods, is often used in map generation.

There have been substantial efforts to develop soil property sensors to avoid the sampling costs and problems (Hummel et al., 1996; Stafford, 1988). The organic matter sensor of Gaultney et al., (1988) has been commercialised.

Needed technical capabilities

No matter what type of mapping is being done, the crucial element always seems to be the measuring of the quantity to be mapped. Accurate, reliable sensors are needed to convert the physical or biological quantity into some sort of electronic data value. The sensor reading needs to be predictably related to the quantity being measured without being affected by other quantities or environmental conditions. The dynamic response, or speed, of the sensors is also important. Slowly reacting sensors will tend to smooth the data and will result in averaging of the dynamic data. The amount of spatial averaging is proportional to the time averaging of the sensor multiplied by the speed the sensor is moved through the field.

Mapping also requires accurate locators to unequivocally establish the geographical location of the quantities being measured. There are a number of methods, such as dead reckoning and radio-wave tri-angulation, to establish location in a field. But the dominant method in spatially-variable crop production is the use of the Global

Positioning System (GPS), more specifically differential GPS (DGPS). Most precision farming guides for farmers in the USA say that GPS should be used.

Finally, there needs to be some sort of computer system to record and store both the data from the sensor measurements and the location data. Theoretically, these two types of data could be stored on different computers and merged when mapping. But usually the merging occurs during data gathering. If the computer has appropriate display capabilities, the map can be displayed to the equipment operator while the data is being gathered. However, if there is post-processing of the sensor data or of the differential correction for GPS, such a real-time map may not be available in its most accurate form.

Use of maps

The use of soil or crop maps can be categorised into:

1. Information Acquisition
2. Strategic Decision making
3. Tactical Spatially-Variable Control

Information Acquisition provides information on the soil or crop variability to the farm operator. It alerts him to the variations which are in the fields and allows him to better understand the spatial variability of his farm.

The greatest mapping benefit for the least cost seems to be from the Strategic Decision making the farm manager is able to make. Perhaps the manager would be able to change the field boundaries so that the optimum crops were planted in the optimum locations. Or mapping may show that a portion of a field should have artificial drainage installed. The first yield map of Searcy et al. (1989) demonstrated the yield loss due to machinery traffic compaction and that the machinery traffic should be routed around the field. Mapping provides the information necessary to evaluate problems and potential opportunities due to management changes.

Tactical Spatially-Variable Control is the map-based Temporally-Separate control which to most agriculturalists exemplifies precision agriculture. It according is dealt with in the next major portion of this paper.

CONTROL PROGRAMS

Maps document the spatial variability within fields. Field operations are often controlled to mitigate or take advantage of the variability by responding to the variability in soils, crops, or pests.

The most common response to variability has been to control fertiliser application. Fertility levels and yield goals vary within fields. Usually based upon maps generated from interpolation of soil tests, fertilisers are often spread in a spatially-variable manner. Commercial equipment to perform this task has been available for over a decade (Lullen, 1985; Elliott, 1987). The variability in soil organic matter has been similarly used to control the application rate of herbicides which are sensitive to organic matter (Gaultney et al., 1988). Soil moisture conditions may be used to control irrigation or planting depth or population. Of course, soil type and topography maps may also be used to control fertiliser application, pesticide application, tillage (Schafer *et al.*, 1984), irrigation (Wall *et al.*, 1996; Sadler *et al.*, 1996), or planting depending upon what makes the most agronomic sense.

Maps of crop yield, size, or colour can be similarly useful. For example, remote sensing maps of crop stress can be used to guide spatially-variable interventions, if the cause of the stress can be determined. Spatially-variable operations such as patch spraying (Stafford and Miller, 1996), variable irrigation, or variable nitrogen side-dressing can be guided with such information.

Pest infestation maps can be used to guide application. Patch sprayers can also use these maps. Herbicides can be added at needed locations while applying fertilisers. There are many questions as to strategy. Will spraying boundary areas contain a pest? Will maintaining a small population in an untreated field portion delay the development of pesticide resistance? More research is needed to develop both the Strategic and Tactical strategies for all these operations.

Needed technical capabilities

The control of field operations in a spatially-variable manner, such as controlling fertilisation, pesticide application, irrigation, tillage, or planting, is essentially the reverse of the mapping operations. A map in a computer is used to control of the actuation of some piece of field equipment.

The field equipment will need the following items:

1. Control Computer
2. Locator
3. Actuator

The Control Computer will co-ordinate the field operation. It has a map of desired activity as a function of geographic location. It receives the equipment's current location from the Locator and decides what to do based upon the map in its memory or data storage. It then issues a command to the Actuator.

Dynamic response is very important. Otherwise, delays from information transfer and actuator and equipment dynamics will result in a control action being performed long after it was desired. For example, good fertiliser applicators mix and transport fertiliser not based upon the current location of the applicator, but rather upon the predicted location of the applicator when that portion of fertiliser hits the soil (Schueller and Kulkarni, 1988; Schueller and Wang, 1992). Modern microprocessors can easily perform these tasks, if they are programmed with the proper algorithms. Although this type of algorithm (referred to in control theory as feed-forward compensation) is useful, care must be taken in mechanical equipment design to speed dynamic response, thereby reducing potential inaccuracies.

The requirements for location determination in control are similar to those in mapping. However, post-processing cannot be done to remove noise or add differential or other corrections. The accurate location must be available immediately.

The command from the Control Computer must be converted into some change in the actions of the equipment. Usually, this means converting an electrical signal into some sort of displacement or force. Electro-hydraulic or electromechanical systems are typically used. For example, a proportional solenoid hydraulic valve can be stroked to vary the hydraulic oil flow to a hydraulic motor turning a centrifugal fertiliser pump, thereby varying liquid fertiliser flow. The Actuator system must act as commanded and must act quickly. Generally getting the desired mechanical output is not the most significant problem if good mechanical design principles are followed. The greater difficulty is insuring that the mechanical output corresponds to the correct agronomic output, such as insuring accurate control of fertiliser flow at varying compositions and temperatures or of planting depth in variable soil conditions.

CONTEMPORARY SITUATION

Spatially-variable crop production has been commercialised in the USA. Although there is no systematic means of evaluating the extent of its adoption, anecdotal evidence and some limited studies have shown it to becoming increasingly popular with farmers. One major grain harvester manufacturer claimed one-third of its new combines were leaving the factory equipped with yield monitors. A survey commissioned last year by Case-IH found that about 19% of the North American farmers with 750 acres (300 hectares) or more are currently using GPS with 90% expecting to adopt the technology in the next five years (Finck, 1996). AgChem has sold a significant number of their SOILECTION units which variably apply fertiliser.

The most common technology is grain harvester yield mapping. Between the units sold by the combine manufacturers and by after-market suppliers, a large percentage of grain harvesters are capable of yield mapping. The advice of university personnel, the availability of equipment from multiple sources, and the inherent utility and aesthetic appeal of the yield map to farmers have contributed to its popularity. It is now an accepted technology, although tests of accuracy have been called for (Schueller, 1995). It is predicted that about 20,000 grain yield monitors will be operational for the 1997 North American harvest, although not all of them have GPS to allow yield mapping (Mangold, 1997).

Yield mapping for other crops is still a research question. The primary difficulty has been the development of accurate sensors. A load-cell system for potato harvester conveyors and other similar applications is commercially available. Other yield mapping systems are close to commercialisation, but more research is needed for improved performance and additional crops.

Soil fertility maps are widely commercially used to guide the application of chemical fertilisers. Generally these are produced from rather intensive soil samples subjected to laboratory analysis. There is currently a healthy debate about how much other knowledge such as topography and soil type should guide sampling and map generation and also debate over what level of sampling intensity can be economically justified.

There has been commercialisation of remote sensing technology. Some have achieved commercial success, although that success might be considered limited. Issues of timeliness, resolution, ground-truthing, and cost have caused difficulties.

AgChem is the current owner of the Ortlip (1986) USA patent on applying fertiliser according to digital maps. They sell various equipment, primarily based upon their new FALCON controller, to apply fertiliser and pesticide in a spatially-variable manner. There is current patent litigation between AgChem and Tyler regarding Tyler's spatially-variable applicator. Others manufacture equipment which can easily be used or modified to perform spatially-variable application, but explicit claims to do so in the USA appear to be affected by the legal situation.

Controllers are commercially available to vary the planting population. Spatially-variable control of planting variety and depth has been discussed but has received minimal implementation.

Many researchers and developers in both the public and private sectors are working on developing additional technologies for spatially-variable crop production. New or more accurate methods of sensing soil, crop, and pest situations are eagerly anticipated. Other machinery to vary seed, fertility, pesticide, and soil tillage is also under development.

It is difficult to predict where the next innovation will come from. However, since agricultural equipment is not isolated from the general advances in society, its advances may occur in areas of general technological advances. Data communications and networking are proliferating. It is likely that all of a farm's mappers, controllers, and computers will be linked in a wireless network. Advances in computational speeds and storage capacities make it likely that vision techniques will be widely used for soil, crop, and pest determination. Competition between various one-meter resolution satellites and the extending of the internet will bring inexpensive remote sensing data quickly to the farmstead. Improving locator accuracy will finally make widespread controlled traffic a reality.

It should be restated here that spatially-variable crop production can use Automatic control as well as Temporally Separate. In those cases, the Locator and several functions of the control computer (e.g., map storage and Locator correction) can be dispensed with. Of course, that demands an accurate, fast sensor and a known consistent relationship between sensed quantity and desired control action. Perhaps a combination Automatic and Temporally Separate system might prove the best. For example, a nitrogen side-dressing rig in maize (corn) could have stored map data on spatially-variable soil type and other limiting factors. Real-time sensors would measure current crop condition and soil nitrate level and then meter just the right amount of nitrogen according to algorithms which consider the map data.

It should be remembered that there are a variety of means of achieving spatially-variable crop production without the use of electronics. For example, reorienting field boundaries or contour strips on hillsides achieves more uniform areas which can be treated in different manners. Wiping applicators which apply herbicides to tall weeds do so only where the herbicides are needed.

CONCLUSIONS

Spatially-variable crop production, often known as precision farming, has become widespread and has generated widespread equipment and management research and development. Various systems have been developed to generate field maps, especially of crop yield. Such yield mapping is very popular in grain and soybean crops. The spatially-variable application of crop inputs has also had some commercial success. The component engineering technologies include sensors, actuators, locators, and computer control. The technologies for precision farming will continue to advance with general technological progress.

REFERENCES

Auernhammer, H. (Ed.) (1994) Special issue: Global positioning systems in agriculture. *Computers and Electronics in Agriculture,* **11(1)**.

Auernhammer, H., Demmel, M., Muhr, T., Rottmeier, J., Wild, K.J. (1994) GPS for yield mapping on combines. *Computers and Electronics in Agriculture,* **11(1),** 53-68.

Birrell, S.J., Sudduth, K.A., Borgelt, S.C. (1996) Comparison of sensors and techniques for crop yield mapping. *Computers and Electronics in Agriculture,* **14(2/3),** 215-234.

Carr, P.M., Carlson, G.R., Jacobsen, J.S., Nielson, G.A., Skogley, E.O. (1991) Farming soils, not fields: a strategy for increased fertiliser profitability. *Journal of Production Agriculture,* **4,** 57-61.

Cassman, K.G., Plant, R.E. (1992) A model to predict crop response to applied fertiliser nutrients in heterogeneous fields. *Fertilizer Research,* **31,** 51-163.

Cox, G. (1997) Yield mapping sugar cane. World-wide-web page: http://neptune.eng.usq.edu.au/~coxg/ymaping.htm.

Elliott, C. (1987) Fertilizing-blending and spreading on-the-go using computerised soil maps and radar guidance. SAE paper 871676.

Finck, C. (1996) On tap for technology. *Farm Journal,* **112(13),** 24.

Gaultney, L.D., Schueller, J.K., Shonk, J.L., Yu, Z. (1988) Automatic soil organic matter mapping. ASAE paper 88-1607.

Hummel, J.W., Gaultney, L.D., Sudduth, K.A. (1996) Soil property sensing for site-specific crop management. *Computers and Electronics in Agriculture,* **14(2/3),** 121-134.

Lullen, W.R. (1985) Fine-tuned fertility: tomorrow's technology here today. *Crops and Soils,* **38(2),** 18-22.

Mangold, G. (1997) Number of yield monitors. Posting on Email discussion group: precise-agri@Soils.Umn.EDU, March 19.

Nielsen, D.R., Bouma, J. (Eds) (1984) *Soil spatial variability. Proc Int Soc Soil Sci and Soil Sci Soc Am workshop.* Wageningen, Netherlands, PUDOC Publishers.

Ortlip, E.W. (1986) Method and apparatus for spreading fertiliser. USA Patent 4,630,774. December 23.

Robert, P.C., Rust, R.H., Larson, W.E. (Eds) (1993) *Proceedings of Soil specific crop management: A workshop on research and development issues*. Minneapolis, April 14-16, 1992.

Robert, P.C., Rust, R.H., Larson, W.E. (Eds) (1995) *Proceedings of the second international conference on site-specific management for agricultural systems. Minneapolis*. March 27-30, 1994.

Robert, P.C., Rust, R.H., Larson, W.E. (Eds) (1996) *Proceedings of the third international conference on precision agriculture*. Minneapolis, June 23-26.

Sadler, E.J., Camp, C.R., Evans, D.E., Usrey, L.J. (1996) A site-specific centre pivot irrigation system for highly-variable coastal plain soils. *Proceedings of the third international conference on precision agriculture*. Minneapolis, June 23-26, P.C. Robert, R.H. Rust, W.E. Larson, (Eds), pp 827-834.

Schafer, R.L., Young, S.C., Hendrick, J.G., Johnson, C.E. (1984) Control concepts for tillage systems. *Soil & Tillage Research,* **4,** 313-320.

Schneider, S.M., Han, S., Campbell, R.H., Evans, R.G., Rawlins, S.L. (1996) Precision agriculture for potatoes in the Pacific Northwest. *Proceedings of the third international conference on precision agriculture*. Minneapolis, June 23-26, P.C. Robert, R.H. Rust, W.E. Larson, (Eds), pp 443-452

Schueller, J.K. (1988) Machinery and systems for spatially-variable crop production. ASAE Paper No. 88-1608.

Schueller, J.K. (Ed-Site Specific portion) (1991) *Proceedings, Automated Agriculture for the 21st Century*. Chicago, Illinois, December 16-17.

Schueller, J.K. (1992) A review and integrating analysis of spatially-variable control of crop production. *Fertilizer Research,* **33,** 1-34.

Schueller, J.K. (1995) We need "Nebraska Tests" for spatially-variable agriculture. *Resource,* **2(5),** 40.

Schueller, J.K., Kulkarni, R.S. (1988) Spatially-variable fluid fertiliser mobile applicator design concepts. SAE paper 88-1290.

Schueller, J.K., Mailander, M.P., Krutz, G.W. (1983) Computer control of combine forward speed and rotor speed. *Proceedings, ASAE Conference on Agricultural Electronics Applications*. Chicago. December 11-13, pp. 99-108.

Schueller, J.K., Wang, M.W. (1992) Spatially-variable fertiliser and pesticide application with GPS and DGPS. *Computers and Electronics in Agriculture,* **11(1),** 69-84.

Searcy, S.W., Schueller, J.K., Bae, Y.H., Borgelt, S.C., Stout, B.A. (1989) Mapping of spatially-variable yield during grain combining. *Transactions of the ASAE,* **32(3),** 826-829.

Stafford, J.V. (1988) Remote, non-contact, and in-situ measurement of soil moisture content: a review. *Journal of Agricultural Engineering Research,* **41,** 151-172.

Stafford, J.V. (Ed) (1996) Special issue: Spatially variable field operations. *Computers and Electronics in Agriculture,* **14(2/3).**

Stafford, J.V., Lebars, J.M., Ambler, B. (1996) A hand-held data logger with integral GPS for producing weed maps by field walking. *Computers and Electronics in Agriculture,* **14(2/3),** 234-247.

Stafford, J.V., Miller, P.C.H. (1996) Spatially variable treatment of weed patches. *Proceedings of the third international conference on precision agriculture.* Minneapolis, June 23-26, P.C. Robert, R.H. Rust, W.E. Larson, (Eds), pp 465-474.

Wall, R.W., King, B.A., McCann, I.R. (1996) Centre-pivot irrigation system control and data communications network for real-time variable water application. *Proceedings of the third international conference on precision agriculture.* Minneapolis, June 23-26, P.C. Robert, R.H. Rust, W.E. Larson, (Eds), pp 757-766.

Walter, J.D., Hofman, V.L., Backer, L.F. (1996) Site-specific sugarbeet yield monitoring. *Proceedings of the third international conference on precision agriculture.* Minneapolis, June 23-26, P.C. Robert, R.H. Rust, W.E. Larson, (Eds), pp 835-844.

MANAGEMENT FOR SPATIAL VARIABILITY

C.J. DAWSON

Chris Dawson & Associates, Westover, Ox Carr Lane, Strensall, York, YO3 5TD, UK

ABSTRACT

Precision Agriculture is defined as more precise farm management. It must continually assess the management requirements of practical agriculture. Simple, cost-effective systems are required despite complexity in their development. Automatically generated yield and quality maps provide the economic and environmental justification for spatially variable management. Non-manual measurement of critical influencing factors is required on farm, both for the identification of areas where yield is limited by controllable factors, and for the determination of intrinsic spatial variability. The opportunity to carry out spatially variable field operations is highlighting the need for better agronomic knowledge, although good husbandry is applicable at all scales and many variable treatments are already being made by farmers. The possibilities being presented by Precision Agriculture are much greater than spatially varied applications, and include re-evaluation of trials procedures, plant breeding programmes and on-farm testing of different cultivation practices. Precision Agriculture offers a potential step-change in productive efficiency.

INTRODUCTION

Precision Farming is essentially more precise farm management, made possible by modern technology, principally through the measurement and consideration of the spatial variability which can exist within the traditional unit of management, the field. No industry can afford to ignore significant variability in a supposed uniform resource.

Precision Farming is a management philosophy or approach, and is not a definable prescriptive system. It does not necessarily demand that specialist machinery is purchased, nor that the farmer has an on-line connection to the local University, although it will probably require special tools and resources or modified equipment in order to take advantage of the opportunities presented. In essence, it is accepting that the more precise the information available, the greater the opportunity for best practice. It can be seen as an agricultural equivalent of the Total Quality Management systems which have been adopted to varying extents in other industries (Searcy, 1994).

The arable agricultural management being practised in Europe today is already relatively precise. Whilst the Conference addresses Precision Agriculture, it should not be inferred that current systems are seriously inaccurate. We are developing more precise systems which will allow both existing and new management expertise to operate at a higher resolution. Existing systems may operate at the farm level, with a resolution of 1000 m, or at the field level of 100 m. We are opening up the opportunity of considering a within-field resolution of 10 m, or even less than 1 m.

However, farmers can be grouped into categories which have differing attitudes to Precision Farming. In the early stages of development of such a concept, farmers who have an enthusiasm for new ideas will be in the forefront. These are valuable to the researcher and developer but are not typical of the majority of farmers and care should be taken not to develop products and services for them without considering the wider market.

A common attitude of a majority of farmers can be illustrated by analogy with the development of the motor car. At the turn of the century there were several workers who were designing motorised vehicles, and alongside them were a number of enthusiastic users. These users were content with the fact that they needed to understand how the motor car worked in order to be able to keep it running, while the majority of the population saw the car as a complex and unreliable novelty which was unlikely to be of any long term use. Now of course the car is considered indispensable by most in the developed world, and no-one considers it necessary to understand the workings of the machine; it is simply required to go from point A to point B.

Perhaps Precision Farming can learn from the experiences with the early motor car! If the time-scale of adoption is to be shorter than that of the car, then the simplicity of the concept should be emphasised. Public discussion of the early technical developments might indicate that Precision Farming is only for the technophiles!

It is essential to remember that the ultimate purpose of current research, much of which is newly reported here at this Conference, is to provide assistance to the agricultural manager - the farmer. This is to help with his agronomic, economic and environmental decision-making. It must be designed to provide this assistance without significantly increasing the management burden, but rather to simplify or decrease it. It should aim to improve and not hinder time management during peak activity periods (Schueller, 1996).

Some large-scale farmers will perhaps have divided the land being farmed into groups of fields which are treated similarly. The justification for this approach is that field averages of variables which might be expected to affect crop performance were such that the manager decided that there was little or no advantage to be gained by treating the fields individually. There can be no argument with that decision, given that it was taken before Precision Farming was an option. The key point is that the decision was made following measurement of the variables. Similarly, the farmer taking a Precision Farming approach might well consider the within-field variability of the measured factors to be of low significance. He could therefore continue to manage the field uniformly. However it can no longer be considered good practice to assume the field to be uniform, without first taking measurements of within-field variability of critical factors.

From the farm manager's point of view the key questions which have to be asked are:

❏ Is there significant spatial variability within field?
❏ Does it affect economic or environmental performance?
❏ Can the cause(s) be identified?
❏ Can quantified measurements be readily made?
❏ Can worthwhile actions be taken?
❏ Will these actions be likely to improve performance?
❏ Can the results of these actions be monitored?

46

The papers and discussions at this Conference address one or other of these questions. They can be considered as the framework for the management of spatial variability, i.e., Precision Farm Management.

VARIABILITY WITHIN-FIELD

The factors which might be considered when investigating variability within field are generally the same as those taken account of when assessing variability between fields. The fundamental difference is that for the latter, the average value of the variable over the whole field is measured; for the former, values for points or relatively small areas are required.

Two particular facts are usually apparent when a series of individual measurements are made within the field:

❑ Almost every measured parameter varies to a considerable extent, and

❑ This variation is usually greater than that observed between the average whole-field values for different fields.

Agronomists and advisors are prepared to make definitive recommendations for crop management which take account of the differences between the average values of a range of parameters measured for separate fields. However there appears to be a general reluctance to make specific within-field variable recommendations according to the measured values of the same parameters. In part this is due to a lack of definitive quantified relationships, but there can be little doubt that the known but unpredictable temporal variability, combined with an expectation that average recommendations might be less risky, plays a significant part.

The factors which can vary within the supposedly uniform field can be divided into two groups:

❑ Resources, including:
    Topography and aspect of the field.
    Soil physical characteristics.
    Soil nutrient status.
    Weeds, pest and diseases.

❑ Outputs, including:
    Yield.
    Quality.
    Environmental impact.

In considering relevant variation within-field and its causes, these two groups should perhaps be in reverse order, since it is primarily necessary to measure the output variation and then to look for relationships with the measured resources. It is considerably less satisfactory to measure the variation in the resources and then, having attributed an assumed effect on output, to take action to address the variation in resource.

47

THE EFFECT ON PERFORMANCE

The prime question is whether there is significant variability in the performance of the crop in the field, in terms of yield, product quality and/or environmental effect. In general, the answer to this question is positive, but it is not always so. If there is little significant variability then there will probably be little justification in any variable treatment or management of the field.

Prior to offering management advice, it is necessary to know how the field is performing. Without a yield map, discussion of the variability of resources relies upon a manager's knowledge. However it is not economically viable for a farmer to address the variability in a potentially influential factor without knowing whether in fact it is actually having a significant effect on the performance of the crop.

Figure 1 illustrates the variability in yield which is frequently seen in fields of winter wheat in the UK - in fact the less productive headlands of this field are already in non-rotational setaside resulting in the variation shown being less than for the whole field. Nevertheless approximately 12.5% of the field is still yielding at least 1.0 t/ha less than the average yield of 9.6 t/ha. As margins become tighter, the precision farm manager is asking what could be causing these lower yields, and whether he can take any remedial or other action which will improve his margin.

In addition to the variability in yield, which can be quantified both in terms of tonnes and financially, significant variation in several quality parameters have also been identified.

Agricultural crops for which yield and quality maps have been generated, include cereals, oilseeds, pulses, potatoes, sugar beet and cane, cotton and forage crops.

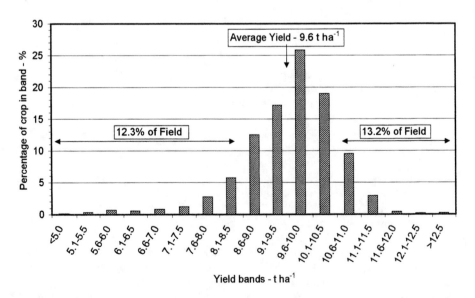

FIGURE 1. Frequency distribution of grain yield bands in a winter wheat crop in 1996. Massey Ferguson yield mapping combine. (Source: Shuttleworth Farms).

From the point of view of the farmer, for whose benefit the tools of Precision Farming are being developed, the illustration of significant variability in yield and crop quality is a powerful inducement to consider further investigation of the opportunities for improved management precision. However he is concerned that the expertise and technology required is very complex and this can have a negative influence on his enthusiasm.

CAUSES OF VARIABLE PERFORMANCE

When confronted with a yield map the grower's first question is to ask for a reason for the yield differences. Unfortunately it is rare that there is a simple answer. Yield variability usually results from a combination of factors. Similarly, any lack of consistency in the pattern of yield from season to season will be queried. It might be expected that high yielding areas would be spatially relatively stable in the field year after year for similar crops; it would be reasonable to assume that they were high yielding because there were few limitations to high productivity. On the other hand, areas of low yield may be more 'mobile' since different potential limitations to yield will occur in different areas in different years. For example, in a wet season, a heavy soil may be affected by waterlogging, resulting in a negative effect on yield. On the other hand in a dry summer, that same part of the field might perform relatively well due to better water retention. There is a need for programs to provide some statistical resource which allows simple objective comparisons to be made between maps from different seasons or data on different measured variables.

The causes of yield variation can be divided into two categories:

❏ Those over which the farmer has some control, such as:
   Soil compaction.
   Soil pH.
   Field drainage.
   Soil nutrient status.
   Weeds, pests and diseases.

❏ Those 'intrinsic' variables over which he does not have control, such as:
   Soil type.
   Soil depth.
   Soil water.
   Slope.
   Aspect.

There are many factors which can affect relative variability of yield and quality. The quantity of available water which is largely a function of soil type and depth probably dominates for non-irrigated crops in Europe. However, before the farmer considers the management decisions appropriate to variability caused by such intrinsic drivers, it is essential to identify and eliminate any effects on performance caused by factors over which he does have control.

From a management point of view, therefore these two categories can be distinctly separated. The objective with the first group, is to remove avoidable limitations on yield and performance. The second group however provides the on-going challenge of managing the influence of intrinsic variability and of optimising gross margin and overall field performance.

The yield map is particularly useful as an aid to directing in-field investigation of reasons for poor performance. Not only are the locations of areas of low yield readily identified, but the reduction in income from those areas is also apparent. The yield map may for this purpose, be termed a 'management' map, since it indicates areas where investigation might most profitably be carried out. To a considerable extent, it can reduce the need for complete field maps of influential variables. In practice, areas of low soil pH, soil compaction and weed infestation have commonly been seen as immediate and correctable causes of yield depression.

The causes of variability in crop performance within-field are unlikely to be any different from those which cause variability on a larger scale between fields. It is unlikely that new influential factors will be found. The farmer or agronomist will thus be aware of the factors which merit investigation as possible causes of yield or quality limitation.

Two primary questions arise in the assessment of a potentially limiting factor:

❑ How can the 'value' of the variable be economically measured, and
❑ Is a critical limiting 'value' for the variable known.

Without the answer to these questions, it will not be possible to define the variable factor as limiting or otherwise at the point of measurement (Dawson and Johnston, 1997; Oliver *et al.*, 1997). This principle is more readily applied to soil chemical characteristics such as nutrient status than to physical factors such as compaction. Nevertheless it is probable that only by attributing quantitative values to variable factors will it be possible to manage their variability without undue time and effort being involved. And time and effort must be considered if Precision Agriculture is to offer benefits without unreasonable extra management input.

It is a considerable problem that few, if any, of the manageable factors which affect crop performance can be measured both quantitatively and automatically by commercially available equipment.

MEASUREMENT OF VARIABILITY

Since the introduction of the yield monitor, which measures crop yield, there has been considerable research activity to develop systems which can measure variable factors and also different aspects of crop output quality. The main questions are what to measure, and how to measure it (Stafford, 1996).

It is difficult to envisage manual data collection of key information for a Precision Agriculture system. The data are expensive to collect and therefore generally limited in quantity, thus requiring that estimates be made of the values at the points or areas not sampled. While the geostatistical interpolation required for such estimation is relatively well established, the variograms required are difficult to develop. Thus although some use of manual measurement is reasonable at this stage in the implementation of precision agriculture, it is unlikely to be acceptable in the long term except for factors which do not change significantly with time, such as soil clay content, depth or other physical characteristics.

Where automatic measurement of a specific variable parameter would be difficult to achieve, the use of a surrogate for that variable may offer an alternative, such as the measurement of chlorophyll in lieu of leaf nitrogen.

Before reliable measurements can be taken, it is essential that the method and frequency of sampling is established as being appropriate for the specific variable, and that the error or variance is both known and acceptable. There can be little doubt that many data sets have been collected which have large and unknown errors. It is to be expected that data on different variables, would require different sampling protocols. It is important that the data, on which management decisions are to be based, are collected using established and standard procedures.

There is considerable interest in the use of remote sensing of the growing crop as a means of identifying areas of differing performance. At the present time, processed images present a useful indication of the pattern of variability at the time of capture, and may thus be of value as a guide to physical inspection of the crop. It may be some years before such images have been sufficiently validated to allow quantitative estimations to be derived from them.

Measurement is thus required of both manageable and intrinsic factors which can cause variation in crop performance. For temporally stable factors, such as soil depth, greater cost may be acceptable in the collection of the data, since it will only be collected once. However even in this category, the cost may only be justified for those variables which are known to have a potentially significant effect. Those responsible for the management of variability require the identification of significant spatial variables, and their classification in order of potential effect.

The current requirement is for researchers to measure potentially influencing factors, in order to establish their relative importance, and to develop correlations with crop performance. This information will indicate where development in automatic data collection is required, and whether the data will be required annually, infrequently or only once by those managing the variability.

It is clear that the management of spatial variability requires an understanding of inter-relationships between different variable influencing factors and also the effects of seasonal differences. The use of the traditional replicated trial as a research tool may no longer be generally appropriate, with other investigative procedures, involving the new sensing and location technologies, becoming more appropriate to the needs of Precision Agriculture.

FEASIBLE ACTIONS

The first action which must be taken in the commercial management of variable crop performance is the removal of manageable limiting factors. Only when these have been identified, and corrected, can management of the intrinsic or long term variability be considered. The areas in which yield may be limited are difficult to identify without the yield map, probably requiring whole-field grid sampling for any causative factors to be found. This may incur an unreasonable cost although it is likely to be worthwhile for the measurement of soil pH. This major factor will require monitoring on a regular ongoing basis, and there is a real need for automatic, or semi-automatic measurement of pH.

The actions which may be taken can be divided into those which are required to correct limitations to performance, and those which would be expected to be carried out on a continuing basis as part of the management of intrinsic variability. This latter group involves simple or automatic measurement and must be readily implemented using systems which are simple to manage.

Fertilisers: phosphorus and potassium

The value of the yield map as a management guide has already been mentioned, and it is evidently useful as an indicator of areas of yield depression possibly caused by manageable factors. This is particularly so for potential limitation by soil phosphorus reserve. It has been found in the UK, where arable soils have generally been fertilised according to average crop removal and soil status for many years, that there is frequently a negative relationship between yield and extractable soil phosphorus. This may be quite readily explained if the areas of higher yield in the field tend to be temporally stable. The removal of phosphorus in the harvested crop in these areas will be greater every year than the rate of replacement for the average yield. In these circumstances, the limitation to crop performance may be found in the areas of highest rather than lowest yield.

A strategy for soil sampling for phosphorus, (and also potassium), may thus be designed to measure initially the values at the highest and lowest yield points in the field. If variability in soil P or K status is being driven by variable crop off-take, then this should become apparent. If low or limiting values are found at sampled points, then further grid sampling will be required to determine the extent of the deficient areas. This strategy can reduce the number of samples which need to be taken to obtain an indication of where P and K are limiting. Such a strategy would apply for soils which are reasonably well buffered for P and K status and where sound whole-field fertiliser policies have been practised for some time.

In these historically well managed soils, there does not appear to be good association between patterns of different soil types and patterns of available P or K reserve, except where there are very significant differences between soil types. This may be contrary to expectation, but it is probable that fertiliser applications to relatively well buffered soils over-ride potential differences over the years.

Another factor which requires consideration is the effect of the depth of ploughing, and of any sub-soil enrichment. Standard soil analysis reports nutrient status in terms of concentration to the depth of sampling. The total depth of nutrient enriched soil must also be known (usually the depth of ploughing) to determine the quantity of nutrients available. Relationships between total available soil nutrient and crop performance may exist which are not apparent if only the concentration is measured.

If limiting nutrient status has been identified and corrected, (probably by specific local application of high rates over a period of a few years), continuing variability in nutrient reserve is not of great significance. The critical issue is whether or not the factor is limiting.

Having removed such areas of limitation, the fertiliser policy can be driven solely by the need to replace nutrients removed in the harvested crop, (in theory without the need for further soil analysis). The yield map gives a direct indication of phosphorus and potassium off-takes and thus maintenance fertiliser applications, designed to sustain the fertility of the soil, can be variably applied. The ratio of P:K in the fertiliser will be constant within the field for the crop being grown, and the variable application can thus be made using a standard fertiliser appropriate to the crop grown, with simply the rate being varied. This fertiliser may be applied retrospectively for the majority of arable crops, although the policy may need adaptation where nutrient responsive crops such as potatoes are grown in the rotation. Also there may be areas of high reserve within the field, possibly associated with areas of low yield, which may justify reduced or no further nutrient applications for both financial and environmental reasons (Dawson, 1996).

Fertilisers: nitrogen

Of all of the variable inputs, nitrogen is perhaps the most critical. It is also one of the most difficult to manage, except in retrospect! There are within-field variations in the quantity of nitrogen used by arable crops, and there is therefore a desire to vary the input according to apparent need. The difficulty is that the poorer yielding areas of a field are not low yielding through shortage of nitrogen, but rather through a productive inefficiency. Thus because of this inefficiency more nitrogen, and other inputs, must be available per unit of output than in the higher yielding areas. Unless it is possible to improve the efficiency of utilisation of resources in areas of low productivity, and to do this cost effectively, the opportunities for reducing the application rate of nitrogen may be limited. The economic response to nitrogen is generally high, and the temporal variability in yield is also relatively high, and unpredictable. Thus, the incentive to reduce nitrogen application in areas of poor performance is not great, although it is seen as an obvious goal.

The corollary is that the efficiency of utilisation of inputs is relatively high in the higher yielding areas. It may be possible that additional yield is economically achievable through the application of extra nitrogen. However, it may be that the amount being applied, (presumably appropriate for the anticipated average yield), is adequate due to above-average efficiency. If these areas of high yield are temporally relatively stable, it is also reasonable to expect that soil organic matter levels may be higher than average, due to the greater production of biomass annually. This could lead to a relatively greater level of mineralised nitrogen in these higher yield areas which would also contribute to the greater quantity of nitrogen removed in the crop at these sites.

It is possible to envisage a scenario in which the most cost effective and risk reducing strategy for nitrogen application is one in which nitrogen application is not varied within the field, despite significant variations in yield, and differences in nitrogen removal of up to 100 kg N ha^{-1} between different parts of the field. More fertiliser than is calculated as being necessary would be applied to the lower yielding areas because of the lower efficiency, and vice versa in the high yield areas, leading to uniform application. The unpredictable difference in actual nitrogen response between seasons may be greater than the spatial differences in off-take within season, further complicating nitrogen fertiliser recommendations and influencing assessment of risk.

The calculation of nitrogen requirement of arable crops frequently includes the quantity of available mineralised soil nitrogen, as estimated from soil analysis. If grid point samples are required, the number of sub-cores at each point must be determined, and a calculation of the number of grid points made. Early indications from current unpublished research suggest that a higher grid density is required than that considered acceptable for soil PK analysis, and this would seem to preclude manual grid sampling due to high cost. There may be a need for an automatic soil N measurement system providing robust data. Alternatively, the growing crop might be used as an indicator of soil N, and measurement be made of its nitrogen status or some related variable. There is also a need to investigate the reasons for any spatial variation in soil mineral nitrogen, and to consider the possibilities of modelling, rather than measuring, the variability.

A strong negative spatial correlation is frequently found between grain yield and grain protein concentration in winter wheat in the UK. If the areas of high yield are seasonally relatively static, as might be expected, then areas of potentially low grain protein concentration can be predicted in advance. It is also known that late applications of fertiliser nitrogen, (probably as a foliar application of aqueous urea), can raise the protein content of wheat grain. Thus the farmer has the opportunity for selective application of additional nitrogen to areas of high yield in order to raise the quality of the grain being produced, without the application being applied to areas of lower yield where the grain already has an adequate protein status.

Soil structure and cultivations

Areas of potential limitation of crop performance caused by soil compaction and other structural damage can be addressed with greater confidence, since the financial effects can be inferred from the yield map. This map assists by identifying areas in which measurement of soil compaction may be worthwhile, in order to determine whether specific cultivations should be carried out. Equally, the financial penalties of soil structural damage are apparent, and strategies can be implemented which will reduce the probability or extent of such structural damage recurring. Measurements of compaction are currently relatively expensive as they are made manually, and again some automatic assessment of soil compaction would appear to be a useful management aid.

It has been said that few of the actions which may be taken in order to implement a Precision Farming policy will in fact be new, but rather that they will be better focused and therefore more effective both financially and environmentally. Husbandry practices which were effective or 'good' on the field scale will also be good when variably applied for specific reason within the field, but the cost effectiveness should be improved. To an extent, such variable within-field actions are already taken, with the limitation being the extent to which such detailed treatments are manageable. Precision Agriculture brings technology to assist with this management, and allows detailed treatments to be managed on a large scale.

Where soil types vary within the field, it can be envisaged that there would be benefits from being able to vary cultivations which are designed to create a uniformly suitable seedbed (Stafford and Ambler, 1990), such as by changing the offset angle of a disc cultivator according to the soil type.

Seeding

Similarly, the quantity of seed sown per hectare, and the depth to which it is drilled are traditionally varied according to the quality of the seedbed which has been prepared. Within Precision Agriculture management systems, the seedrates of winter wheat and other crops have already been automatically varied to apply extra seed in areas of predetermined poorer seedbed within the field, with apparently beneficial results, using a variable rate drill linked to DGPS (Unpublished, Shuttleworth Farms, 1994).

As different varieties of the same species of a crop perform differently depending on the fertility and yield potential of a soil, it would be feasible to sow more than one variety of a crop in a field with known variable performance. It would be essential that all varieties can be managed together and will mature at similar rates. A variety whose full potential can be realised in the highest yielding situations may well be justifiably accompanied by one which gives a better potential performance in poorer areas. Similarly seedrates may be varied in order to optimise performance from areas of different yield potential, and also to take account of areas in which limiting water supply could justify reduced crop density.

Agrochemical use

The significant difference between agrochemicals and fertilisers is that a precision agriculture system seeks to minimise agrochemical use but to optimise nutrient application rates. The broad objectives for precision application would therefore be whether or not to apply an agrochemical, whereas for fertilisers the decision would concern an appropriate rate.

The clearly evident patchy nature of weed infestations has already led to the variable application of herbicides (Stafford and Miller, 1996). Optical sensors have been developed which are intended to assist in the identification of weed infestations, with the objective of simplifying the management of variable weed control. In general, variable application of agrochemicals is restricted to the control of spatially dependent pests such as weeds, cyst nematodes and soil-borne fungi, rather than to pests which are more temporally variable, such as aphids and foliar fungal disease.

Since the economic and environmental objectives of reducing agrochemical use are similar, it is anticipated that considerable progress will be possible in this area. The limitation is again likely to be the absence of automatic quantitative data collection, although the economic benefits of precision application of nematicides already justify manual sampling for measurement of levels of infestation.

General management

The extent to which general management decisions are influenced by Precision Agriculture will depend upon the management style and objectives of the grower. Nevertheless, considerable financial advantage has already been gained by some who have allocated their non-rotational setaside with the assistance of the yield maps, and thereby have improved their overall yield per cropped hectare. Some have noted that, even without setaside, there are identifiable areas which would be unlikely to justify cropping, and that the presence of such uncropped areas, (usually certain field headlands), also improves the efficiency of management activities within the field. The identification of areas of potentially good crop performance allows more precise and appropriate cropping plans, particularly when responsive high-value crops are included in the rotation.

IMPROVEMENT IN PERFORMANCE

The justification for adopting a Precision Agriculture approach for the individual grower is that it offers the potential for improved production efficiency and overall profitability. Unless the adoption of such an approach provides these results, and in a way which is relatively simple to manage, then it will not become common practice.

Some of the possible actions which the commercial farmer might take have been discussed in previous sections, but these were based mainly on the use of existing management expertise, and not on the use of models. There is no doubt that models of probable performance, based on measurement and experience, will become a major feature of a developed use of Precision Agriculture. Unless such models are available, the management decision making that will be required will be unreasonable if not impossible. For a given set of variables and conditions, certain decisions will be made, and these may be specific for a farmer. At its simplest, these decisions will be recorded and may be offered as the default decision when the set of conditions recur.

As the principal drivers of intrinsic variability are confirmed within field, and the annual performance of the crop monitored, then relationships between these can be developed. The key question for the practical farmer is to know what needs to be measured, and how, and to what extent these measured variables affect crop performance, to assist him in the initial planning and subsequent day-to-day management of the crop. A model can be envisaged in which the main drivers of spatial variability are perhaps soil type and depth, and from which a 'standard' performance map can be constructed for given weather conditions, assisted by a series of yield maps. The temporal variability in performance, both within season and between seasons, will be driven by the weather, which can be measured continually. Thus the model can be run at any time during the season to predict the state of the crop at that time. This model can be compared with a remotely captured image of the crop, illustrating perhaps NDVI or some other indicator of crop performance. By comparing these two information sources and identifying any areas which appear to show differences, the farmer will be in a position to investigate these areas which are not performing to expectation and thus focus his management effort.

Agronomists are frequently asked whether trials have been carried out on variable within-field treatments, despite the fact that the treatments are already standard practice on a whole-field basis. There seems little logic to this question as the benefits of good husbandry should not need to be reassessed simply because the resolution changes. Indeed much of the initial research, for example into the beneficial effects of appropriate levels of nutrients or soil pH status, was carried out at plot level. If it is known that a low average pH in a field is unsuitable for a crop, then it is equally unsuitable in patches in the field. The question as to whether there is a financial or environmental advantage is more difficult to demonstrate by means of trials, and may require the monitoring of trends in order to accommodate seasonal variability.

56

EVALUATING THE RESULTS

The measurement of improvement in performance resulting from a change in management systems is often difficult in the short term, and this is particularly so for agricultural systems which are so affected by seasonal weather differences. Furthermore, there are often several unexpected, and usually unquantifiable, benefits which are generated by a more focused management approach. Recent estimates suggest that net profit could be improved by about 10% for a range of arable and forage crops, excluding any additional costs for special equipment (van Kraalingen, 1997).

The assessment of an improvement in economic performance through the introduction of Precision Agricultural practices is sought by those demonstrating the use of special techniques, services or equipment, and this frequently involves the use of alternate strips of variable and uniform treatments across a field. This demonstration of a possible effect within a field may be readily understood and perhaps helpful, but it cannot demonstrate or evaluate the potential benefits from the whole-farm adoption of Precision Farming. There is a need for multi-season whole-farm effects to be recorded, from which economic and environmental changes could be measured.

Those farmers who have yield mapping, and who have introduced some aspects of Precision Farming, are able to use the annual yield map to record changes in performance of yield, quality and gross margin of the individual field. These growers have also, in many cases, used the yield mapping/recording facility to investigate for themselves different treatments, for example of cultivations and agrochemical use, within the field, by treating alternate tramline-separated strips differently. From this experimentation they are building a resource of particular information upon which they are or will be able to base their variable decision-making. They are, in effect, building their set of defaults for a specific management model.

However, in addition to the measurement of the effects in terms of business performance, there are significant potential benefits which can result from a more precise measurement of the inputs and outputs. Alongside the development of automatic or managed input instructions for application equipment, there is the opportunity to record in detail the actual applications made, and for these to form part of a record of the crop production. This information may be of value in demonstrating conformance with legislative or environmental requirements, and may equally be of value in the context of traceability of the crop product.

Automatic measurement of variations in crop quality, in addition to the yield measurement, will allow not only the ability to keep records for traceability, but also to select or separate product with specific quality criteria from within the field as it is harvested. As consistency of products manufactured from agricultural raw materials becomes increasingly important, the ability to offer consistent raw materials of known specification becomes a valuable marketing advantage.

CONCLUSIONS

The concept of Precision Agriculture presents a considerable challenge to those involved in its development, and apparently a considerable opportunity to those prepared to implement it. There is a widespread belief that it offers developed and developing agriculture the prospect of a step-change in productive efficiency, combining both a potential reduction in unit cost with an improvement in product quality. The underlying requirements for measurement and recording have demonstrated themselves to be the basis for major improvements in other industries, and it is not unreasonable to consider Precision Agriculture to be a revolutionary opportunity for farming.

REFERENCES

Dawson, C.J. (1996) Implications of precision farming for fertiliser application policies. *Proceedings of the Fertiliser Society*, **391**, 43 pp, Peterborough, UK.

Dawson, C.J., Johnston, A.E. (1997) Aspects of soil fertility in the interpretation of yield maps as an aid to precision farming. *Proceedings of the First European Conference on Precision Agriculture*, J.V. Stafford (Ed.), London, SCI.

Oliver, M.A., Frogbrook, Z., Webster, R., Dawson, C.J. (1997) A rational strategy for determining the number of cores for bulked sampling of soil. *Proceedings of the First European Conference on Precision Agriculture*, J.V. Stafford (Ed.), London, SCI.

Schueller, J.K. (1996) Short communication: Impediments to spatially-variable field operations. *Computers and Electronics in Agriculture*, **14**, 249-253.

Searcy, S.W. (1994) Engineering systems for site-specific management: opportunities and limitations. *Proceedings of Site-specific Management for Agricultural Systems - 2nd International Conference*, Madison, WI, USA, American Society of Agronomy, pp 603-612.

Stafford, J.V., Ambler, B. (1990) Computer vision as a sensing system for soil cultivator control. *Proceedings of the Institution of Mechanical Engineers*, **C419**, 123-129.

Stafford, J.V., Miller, P.C.H. (1996) Spatially variable treatment of weed patches. *Proceedings of the 3rd International Conference on Precision Agriculture*, Minneapolis, USA, June 23-26. Madison, WI, USA, American Society of Agronomy, (in press).

Stafford, J.V. (1996) Essential technology for precision agriculture. *Proceedings of the 3rd International Conference on Precision Agriculture*, Minneapolis, USA, June 23-26, Madison, WI, USA, American Society of Agronomy, (in press).

Van Kraalingen, D.W.G. (1997) Opportunities for Precision Agriculture, Nota 43, AB-DLO, Wageningen, The Netherlands.

SPATIAL VARIABILITY
IN SOIL AND CROP

FROM SOIL SURVEY TO A SOIL DATABASE FOR PRECISION AGRICULTURE

J. BOUMA

Department of Soils, Agricultural University, Box 37, 6700 AA Wageningen, The Netherlands

ABSTRACT

Existing soil databases derived from published soil surveys are not directly suitable for application in the context of precision agriculture because 'representative' soils for mapping units are not georeferenced. Soil behavior in space and time is important for precision agriculture and the soil water regime plays a central role. Soil data for precision agriculture is to be derived from a specific survey based on point observations, the spacing of which are determined by geostatistical analyses. A database with such primary soil data is used to obtain basic input data for simulation modeling by tapping into a general database with pedotransfer functions. Grouping of simulated results for crop yield and solute fluxes for any given soil series is advocated, to allow a preliminary assessment of possibilities as soon as the different soil series occurring within a farm are known.

INTRODUCTION

Soil surveys are widely available in many countries. Soil maps are accompanied by soil survey reports with analytical data for 'representative' soil profiles for each soil unit on the map. In addition, interpretations are provided for different types of land use in terms of relative suitabilities or limitations. Such interpretations are useful when made in a regional context when broad evaluations are needed covering a wide variety of soils. However, more specific interpretations are needed when dealing with precision agriculture where applications of agrochemicals and proper timing of tillage, soil traffic and sowing and planting are of prime interest. Clearly, statements as to the relative suitability of a given soil for a given form of land use are less relevant in this context. Even though information on land suitability from soil surveys is irrelevant for precision agriculture, soil data as such are very important because soil conditions determine to a large extent the type of management measures to be made in precision agriculture.

One additional problem of 'representative' soil profiles is the lack of georeferencing. A soil is considered to be representative for areas on the soil map, ignoring internal variability. This certainly may be a problem on small-scale maps (1:50000 and smaller) but also at large scales, such as 1:5000. Aside from the problem that spatial variability is poorly represented by 'representative' soil profiles, we have to deal with the fact that data traditionally provided in soil surveys cannot directly be used when dealing with questions on precision agriculture. For instance, soil texture and organic matter content have implications for soil behavior but they cannot directly be used in models predicting such behavior. Aside from making soil observations, we have to make a 'translation' towards a functional rather than a pedological characterization. In dealing with these problems, we have defined a series of procedures that allow distinction and use of soil data in the context of precision agriculture to be focused on agricultural fields, as follows:

(i) make a series of point observations with spacings that reflect the variability of the area to be characterized. Sometimes clearly different subareas, based on depositional or soil-forming factors, can be distinguished within a field. Then, point observations should be defined for each subarea;

(ii) make soil observations for each point and transform these observations into data that can be used for feeding expert systems or simulation models of crop growth and solute fluxes;

(iii) run the models in both a retrospective and a predictive mode, focusing on both historic weather data for periods of thirty years or more and on weather conditions for the season to come, using weather generators or stochastic procedures;

(iv) interpolate point data to obtain expressions for subareas of the field, taking into account occurrence of natural soil patterns, and:

(v) define a database for primary and secondary soil-data to be used in decision support systems.

Each of these elements will now be briefly discussed. The reader will be referred to more detailed source publications.

DETAIL OF WORK

In the introduction the suitability of traditional land-use interpretations based on soil surveys for applications in the context of precision agriculture were discussed. This, however, does not mean that soil survey data would not be useful. The level of detail by which the data are to be generated has to be considered. This level will strongly depend on the type of problem to be studied. The diagram in Figure 1 illustrates the choices to be made (Hoosbeek and Bryant, 1992; Bouma and Hoosbeek, 1996; Bouma, 1997). Approaches range from qualitative to quantitative on the one hand and from empirical to mechanistic on the other. Different scale hierarchies are distinguished, ranging from the molecular level to the world level. Hierarchies are indicated in terms of an i code, the i level being the level of the individual soil (the pedon). Smaller scales are indicated with an i- code, higher scales with an i+ code (see Figure 1). The numbers in the figure, which are distinguished at each hierarchial scale, indicate users' expertise (K1); expert knowledge (K2); use of simple models (K3) in which soil processes are not explicitly represented; use of more complex models in which processes are mechanistically expressed (K4) and complex models in which certain aspects of soil behavior are studied in detail (K5). When dealing with general land use questions in a large area, traditional soil survey interpretations in terms of, e.g., relative suitabilities for different crops at level K2 or K3, are very suitable. However, precision agriculture requires a more quantitative approach at level K4 or K5. In fact, procedures being followed in practice are more complex, because activities at different hierarchial scale levels have to be distinguished as well (see Figure 1). At field (i+1) level a K2 approach will be valuable in distinguishing natural soil patterns on the basis of which point observations can be made most efficiently. At pedon (i) level a selection has to be made about necessary detailed physical or chemical measurements (K5) considering the amount of already available data and the field experience of the soil scientist which will result in 'smart' sampling (K2). After making the measurements, results are scaled up again to the field level where the effect can be equivalent to the K4 level if sufficient representative samples have been taken (Figure 1).

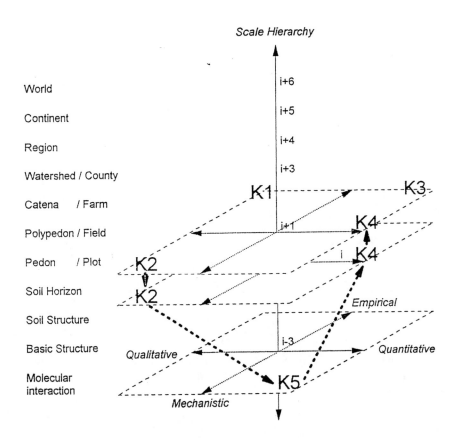

FIGURE 1. Scale hierarchy diagram according to Bouma and Hoosbeek (1996). Three hierarchy levels are distinguished: the field (i+1), pedon (i) and soil structure level (i-3). The diagram illustrates how methodology decisions are made at different scales and how they are interrelated.

OBTAINING POINT DATA

Geostatistical methods can be used to estimate the optimal distance of soil observations in an area taking into account spatial variability. Clearly, more observations are needed in areas that are heterogeneous than in areas that are homogeneous. An exploratory survey is made to establish heterogeneity (Finke and Bosma, 1993; Verhagen et al., 1995; Groenigen et al., 1997). Then, the real survey follows. It is often advisable to use remote sensing techniques or aerial photographs to establish soil patterns originating from sedimention or erosion processes. In fact, this represents K2 expertise, which is followed by the already mentioned geostatistical analysis within the major landscape units (K5 expertise). To mix clearly different land units into one geostatistical analysis is inefficient (Stein et al., 1988).

SOIL OBSERVATIONS

The discipline of soil science has a good tradition in standardizing data gathering techniques in the context of soil surveys. However, increasingly we are less interested in gathering general sets of data without really knowing how they will be used. In the context of precision agriculture, a number of questions are raised, procedures are defined to answer these questions and data gathering techniques are devised to obtain the necessary data, assuming that they cannot be obtained from existing databanks. So before defining data gathering procedures, we should define our methodology realizing that all these procedures are governed by the questions to be answered.

WHAT IS THE QUESTION?

Precision agriculture deals with 'where' and 'when' questions. Many, but not all of those questions, are related to soil conditions in space and time which are the focus of this paper. It is useful to use an activity record during the year to systematically list management activities that are needed for growing a particular crop and to see when these activities occur, how timing could be improved and, particularly, whether a varying pattern of application over the field is profitable (e.g., Bouma *et al.*, 1993; Bouma *et al.*, 1997).

For growing most crops in the world, we deal with a number of questions that relate to soil conditions. Examples are when to: (i) sow or plant; (ii) fertilize; (iii) apply biocides to combat pests and diseases; (iv) allow soil traffic; and (v) apply tillage. Not only the activity itself but also its mode have to be defined. Two considerations are important here:

1. All activities are to be planned in an uncertain future in terms of weather conditions. In research we are used to working with and interpreting historic data. At the beginning of a growing season, a farmer faces the unknown. He has to make choices in which he needs support.

2. The soil moisture regime is an important connecting link in many of the above mentioned activities: sowing, planting, soil traffic and tillage cannot be done when the soil is too wet, nor are sowing and planting advisable when the soil is likely to remain very dry for some time. Fertilization and application of biocides under wet conditions may result in excessive losses by leaching. So when considering the question as to what is important when dealing with precision agriculture, the focus should be on the characterization of the soil moisture regime as a central issue.

The earlier question raised in this section can now be specified by asking how the soil moisture regime can be predicted and which moisture contents at any given time are important as critical values for the various management procedures discussed above. Critical values are often referred to as 'threshold values' and this term will be used from here on.

SOIL MOISTURE REGIMES AND THRESHOLD VALUES

To adequately represent soil moisture regimes in the context of precision agriculture we need a K4/K5 approach which is quantitative and either empirical (for simple flow systems with deep groundwater tables) or mechanistic for more complicated flow systems.

Such models require hydraulic conductivity (K-θ) and moisture retention (h-θ) data which can be measured with modern techniques (Verhagen *et al.*, 1995). Coefficients can be defined to express these data, in terms of, e.g., the Van Genuchten parameters. However, in soils with larger pores, it may be advisable to use tables listing K and h as a function of the moisture content, as input in the flow models (Leummens *et al.*, 1995). Also, soil parameters can be related to the hydraulic parameters by pedo-transferfunctions (PTFs) which relate existing soil data (such as texture and organic matter content) to data which are not available and often difficult to measure (such as the hydraulic characteristics) (Bouma and van Lanen, 1987; Tietje and Tapkenhinrichs, 1993).

In our work, we have used 'functional' soil horizons to estimate hydraulic characteristics (Verhagen *et al.*, 1995). First, measurements are made in soil horizons, as distinguished in soil surveys, in any given soil occurring within the fields of a farm. When identical results are obtained in different pedological horizons, the pedological horizons are put together to form one 'functional' horizon or 'building block'. For further details, the reader is referred to Finke and Bosma (1993) and Verhagen *et al.* (1995). In the Netherlands, there is a national database which links soil characteristics, such as texture, organic matter content and bulk density, to the hydraulic characteristics mentioned here (Wosten *et al.*, 1994; Wosten *et al.*, 1995). The establishment of international databases along these lines is advocated as well. Efforts are in progress within the European Union to establish such databases.

Aside from the soil moisture regime, threshold values are needed for soil tillage, traffic, etc. (Bouma and van Lanen, 1987). Results for trafficability and workability in Dutch soils have recently been reported by Droogers *et al.* (1996); Droogers and Bouma, (1996). Other threshold values are the critical nitrate concentration in the groundwater (50 mg/liter), which is important in studies which balance the need for agricultural production with acceptable leaching losses.

HOW MODELING CAN ANSWER KEY QUESTIONS

Booltink and Verhagen (1996) have demonstrated how predictive modeling, applying a weather generator, can be used to assist the farmer in proper timing of his fertilizer application and in varying the application rate within a single field with different soil units. In these and related studies by the research group of the Department of Soil Science of the Agricultural University, we used point data, obtained by simulation, and interpolation techniques to obtain expressions for areas of land. Verhagen and Bouma (1997) showed that leaching of nitrate occurred mainly during the period of the year with a precipitation surplus. They defined critical N contents at the beginning of the surplus season which are likely to result in acceptable leaching of nitrates in most years. They used model results for a 30 year period to obtain probabilistic expressions. Again, as in the Booltink and Verhagen (1996) study, critical N contents varied significantly over the field for which calculations were made, providing a specific entry point for precision agriculture.

A FUTURE SOIL DATABASE

Traditional soil databases based on soil surveys have a limited applicability for precision agriculture as has been discussed above. Demands are better met with two types of databases in the future:

(i) A specific soil database for individual farms consisting of primary and secondary point data. Primary data consist of texture, organic matter content, soil structure and bulk density and occurrence of soil horizons. In addition, each point observation should be classified into a taxonomic unit of the local soil classification. This could be the soil series of the USDA classification (Soil Survey Staff, 1951). Secondary data refer to (i) hydraulic characteristics, such as K-θ and h-θ curves which are needed for simulation of the soil moisture regime; and (ii) basic parameters such as rate constants to describe N and biocide transformations, and threshold values for important land qualities being considered. Secondary data can sometimes be measured on-site but often there will not be funds available to do so. Then, secondary data will have to be derived from:

(ii) the second type of database which contains pedo-transfer functions that allow reasonable estimates of parameters needed for simulating crop growth and solute fluxes (see discussion above).

The second type of database should also contain all data from a given soil series as a separate entity. We find that different types of management can change the properties of a specific type of soil in a significant manner (Droogers and Bouma, 1997). These authors compared the behavior of a conventionally cropped soil (conv) with the same soil under permanent pasture (perm) and with a soil that had been managed with biological methods for 70 years (bio). This includes lack of application of agrochemicals and emphasis on organic and green manure. Some modeling results are shown in Figure 2, which shows the probabilities that yields of wheat exceed certain limits as a function of leaching of nitrates. Potential yield is 7500 kg ha^{-1} in all cases, but real yields are particularly low for the bio treatment which avoids chemical fertilization. The diagram offers the possibility of estimating yields of wheat in relation to the probability that the yearly threshold value for nitrate leaching is exceeded. Big yield differences are observed when this does not occur at any time as compared with a 3% and a 10% probability. This type of graph is characteristic for the particular soil type being considered. Rather than present one set of curves for the soil type, we prefer to give a series of curves, each series corresponding to a given type of land use. Such data could be part of the general database (type iii).

CONCLUSION

In summary, soil data for precision agriculture can be grouped into three categories: (i) point data consisting of soil characteristics for the fields of a farm; (ii) secondary data for these points to be derived from pedo-transferfunctions being assembled elsewhere; and (iii) results from simulations of crop yields and solute fluxes for point locations and for areas of land using interpolation techniques. Results may be grouped according to well defined soil series that have been obtained by running the model for a large series of years with different weather conditions. Use of the database can start with data from the third category, after establishing which soil series occur within the farmer's fields. A soil survey will always be necessary. If results are considered unsatisfactory, secondary data of category (ii) can be specifically derived from pedo-transferfunctions to run the models, using primary soil data gathered for each observation point during the soil survey.

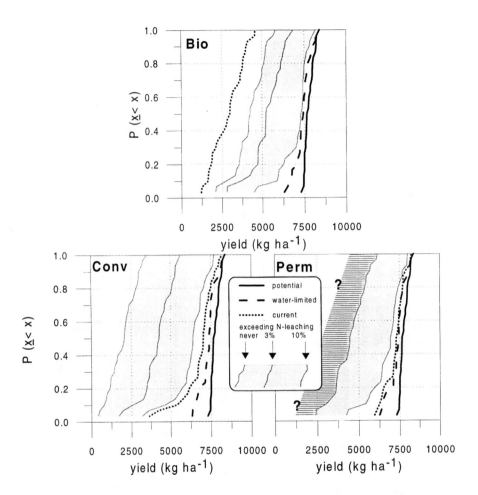

FIGURE 2. Diagram from Droogers and Bouma (1997) in which the relation is presented between yield and three probabilities that a threshold value for nitrate leaching is exceeded. Data are based on simulations for a twenty year period, reflecting temporal variability. One Dutch loamy soil type is characterized with three types of management (Perm = permanent meadow; Conv = conventional arable farming and Bio = biodynamic farming without applying agrochemicals).

ACKNOWLEDGEMENTS

The author acknowledges contributions by J. Verhagen and H.W.G. Booltink. This work was partly funded by the EU-AIR 'IN-SPACE' project 921204: 'Reduced fertilizer input by an integrated location specific monitoring and application system', coordinated by Dr. D. Goense, Wageningen, Netherlands.

REFERENCES

Booltink, H.W.G., Verhagen, J. (1996) Using decision support systems to optimize barley management on spatial variable soil. *Applications of systems approaches at the field level,* M.J. Kropff, P.S. Teng, P.K. Aggerwal, J. Bouma, B.A.M. Bouman, J.W. Jones and H.H. van Laar (Eds), Dordrecht, The Netherlands, Kluwer Academic Publishers.

Bouma, J. (1997) Role of quantitative approaches in soil science when interacting with stakeholders. *Geoderma* (in press).

Bouma, J., van Lanen, H.A.J. (1987) Transfer functions and threshold values: from soil characteristics to land qualities. *Quantified Land Evaluation, ITC Publ. no.6.,* The Netherlands, ITC Enschede, pp 106-111.

Bouma, J. Hoosbeek, M.R. (1996) The contribution and importance of soil scientists in interdisciplinary studies dealing with land. *The Role of Soil Science in Interdisciplinary Research,* R.J. Wagenet and J. Bouma (Eds), Soil Sci. Soc. Amer. Spec. Publ. 45, 1-15.

Bouma, J., Verhagen, J., Brouwer, J., Powell, J.M. (1997) Using systems approaches for targetting site specific management on field level. *Applications of Systems Approaches at the Field Level,* M.J. Kropff, P.S. Teng, P.K. Aggerwal, J. Bouma, B.A.M. Bouman, J.W. Jones and H.H. van Laar (Eds), Dordrecht, The Netherlands, Kluwer Academic Publishers.

Bouma, J., Wagenet, R.J., Hoosbeek, M.R., Hutson, J.L. (1993) Using expert systems and simulation modeling for land evaluation on farm level: a case study from New York State. *Soil Use and Management,* **9(4),** 131-139.

Droogers, P., Bouma, J. (1996) Effects of ecological and conventional farming on soil structure as expressed by water-limited potato yield in a loamy soil in the Netherlands. *Soil Sci. Soc. Amer. J.,* **60,** 1552-1558.

Droogers, P., Fermont, A., Bouma, J. (1996) Effects of ecological soil management on workability and trafficability of a loamy soil in the Netherlands. *Geoderma,* **73,** 131-145.

Droogers, P., Bouma, J. (1997) Soil survey input in exploratory modeling of sustainable soil management practices. *Soil Sci. Soc. Amer. J.* (in press).

Finke, P.A., Bosma, W.P.J. (1993) Obtaining basic simulation data for a heterogeneous field with stratified marine soils. *Hydrological Processes,* **7(2),** 63-75.

Groenigen, J.W., Stein, A., Zuurbier, R. (1997) Optimization of environmental sampling using interactive GIS. *Soil Technology,* **10,** 83-97.

Hoosbeek, M.R., Bryant, R.B. (1992) Towards the quantitative modeling of pedogenesis - A review. *Geoderma,* **55,** 183-210.

Leummens, H.J.L., Bouma, J., Booltink, H.W.G. (1995) Interpreting differences among hydraulic parameters for different soil series by functional characterization. *Soil Sci. Soc. Amer. J.,* **59,** 344-351.

Soil Survey Staff (1951) *Soil Survey Manual. USDA Handbook 18,* Washington DC USA.

Stein, A., Hoogerwerf, M., Bouma, J. (1988) Use of soil map delineations to improve (co) kriging of point data on moisture deficits. *Geoderma,* **43,** 163-177.

Tietje, O., Tapkenhinrichs, M. (1993) Evaluation of pedo-transfer functions. *Soil Sci. Soc. Amer. J.,* **57,** 1088-1995.

Verhagen, J., Booltink, H.W.G., Bouma, J. (1995) Site specific management: balancing production and environmental requirements at farm level. *Agricultural Systems,* **49,** 369-384.

Verhagen, J., Bouma, J. (1997) Defining threshold values at field level: a matter of space and time. *Geoderma* special issue (in press).

Wosten, J.N.M., Veerman, G.L., Scholte, J. (1994) Water retentie en doorlatendheidskarakteristieken van boven- en ondergronden in Nederland: De Staringreeks. Wageningen. DLO- Staring- Center. technical Document, 18, 66 pp.

Wosten, J.H.M., Finke, P.A., Jansen, M.J.W. (1995) Comparison of class and continuous pedotransfer functions to generate hydraulic characteristics. *Geoderma,* **66,** 227-237.

SPATIAL VARIABILITY OF SOIL AGROCHEMICAL PROPERTIES AND CROP YIELD IN LITHUANIA

A. BUCIENE, A. SVEDAS

Lithuanian Institute of Agriculture (LIA), Department of Agrochemistry, Instituto al. 1, 5051 Dotnuva-Akademija, Kedainiai distr., Lithuania

ABSTRACT

Lithuanian soils are generally derived from moraine. In the Middle Plain, the region of the most fertile and comparatively homogeneous soils on a large scale, spatial variability of soil types, texture, physical and agrochemical properties was considerable even in small arable plots of 5-10 ha size.

The prevailing soil type in the 50 fields studied in the Middle Plain was sod-gleic sandy-loam, however smaller outliers of sod-calcareous and sod-podzolized bogy loam and silt loamy soils were observed. Soil pH_{KCl} in the top-layer varied on average from 5.9 (minimum value) to 7.7 (maximum value), humus content varied from 2.1 to 6.5 %, total N from 0.14 to 0.37 %, available P_2O_5 from 57 to 357 mg/kg and available K_2O from 47 to 225 mg/kg. Higher variations in these factors have been observed in the hilly regions of Lithuania.

Crop yield variation depends significantly on the variability of soil agrochemical properties on the field scale.

INTRODUCTION

Lithuania is situated on the south eastern shore of the Baltic Sea. It has a more severe climate than Western Europe, because it is located in the transition zone between the Atlantic marine climate and the continental climate. Comparatively mild winters with many thaws, cool springs, moderately warm summers and warm rainy autumns are typical seasons in Lithuania. The average annual air temperature is 6 ± 1.6 °C. The average annual precipitation ranges from 700 mm in the far west to 559 mm in the east. The land area of Lithuania is 6.520 million ha and is distributed as follows (on 01.01.1996): 54 % agricultural land (3.5 million ha); of which 70.5 % is arable land (2.5 million ha); 27.5 % grassland; 2.0 % shrubs, sands and slopes. Forests cover 27.6 % of area (61.0 % coniferous and 39.0 % deciduous). The remainder 18.4 % belongs to the cities, towns, villages, lakes, rivers, roads, etc.

The main soil and climatic zones are shown in Figure 1 (Lietuvos TSR fizine geografija, 1958). The figure illustrates the direct and long-term relationship between climatic factors and soil formation processes. From both maps three main soil-climatic zones can be distinguished: Western, Middle and South-Eastern. These reflect common features of climate, organic matter accumulation - mineralization rate, matter migration character, etc.

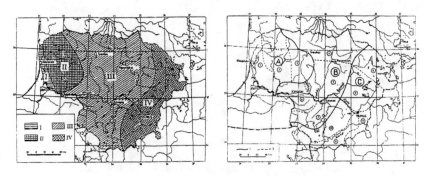

a   b

FIGURE 1. Climatic zones (a) and main soil regions (b) in Lithuania: I - Sea shore; II - Zemaiciai; III - Middle; IV - Eastern; A - Western; B - Middle; C - South-Eastern.

SPATIAL VARIATION OF SOIL PROPERTIES AS A CONSEQUENCE OF NATURAL PROCESSES AND HUMAN-RELATED ACTIVITY

General features of soil cover in Lithuania

Lithologic-genetic features of the topography and moisture distribution have caused variations in the structure of the soil cover. In Lithuania six principal genetic types of soils can be distinguished: podzol, podzol-bog, sod-calcareous, sod-gley, bog and alluvial soils. In addition, soils formed primarily under the influence of anthropogenous impact can also be defined: alluvial-deluvial, deluvial, eroded soils and disturbed land. This soil classification was developed in accordance with the Russian soil classification system in 1959. Later it was improved by adapting it to Lithuanian conditions. However, this work remains unfinished. Accumulation of abundant data from investigations continues to reveal a more detailed picture of soil gleyzation and to identify some additional soil types such as brown soil and Lessives (Luvisols) (Vaicys,1996).

Soil properties depend not only on the texture of the topsoil, but also on the subsoil. In Lithuania seven groups of soil texture in both top and subsoil are distinguished: sand, loamy-sand, sandy-loam and loam, clay-loam and clay, muck sod, shallow peat, peat. Loamy-sand and different loams make up almost 60 % of all soils (Zemes kadastras, 1989). Soddy-podzolic soils are prevalent (45.3 %).

The two western soil regions (Figure 1) are mainly formed from medium loam parent rock of moraine origin. The occurrence of carbonate is influenced by the texture and the intensity of the soil - moisture regime. It was found that the spatial distribution of carbonate content in the parent rocks was determined by geomorphological elements and their age (Eidukeviciene, 1993).

The Eastern soil region is part of the Baltic Range. It consists of highland divided in two parts: northern (Aukstaiciai highland) and southern (Medininkai). Contrary to the individual isolated hills which dominate the West, the Eastern region hills are composites of many small hills with short and irregularly shaped slopes. Variations in soil texture are also complicated: gravel drifts are usually found inside the hill which is then covered with different thickness of loam, sandy-loam or clay-loam. Both Western and Eastern soils are more heterogeneous on a large scale than the soils formed in the plains. However, podzolization, sodding and bogging are more common in the Western region, than in the Eastern one. The morainic plains of the Seashore and the Middle Plain are of smooth or wavy relief and are formed mainly from the ground moraine drifts (Svedas, 1993).

Historically, the Middle soil region was the most important agricultural area, and was distinguished by its fertile and comparatively homogeneous soils on a large scale. Sod gley and sod - calcareous soil types prevail in this region (Zemes kadastras, 1989). However in some places the soil texture and microclimate vary significantly and create considerable local soil diversity.

The range of main soil agrochemical indices in the topsoil (regional level)

Data obtained from experimental sites and Land survey expeditions in 1950-1961 was analysed and grouped into different soil regions and textural classes (Table 1).

TABLE 1. The range of agrochemical indices of the topsoil in Lithuania (1950-1961).

Indices	Sod-podzolic				Sod-podzolic gley			Sod-gley
	sand	loam				loam		loam
		Zemaiciai highland	Baltija highland	Old landscape	Western zone	Middle zone	Eastern zone	
pH_{KCl}	4.5-5.5	4.0-5.5	4.1-6.5	4.1-6.0	4.1-6.0	4.1-6.5	4.1-5.0	6.6-7.5
Exchange Al, mekv/kg	0.5-4.2	0.6-4.0	2.7-4.0	2.9-10.7	0.3-0.9	-	0.5-2.0	-
Hydrolytic acidity, mekv/kg	20-36	21-60	10-35	26-45	21-70	16-40	26-50	4-11
Cation exchange capacity, mekv/kg	12-35	20-70	26-70	5-40	26-88	36-69	26-50	11-35
Humus, %	1.1-2.3	1.5-3.0	1.1-1.8	0.9-1.8	2.3-4.0	2.1-3.4	2.1-3.4	1.7-4.5
Total N, %	0.05-0.13	0.08-0.19	0.05-0.15	0.06-0.13	0.12-0.27	0.12-0.19	0.12-0.18	0.12-0.25
Total P_2O_5, %	0.05-0.10	0.04-0.16	0.04-0.16	-	0.11	0.08	0.11	0.07-0.12
Avail.P_2O_5, mg/kg	50	10-50	15-100	40-90	10-90	20-70	10-100	50-100
Total K_2O, %	1.12	1.4-1.9	1.4-1.9	-	1.8	1.5	2.0	-
Avail. K_2O,	70-210	50-300	160-350	180-230	110-300	100-250	50-250	40-170

Note: the analysis methods used were: pH $_{KCl}$ - potenciometric; exch. Al - by A.V.Sokolov; hydrolytic acidity and cation exchange capacity - by Kappen-Hilkovic; humus - by Tiurin; total N - by Kjeldal; total P_2O_5 - by Lebediancev; available P_2O_5 - by Kirsanov; total K_2O by Smith; available K_2O by J.V.Peive (0.2 N HCl).

The data reflect the natural distribution of the main soil agrochemical indices, since the agriculture of that period was very extensive and low-input. The highest humus and nitrogen contents are typical of the Western soil region where there is a higher precipitation rate and a milder temperature regime. Higher humus content was also found for sod-gley or gleic soil types. More acid soils were found in the Western region and in the Old landscape district of Medininkai highland. The age of the rock was directly related to the soil pH in profile (Eidukeviciene, 1993).

During the last 30 years, comparatively intensive land use and reclamation, extended soil liming and fertilising activities, have added significantly to the microdiversity of the soil agrochemical properties even in small separate fields (Mazvila et al., 1995).

Spatial variation of soil agrochemical properties in the topsoil (field/hill level)

Investigations on the local spatial variation of soil properties started in the Kedainiai and Radviliskis districts (Middle Plain) in 1977 (Table 2). Coverage is not yet complete.

Both the topsoil and the subsoil were tested. In addition to soil sampling, crop yield was analysed from the same small plots (Buciene, 1984). The investigations lasted from 1977 until 1983 and were renewed in 1991-1995 on watershed and field levels (Buciene et al.,1996). The fields examined during the first phase of study (1977-1983) were of different sizes, but most were between 5 and 10 ha. Soil and plant sampling was performed by moving in straight lines across the field at distances of 35-400 m from the field margin with samples taken at equal intervals (Figure 2).

TABLE 2. Variation of topsoil agrochemical properties within several fields (5-10 ha) in the Middle Plain of Lithuania in 1977-1991.

Field address, year, crop, number of soil samples	Variable	Average	Minimum	Maximum	Coefficient of variation %
Experimental farm of LIA,. Kedainiai distr.,1977, ley; 28	Available P_2O_5, mg/kg	350	110	970	42.3
Zemaite kolxoz, Kedainiai distr.,1977, sugar beet; 40.	Available K_2O, mg/kg	100	50	360	56.1
Ateitis kolhoz, Radviliskis distr.,1977, potatoes; 30.	humus content, %	2.1	1.2	3.1	56.3
Melnikaite kolhoz,Kedainiai distr.,1979, barley; 38.	Available P_2O_5, mg/kg	140	60	340	73.0
Tiesa kolhoz, Kedainiai distr. 1977, winter wheat; 42.	Available P_2O_5, mg/kg	110	30	260	72.4
Krakes kolhoz, Kedainiai distr.,1982, sugar beet; 40.	Available P_2O_5, mg/kg	330	170	1200	61.3
Experimental farm of LIA, Kedainiai distr.,1991, ley;23.	humus content, %	2.9	2.0	4.6	20.8
Experimental farm of LIA, Kedainiai distr.,1991, ley;25.	Available P_2O_5, mg/kg	163	79	363	42.4

Note: the available P_2O_5 and K_2O were determined by A-L method; humus - by Tiurin method.

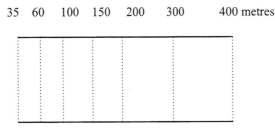

35 60 100 150 200 300   400 metres

FIGURE 2. Soil - plant sampling scheme in the field.

The prevailing soil type in the 50 fields studied in the Middle Plain was sod-gleic sandy-loam, however smaller outliers of sod-calcareous and sod-podzolized bogy loam and silt loamy soils were observed. Soil pH_{KCl} in the top-layer varied on average from 5.9 (minimum value) to 7.7 (maximum value), humus content varied from 2.1 to 6.5 %, total N from 0.14 to 0.37 %, available P_2O_5 from 57 to 357 mg/kg and available K_2O from 47 to 225 mg/kg.

Figure 3 shows the spatial variation of soil texture and pedons in one of the experimental sites of LIA. The prevailing soil type (white contours) is sod gley slightly podzolized sandy loam, however other types with different soil texture also occur in the area.

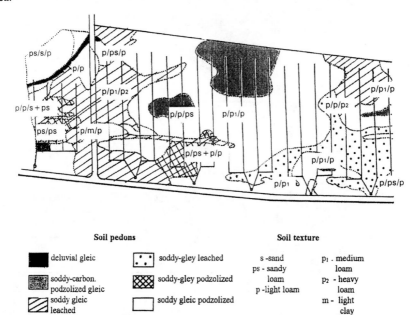

FIGURE 3. Variation of soil texture and pedons within an experimental site (4.5 ha) in Valinava, LIA, 1991.

The Middle Plain soils are comparatively rich in available phosphorus since over the last few decades they have received higher amounts of phosphorus fertilisers than other regions. Figure 4 shows the spatial distribution of available P_2O_5 in the topsoil of one demonstration farm in the Kedainiai district (Sileika *et al.*, 1996). The farm has inherited from the previous Melnikaite kolhoz a rather high supply of available phosphorus in both the topsoil and the subsoil. The prevailing soil type on this farm is sod gley sandy loam.

The spatial variation of soil agrochemical properties was even higher in the hilly regions (Table 3).

THE RELATIONSHIP BETWEEN AGRICULTURAL CROP YIELD AND SOIL AGROCHEMICAL PROPERTIES

Different data obtained from field experiments and expeditions over the period from 1929 to 1993 was compiled and analysed using methods of correlation and regression

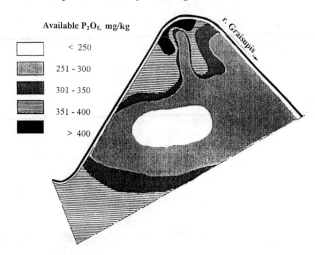

Available P_2O_5, mg/kg

< 250

251 - 300

301 - 350

351 - 400

> 400

r. Graisupis

FIGURE 4. Distribution of available P_2O_5 in the topsoil of V. Liutkevicius demonstration farm (16.7 ha) 1995 (based on analysis of 35 soil samples).

TABLE 3. Variability of agrochemical properties in the topsoil of a single hill (Svedas, 1974).

Place on the slope	Hydrolytic acidity mekv/kg			Humus %			Available P_2O_5 mg/kg		
	n	average	range	n	average	range	n	average	range
Top	12	23±1.4	16-30	26	3.5+0.14	2.6-5.6	8	146±27.3	84-275
Slope	12	24±1.2	19-31	26	2.1±0.12	0.3-2.9	8	75±5.5	56-108
Foot	12	30±1.3	22-36	26	2.4±0.15	1.0-3.6	8	40±3.3	25-50

Note: n - number of soil samples.

analysis (Buciene, 1984; Buciene et al.,1996; Svedas, 1993; 1996). On the basis of these findings relationships between the crop yield and pH value, humus, total nitrogen, available phosphorus and potassium, and amounts of fertiliser used were determined. The relationship between the yield of different crops and the pH value can be expressed by a quadratic equation:

$$y = y_{pH} (-2.167 + 1.018 \text{ pH} - 0.0813 (\text{pH})^2); \eta = 0.42 \qquad (1)$$
$$\eta_{95} = 0.35$$

where y = crop yield, t/ha

y_{pH} = average crop yield at optimal pH level, t/ha

η = the correlation ratio

η_{95} = the correlation ratio at 5 % probability

Regression coefficients are different for different crops (Svedas, 1996).

Table 4 shows the relationships between crop yield and other agrochemical indices.

It was found, that by increasing the humus content to 3.5-4 % and the nitrogen content to 0.28-0.32 % the yield of all examined crops could be increased significantly. The optimal amounts of available phosphorus and potassium for almost all the crops were between 150 and 200 mg/kg.

CONCLUSION

On the basis of the relationships between yield and soil properties, equations were derived which could then be used to calculate the expected yield of specified plots for certain amounts of applied fertilisers. They were used in the preparing of a computer based programme on fertilising plans, worked out in LIA and now verified.

TABLE 4. Relationships between crop yield and soil agrochemical properties in Lithuania.

Crop	Regression equation	η	η_{95}
Winter wheat (grain)	$Y=4.169*X_1*(0.45+X_1)^{-1}$	0.99	0.81
	$Y=4.006*X_2*(89.2+X_2)^{-1}$	0.70	0.14
	$Y=4.15*X_3*(120.8+X_3)^{-1}$	0.64	0.14
Barley	$Y=3.644*X_1*(0.29+X_1)^{-1}$	0.99	0.75
	$Y=3.687*X_2*(113.2+X_2)^{-1}$	0.80	0.11
	$Y=3.81*X_3*(100.4+X_3)^{-1}$	0.67	0.11
Ley (hay)	$Y=7.423*X_1*(1.001+X_1)^{-1}$	0.93	0.75
	$Y=7.352*X_2*(250+X_2)^{-1}$	0.22	0.09
	$Y=5.371*X_3*(39.2+X_3)^{-1}$	0.55	0.09
Sugar beet (roots)	$Y=30.3*X_1*(1.02+X_1)^{-1}$	0.99	0.88
	$Y=57.2*X_2*(121.3+X_2)^{-1}$	0.65	0.25
	$Y=30*X_3*(200+X_3)^{-1}$	0.35	0.29
Potatoes (tubes)	$Y=21.3*X_1*(1.2+X_3)^{-1}$	0.94	0.88
	$Y=42.1*X_2*(104+X_2)^{-1}$	0.70	0.20
	$Y=41*X_3*(110+X_3)^{-1}$	0.32	0.29

Note: Y - (t/ha); X_1 - humus content in topsoil, %; X_2 - available P_2O_5, mg/kg; X_3 - available K_2O, mg/kg.

ACKNOWLEDGEMENTS

The authors express thanks to the Mathematic-statistical group in LIA and particularly to E. Ivaskeviciute for their technical assistance in working with the numerous data.

REFERENCES

Buciene, A. (1984) Sviazj urozaja selskochoziaistvennych kultur so svoistvami pocvi i rasstojanijem ot lesa v uslovijach Sredne-Litovskoj nizmennosti, *Avtoreferat dissertacii, Ministerstvo Selskogo Hoziaistva SSSR, Litovskaja Selskochoziaistvenaja Akademija,* Kaunas, 20 pp.

Buciene, A., Masauskiene, A., Svirskiene, A., Slepetiene, A. (1996) Ivairaus chemizavimo lygio intensyviu ir organines-biologines zemdirbystes sistemos veleniniuose glejiskuose Lietuvos Vidurio Lygumos dirvozemiuose kompleksinis ivertinimas *Zemdirbystes instituto uzbaigtu tiriamuju darbu konferencijos pranesimai. No. 28,* Dotnuva-Akademija, pp 49-53.

Eidukeviciene, M. (1993) Geochemical and geographical validation of optimalisation of liming acid soils in Lithuania. *The work of doctor habilitatis,* Vilnius University, Vilnius, 99 pp.

Lietuvos TSR fizine geografija (1958) ats.red.doc.A.Basalykas, **1**, Valstybine politines ir mokslines literaturos leidykla, Vilnius, pp 201-204, 294-336.

Mazvila, J., Vaisvila, Z., Eitminavicius, L. (1996) Lietuvos dirvozemiu agrocheminiu tyrimu rezultatu apibendrinimas. *Zemdirbystes instituto uzbaigtu tiriamuju darbu konferencijos pranesimai. No. 28,* Dotnuva-Akademija, pp 3-5.

Sileika, A.S., Kutra, G.J., Buciene, A., Gaigalis, K., Strusevicius, Z.(1996) BEAROP in Lithuania. *Report for 1995, Lithuanian Institute of Water Management, Vilainiai,* Kedainiai, 46 pp.

Svedas, A. (1993) Soil - fertilizer - yield relationships. *The work of doctor habilitatis,* Lithuanian Institute of Agriculture, Akademija, 78 pp.

Svedas, A. (1996) Zemes ukio augalu derliaus priklausomumas nuo dirvozemio agrocheminiu savybiu, *Zemdirbystes instituto uzbaigtu tiriamuju darbu konferencijos pranesimai. No. 28,* Dotnuva-Akademija, pp 45-49.

Svedas, A. (1974) *Zakreplenije pocv na sklonach, Kolos, Leningradskoje Otdelenije,* Leningrad, pp 84-93.

Vaicys, M. (1996) Contemporary problems of Lithuanian soils genesis investigation and classification. *Soil classification and land evaluation in the Baltic States, Intern. workshop of soil scientists from the United States of America and Baltic countries,* pp 72-75.

Zemes kadastras (1989), Red. J.Juodis ir J.Pakutinskas, Mokslo leidykla, Vilnius, pp 45-51.

THE VARIABILITY OF pH AND AVAILABLE PHOSPHORUS, POTASSIUM AND MAGNESIUM IN SOILS WITHIN ARABLE FIELDS IN ENGLAND

P.M.R. DAMPNEY

ADAS Boxworth, Boxworth, Cambridge, CB3 8NN, UK

M.A. FROMENT

ADAS Bridgets, Martyr Worthy, Winchester, Hampshire, SO21 1AP, UK

C.J. DAWSON

Chris Dawson and Associates, Westover, Ox Carr Lane, Strensall, York, YO3 5TD, UK

ABSTRACT

Data are presented showing the variability of soil pH, available phosphorus (P), potassium (K) and magnesium (Mg) within 78 individual arable fields in England (mean field size 20.5 ha). On average, 31 composite soil samples were taken from the 0-15 cm soil depth in each field using a grid pattern, and analysed using standard laboratory procedures. Mean Coefficient of Variation (CV) values were 5.1% for soil pH, 36.1% for soil P, 27.2% for soil K and 28.5% for soil Mg. However, there was a wide variation in CV values across the dataset. Variation was greatest in the larger fields. Fields with a high CV for one nutrient tended to have a high CV for other nutrients. This relationship was strongest between soil P and soil K. On average the variation within fields shown by this dataset is less than the variation between arable fields as shown by a national representative soil sampling scheme. The implications of within-field variation in soil pH and nutrient contents are discussed.

INTRODUCTION

Soil analysis is a widely accepted technique for determining soil pH and the potential supply of soil nutrients available for plant uptake. In England and Wales, many tens of thousands of soil samples are routinely taken each year from individual fields, and the results used for determining optimum application rates of lime, phosphate, potash and magnesium fertilisers. In England and Wales, nutrient Indices are widely used as a primary basis of fertiliser recommendations (MAFF, 1994). For combinable arable crops, Index 2 is recommended as the optimum level to maintain soil reserves of phosphorus (P), potassium (K) and magnesium (Mg). The recommended soil pH for mineral soils is pH 6.5 (MAFF, 1994). Current standard methods of soil sampling are designed to obtain a representative composite sample from an area that is considered to be uniform with respect to soil type, previous cropping and fertiliser/lime/organic manure use. In practice, a single field is normally the sample area, although some large fields may be divided into more than one sample.

In many agricultural areas, it is recognised that there is short range spatial variation in terrain, soil type and soil chemical fertility within fields, although this has not been generally quantified. Until recently there has been no practical means of managing this variability, although for many years farmers on acid-prone soils have used *in-situ* measurement of soil pH to identify variation within fields. Standard soil sampling and analysis procedures, and fertiliser recommendations have thus been based on the average condition of the sampled area, which is normally a complete individual field.

GPS (Global Positioning System) technologies, together with the development of yield mapping combines and fertiliser spreaders capable of variable rate application, now allows the opportunity for farmers to apply lime and fertilisers more precisely to individual field areas. Information on how soils vary spatially in their nutrient supply capabilities will be necessary in order to identify those situations where cost-effective variation of lime and fertiliser applications could be beneficial.

This paper presents some available data on the variability of soil pH, P, K and Mg in arable soils in England, and discusses some of the opportunities and difficulties of measuring and managing the soil nutrient status within individual fields.

METHODS

The data reported in this paper have been obtained from laboratory analysis of soil samples taken from 78 individual arable fields in England between 1988 and 1994 (Table 1). The fields used for sampling reflected local farmer interest, and the dataset should not therefore be regarded as a strictly balanced representation of the range of soil/climatic conditions that exist. Fields were sampled on a 60 x 60 or 100 x 100 metre grid basis. An average of 31 individual samples were taken from each field. The average field size was 20.5 hectares. Each sample was a composite sample, usually of five sub-samples, taken from the 0-15 cm soil layer, and from within a small area no greater than 5 x 5 metres around each grid point. Detailed information on the soil type or cropping in these fields is not available, though most fields were in combinable crop rotations in England. Due to the insufficient number of samples per field, no attempt has been made to carry out any geostatistical analysis for spatial dependency of the data.

Data are also presented from the MAFF Representative Soil Sampling Scheme (RSSS). This annual survey provides unbiased data on the mean soil pH and nutrient status of individual arable and grassland fields in England and Wales. Each year approximately 180 farms are selected randomly from a stratified sample of farms. On each farm, a single composite soil sample is taken from each of four fields, composed of 25

TABLE 1. Summary of location, field sizes and sampling frequency.

Number of fields sampled	78
Average field size (ha)	20.5 (range 4.5 - 51.8)
Total area sampled (ha)	1602
Grid spacing (m)	60 x 60 or 100 x 100
Average number of composite samples per field	31
Total number of soil samples analysed	2445
Average field area per soil sample (ha)	0.65

sub-samples collected by walking in a 'W' pattern across the field. Fields are resampled in the same way on a 4-5 yearly cycle.

Analysis of all soil samples is carried out using standard analytical methods on air-dried samples, ground through a 2 mm sieve (MAFF, 1986). In summary, the analysis methods used were:

Soil pH   1 soil : 2.5 water suspension,
Phosphorus (P)   extracted by sodium bicarbonate at pH 8.5 (Olsen's method),
Potassium (K)   extracted by M ammonium nitrate,
Magnesium (Mg) extracted by M ammonium nitrate.

Laboratory results for P, K and Mg are reported as mg/l, but may also be classified into a nutrient Index (Table 2). In this paper, the nutrient Index is used as one basis for assessing the practical significance of the data.

RESULTS

Table 3 gives an overall summary of the soil analysis data. The mean soil nutrient concentrations in the sampled fields (P Index 3, K Index 2 and Mg Index 2) are typical of arable soils in England. The mean soil pH value is well above the recommended optimum level for mineral soils of pH 6.5, which indicates that a high proportion of the soils sampled were naturally calcareous.

For all nutrients, the mean range of analysis values was greater than the mean value itself. The mean range and Coefficient of Variation (CV) was higher for soil P than for soil K or Mg. Assuming a normal distribution, the average 95% range (mean +/- 2 sd) of soil analysis values spanned 3 or 4 Indices, and was greatest for soil P. These data indicate greater variation within fields for phosphorus, which probably reflects the lower mobility of this nutrient in soils compared to potassium or magnesium.

Table 3 also shows the wide range in CV values between the fields. It is clear that some fields are very variable whilst others are relatively uniform. The causes of these

TABLE 2. Classification of soil analysis results into nutrient Indices.

Nutrient Index	Phosphorus (mg/l)	Potassium (mg/l)	Magnesium (mg/l)
0	0 - 9	0 - 60	0 - 25
1	10 - 15	61 - 120	26 - 50
2	16 - 25	121 - 240	51 - 100
3	26 - 45	241 - 400	101 - 175
4	46 - 70	401 - 600	176 - 250
5	71 - 100	601 - 900	251 - 350
6	101 - 140	901 - 1500	351 - 600
7	141 - 200	1501 - 2400	601 - 1000
8	201 - 280	2401 - 3600	1001 - 1500
9	over 280	over 3600	over 1500

TABLE 3. Summary of statistics of within-field variability of soil pH and nutrient concentrations (mean values for 78 sampled fields).

	pH	Phosphorus (mg/l)	Potassium (mg/l)	Magnesium (mg/l)
Overall mean	7.5	27	200	95
Overall median	7.4	23	171	58
Overall range (max-min)	1.3	38	213	99
Mean standard deviation (sd)	0.38	9.8	54.4	27.1
Mean CV (%) - range in brackets	5.1 (0.4-18.1)	36.1 (13.1-99.7)	27.2 (4.9-100.5)	28.5 (7.1-133.2)
Mean 95% range	6.7-8.3	8-47 (Index 0-3)	91-309 (Index 1-3)	41-149 (Index 1-3)

differences cannot be identified in this dataset, but are likely to be a combination of natural factors (e.g., soil type and terrain) and anthropogenic factors (e.g., fertiliser spreading inaccuracy, yield/nutrient balance variation) probably accumulated over many years.

Linear regression analysis was used to examine the relationships between the variability (CV%) found for different parameters (Table 4). Within-field variation was significantly greater in larger than in smaller fields for soil pH, P, K and Mg concentrations. Soil P, K and Mg concentrations were all significantly co-related such that a large CV for one nutrient was reflected by a large CV for other nutrients. This relationship was particularly strong between soil P and soil K ($p < 0.001$).

Figure 1 shows data for individual fields for soil pH, P and K. The recommended range of values for maintaining soil nutrient contents in arable rotations are also shown - soil pH 6.5, 16-25 mg/l P; 120-240 mg/l K. Fields were most variable where the mean soil pH was low. This is likely to be a reflection of the high proportion of calcareous soils sampled, which would be expected to be more uniform than naturally acid soils. About 27% of the fields contained areas where the pH was below 6.0, and damage to many arable crops would be expected at these soil pH levels. In about 8% of fields, a mean pH of over 7.0 'masked' the fact that part of the field had a pH of below 6.0.

For soil P and K there was no relationship between nutrient content and within-field variation. However, for the majority of fields sampled, the range in soil nutrient contents found was greater than the target range for maintenance using fertilisers or other

TABLE 4. Linear regressions coefficient (r^2 adj) between the mean CV (%) values for all combinations of soil pH, nutrient concentrations and field size.

	pH	Phosphorus	Potassium	Magnesium
Phosphorus (P)	8.9	-	-	-
Potassium (K)	2.6	28.8***	-	-
Magnesium (Mg)	2.0	7.5**	10.9**	-
Field size	12.6**	18.0***	11.1**	9.8**

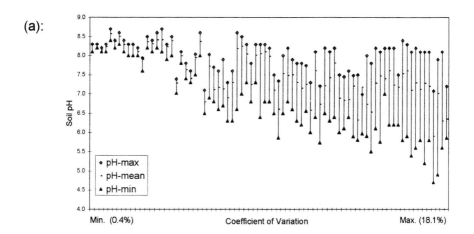

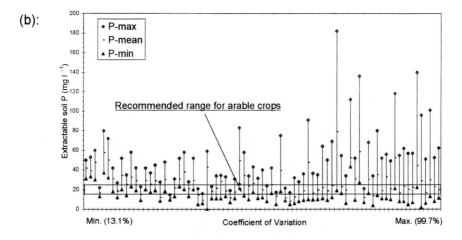

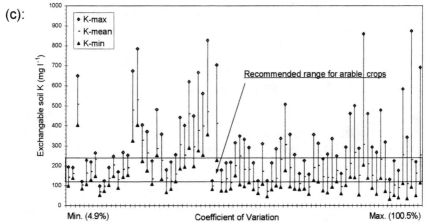

FIGURE 1 (a-c). Maximum, minimum and mean values of (a) soil pH, (b) extractable soil phosphorus, and (c) exchangeable soil potassium in sampled fields, according to Coefficients of Variation.

manures. For most fields, multiple soil sampling identified areas that were both above and below the optimum maintenance levels.

Although fertiliser recommendations are frequently based on the mg/l analysis value, at the simplest level it is the nutrient Index value which is commonly used by farmers. It is appropriate therefore to consider the proportion of individual fields that fall into the mean Index value for that field. On average, for all fields studied, the percentage of a field area within the mean Index was 53% for soil P, 69% for soil K and 63% for soil Mg content. The balance of each field would thus have nutrient Indices both above and below the mean Index value. For some fields, only a very low proportion of the field was in the mean Index value, especially for soil P (e.g., less than 30%).

On a national basis however, the variation within arable fields is less than the variation between arable fields using a single mean value of nutrient status. Table 5 summarises data for soil analysis results from the MAFF Representative Soil Sampling Scheme (1989-1991) for an unbiased, representative group of arable fields in England and Wales. This survey showed that the percentages of fields within the national mean Index were 37% for P (Index 3), 55% for K (Index 2) and 36% for Mg (Index 2). Although Index 3 is the mean national soil P level, the optimum Index for combinable crops is Index 2 for all nutrients.

DISCUSSION

Variation of soil analysis results *between* fields based on a single composite sample per field is well known. For many years, farmers have readily accepted advice to carry out soil analysis every 4-5 years, and to use the results for decisions on lime and fertiliser use. The data presented here indicates that significant variation also exists for soil pH and available nutrient levels *within* arable fields in England. However, although GPS and precision machinery now allow variable rate lime and fertiliser application, it is important that there is a clear identification of those situations where commercial benefits might be expected. A technical ability to carry out variable rate application does not necessarily mean that it will always be a cost-effective management tool for use by farmers.

It is noteworthy that the concept of within-field variable rate application is not new. For many years, farmers on acid land have been applying lime at different rates in different parts of individual fields usually based on *in situ* soil testing. Although the methods of

TABLE 5. Distribution of soil nutrient Indices in arable fields in England and Wales based on a single composite sample per field (MAFF RSSS, 1989-91) - % of samples in each Index.

Soil nutrient	Soil Index					
	0	1	2	3	4	Over 5
Phosphorus (P)	4	14	31	37	11	4
Potassium (K)	1	24	55	17	2	0
Magnesium (Mg)	2	24	36	18	8	12

in situ soil testing used are generally less precise than those used in a laboratory, and the method of lime application has lacked small scale precision, the technique has still been regarded as worthwhile by farmers.

The data presented in this paper show that a large proportion of the arable fields studied contain areas which are significantly different to the mean value for the field. Conventional soil sampling methods based on a single composite sample per field would not identify this variation. There are two potential consequences of this variation. Firstly crop growth and quality may be reduced in field areas where soil pH or nutrient levels are much lower or more deficient than the mean value for the field. In England, the effect of soil acidity is likely to be of more practical significance than low soil P or K contents, since this is a more common problem. Crop responses to low soil P or K contents are rare, and for cereal and other combinable crops, the magnitude of any responses are generally very small. Spatial variation in soil nitrogen is likely to have a larger effect but is not considered in this paper. The second potential consequence is that unnecessary lime and fertiliser may be used in areas where pH or nutrient contents are above the field mean. This is financially wasteful and, for lime, may aggravate trace element deficiency problems.

Although the potential benefits of identifying and managing within-field variation look attractive, there are dangers of attempting spurious precision based on multiple soil analyses, since many sources of error exist. Errors due to sampling are well known but are not well quantified. Soils in England are more variable than in many other parts of the world, and there is little information on the balance between random and systematic variation within fields, or the scale at which variation exists. Beckett and Webster (1971) suggested that up to half of the variance within a field may already be present within any square metre. Whereas systematic variation might be managed if on a large enough scale, random variation could not be practically managed. Even if a high proportion of systematic variation exists, the variation would need to reflect sufficiently large 'uniform' areas to allow practical management. Due to these uncertainties, preparation and interpretation of soil nutrient maps must be carried out with great care.

Sampling methods to accurately identify spatial variability within a field are not nationally agreed, and are the subject of current research. Questions exist concerning the merits of grid sampling versus sampling of zoned areas based on factors like crop yield, soil type or terrain. Optimum sampling densities are not well established, and will need to reflect the scale of spatial dependency of soil properties. Whatever sampling procedure is considered best, errors will exist which must limit the extent to which the information can be applied in practice.

There are also questions concerning the temporal stability of soil analysis data particularly following recent fertiliser and manure applications. Although usually small with modern sophisticated equipment, errors also exist that are associated with laboratory analysis procedures. Although improving, farm fertiliser application machinery is commonly poorly maintained and inaccurate, and even new equipment usually has a CV of between 8 and 12%.

Another important consideration is the relationship between soil nutrient status and the need for fertiliser. Improved precision in recommendation systems may need to be considered to accompany an improved precision in methods of fertiliser application. For instance, current recommendation systems take no account of the depth of cultivated topsoil; the total quantity of available nutrient will be greater from a 25 cm topsoil than a 15 cm topsoil even though both soils may have identical soil analysis values on a 0-15 cm sample depth. No account is currently taken of P or K supply from subsoil reserves which can vary significantly between soil types. Taking all sources of error together, it has been suggested (ADAS, 1995) that minimum differences in soil analysis values for practical decision-making purposes might be 0.2 pH unit, 5 mg/l P, 25 mg/l K and 10 mg/l Mg.

CONCLUSIONS

The data presented here shows that there is significant variation in soil pH and nutrient contents within arable fields. However, there are many unanswered questions concerning methods for accurately quantifying and mapping this variation, and for interpreting the results into meaningful and cost-effective lime and fertiliser application practices. However, it is considered likely that in many situations, there will be benefits to variable rate application, particularly for responsive crops, but that most of the benefit will come from only a small number of management units in individual fields.

ACKNOWLEDGEMENTS

The data presented in this paper are largely based on a review (Project CE 0139) funded by the Ministry of Agriculture, Fisheries and Food (MAFF), whose support is gratefully acknowledged. Data provided by ICI Fertilisers Ltd of Wilton, Cleveland, and SOYL Ltd of Hambledon, Hampshire is also gratefully acknowledged.

REFERENCES

ADAS (1995) A review of spatial variation of nutrients in the soil. Report of Project CE0139 to MAFF CSD, London, UK.

Beckett, P.H.T., Webster, R. (1971) Soil variability: A review. *Soil and fertilisers,* **34(1),** 1-15.

MAFF (1986) *The analysis of agricultural materials (Third edition).* Reference Book 427, London, UK, HMSO.

MAFF (1994) *Fertiliser recommendations for agricultural and horticultural crops (Sixth edition).* London, UK, HMSO.

ASPECTS OF SOIL FERTILITY IN THE INTERPRETATION OF YIELD MAPS AS AN AID TO PRECISION FARMING

C.J. DAWSON

Chris Dawson & Associates, Westover, Ox Carr Lane, Strensall, York, YO3 5TD, UK

A.E. JOHNSTON

Rothamsted Experimental Station, Harpenden, Hertfordshire, AL5 2JQ, UK

ABSTRACT

Mapping crops during the growing season on an individual field basis is now possible using a variety of techniques and yield maps for cereals and root crops can be obtained at harvest. Such maps invariably show differences in both growth and yield but they have little value unless they serve as a stimulus to find the reasons for the differences. Field yield variation can be caused by a wide range of biological, chemical and physical soil conditions and external factors like diseases, pests and weeds. Among soil chemical and physical properties which can cause yield variation are soil pH, readily soluble phosphorus and potassium status, soil organic matter levels and limited rooting depth due to compacted layers of soil. Critical levels of readily soluble phosphorus and potassium for a range of arable crops have been determined for some soils. The high percentage variance in yield accounted for by Olsen P in field experiments suggests that it can be used to help identify causes in yield variation within fields. Thus soil analysis has a major role to play in spatially variable management, i.e., in precision farming, within a field to optimise yield.

INTRODUCTION

For many decades it has been accepted that yield varies within a field. The few early attempts to produce detailed field yield maps were very labour intensive (Mercer and Hall, 1911; RES 1926, 1928 and 1930) and varying husbandry to allow for variable yield potential was too complicated. This has now changed dramatically. Fully automatic and integrated mapping systems can now generate maps showing the variation in performance and yield within individual fields and this now adds a new dimension to the management opportunities for arable agriculture. These new opportunities will allow precision farming, defined here as spatially variable management within a field, to allow more efficient use of inputs, greater profitability and more environmentally benign husbandry. But optimising the opportunities now available will require the causes of yield variation to be ascertained and appropriate remedial measures to be available.

On farm field-yield maps are being produced in Europe, the United States, South Africa and Australia, especially for cereals, pulses and oilseeds but also for crops as diverse as sugar beets, sugar cane and forage. It would be surprising if the causes of yield variation within fields differed greatly from those identified as causing variation between fields, some of which have been investigated. Therefore, it is reasonable to assume that no new causes of variation need be sought.

This paper discusses some of the causes of yield variation and the role of growth and yield maps combined with soil examination and analysis as a management tool to help minimise within field yield variation.

SOIL FERTILITY AND CROP PRODUCTION

There are many possible causes of variation in crop performance within fields but essentially they fall into two categories: those which potentially could be controlled once identified and those which cannot be controlled. Those in the second category, the intrinsic soil properties like soil type and depth, are likely to exert the greater influence on crop performance and some of these properties may be identified through soil mapping. An excellent example was the data from the ICI Ten-tonne Club Competition in 1979 and 1980 (Weir et al., 1984). Farmers who entered the competition presumably considered that they could achieve 10 t ha^{-1} winter wheat grain but in both years the mean and range of yields obtained on some soil series were always larger than those on other soils. Thus it may be impossible to eliminate within field variability if fields contain widely contrasted soil types.

The principles of good management require that, if factors affecting crop performance can be identified and controlled, appropriate strategies should be adopted to do so. Such factors fall into two groups: those which directly affect productivity and those which also affect soil fertility.

Crop productivity

Factors directly controlling crop productivity are those inputs which are applied for each crop like nitrogen (N) fertiliser and agrochemicals to control weeds, pests and diseases. Uneven application of N fertilisers or agrochemicals can be a major cause of yield variation but the remedy lies in careful application. If these inputs are applied at the most appropriate amount and time they benefit principally the crop to which they are applied and leave very little residue. Such inputs are not considered further here except to note that there is a well recognised need to target N and agrochemical applications more efficiently to take account of the possible environmental impacts of these inputs.

Soil fertility

Soil fertility arises from complex and often little understood interactions between the biological, chemical and physical properties of a soil. Here some chemical properties are considered and inputs which affect them often have benefits which persist for more than one year. For example, applications of lime (CaCO$_3$) often maintain a suitable pH, 6.5 for arable crops, for three to five years. The return of crop residues and the use of organic manures benefits soil organic matter (SOM). Johnston (1986a; 1991) has given examples where increased yields were obtained when crops were grown on soils with more organic matter in experiments on the same soil type. The results highlight the fact that the yield potential of many cultivars available today was only achieved when all conditions controlling growth were optimum. These include soil structure which is affected by SOM. Applying phosphorus (P) and potassium (K) both in fertilisers and organic amendments improves and/or maintains the P and K status of the soil and this is discussed in more detail below.

SOIL SAMPLING AND SOIL ANALYSIS

Currently there are wide ranging opinions both about the usefulness of soil analysis generally and about the applicability of specific methods of analysis. For the examples given here, readily soluble P (Olsen P) was determined using Olsen's method, (Olsen et al., 1954) and exchangeable K (exch. K) by equilibrium with 1M ammonium acetate, which gives values strongly correlated with those obtained by extraction with ammonium nitrate.

When soil analysis data are used as the basis for recommending applications of liming materials (e.g., $CaCO_3$) and fertilisers, a number of soil samples are taken, usually to a given depth, at random positions throughout the field and then bulked prior to analysis. On the basis of this average analytical value, a single rate of fertiliser is recommended for the whole field, based on the likely average crop response within defined ranges of analytical values (MAFF, 1994). Boyd (1965) showed that Olsen P either alone or when allowance was also made for soil type, accounted for between 37% and 50% of the variance in the response to P, and Olsen P was by far the largest single factor. Other examples are given by Williams and Cooke, 1965; Webber et al., 1976 and Draycott et al., 1971.

Precision farming, defined as spatially variable management within a field, however, will require a greater use of soil analysis and a better understanding of how soil nutrient status can affect yield. Soils supporting good and poor growth and yield should be examined separately to determine the factors controlling yield. Studies may be made in situ or the soil can be sampled for laboratory tests. Protocols for sampling soil to provide a representative bulked sample have been discussed by Oliver et al. (1997).

In situ, soil texture can be assessed and its likely effect on droughtiness or water logging estimated. Compacted layers can be detected by using a suitable logging penetrometer (Anderson et al., 1980) or by digging to expose a profile. Benefits from relieving soil compaction have been shown in a number of studies (see, for example, Johnston and McEwen, 1984 and the references therein). Benefits were usually largest in the drier years suggesting that once the compacted layer was broken up the roots grew more readily into the subsoil to exploit the water reserves held there.

Critical values for Olsen P and exchangeable K

For soils in which P and K reserves accumulate in plant available forms it is reasonable to assume that if all other growth factors are optimum, yield will increase as the availability of P or K in the soil solution increases until yield approaches an asymptote or plateau yield. Olsen P or exch. K at which the yield closely approaches the asymptote is the critical value and it may vary with crop and soil. Below the critical value yield decreases appreciably, resulting in financial loss to the farmer, but there is no financial incentive for the farmer to increase Olsen P or exch. K appreciably above the critical value. Thus, provided there is no unnecessary environmental risk, there are good agronomic reasons to maintain soils just above the critical value for Olsen P and exch. K. Experiments on the silty clay loam at Rothamsted, the sandy loam at Woburn and the sandy clay loam at Saxmundham have sought to determine such critical values for a range of arable crops.

Critical values for Olsen P

Figure 1 shows yields of potatoes (a) and sugar from sugar beet (b) related to Olsen P on a sandy clay loam soil. For both crops yields in the different years varied by a factor of two due to differences in rainfall but the Olsen P at which the asymptote was approached was essentially the same in the different groups of years for both crops. The Olsen P averaged 25 mg P kg^{-1} for potatoes and 20 mg P kg^{-1} for sugar beet (Johnston et al., 1986).

Subsequent experiments on winter wheat grown on a sandy clay loam showed that the critical Olsen P was not affected by increasing amounts of N which gave larger yields (Table 1). When wheat was given only 192 kg N ha^{-1} in four successive years, yields in pairs of years differed by more than 1.5 t ha^{-1} but the critical Olsen P was 14 and 19 mg kg^{-1} in the high and low yielding years respectively.

Various soil and management factors may affect the relationship between yield and Olsen P. For example, Figure 2 shows the relationship between potato yields and Olsen P in the same experiment but on soils with 2.4 and 1.5% organic matter respectively. On the soil with 2.4% organic matter the plateau yield was approached at about 35 mg P kg^{-1} but on the soil with 1.5% organic matter the same yield was only attained on soils with 60 mg P kg^{-1}. In this experiment, on a poorly structured, difficult to cultivate soil, the organic matter appears to have improved soil structure and therefore

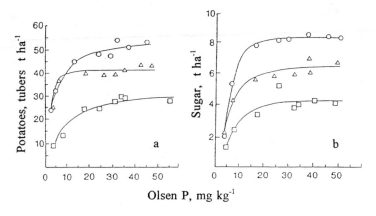

FIGURE 1. Relationship between yields of potatoes (a) and sugar from sugar beet (b) and Olsen P on a sandy clay loam soil. The upper, middle and lower curves were means of years or single year values. (a) Upper 1971, 1974; Middle 1969, 1972, 1973; Lower 1970. (b) Upper 1969, 1971, 1973; Middle 1972; Lower 1970, 1974.

TABLE 1. Yields of winter wheat related to Olsen P.

| | N rate, kg ha^{-1} | | | |
	80	120	160	200
Grain, t ha^{-1}, at 99% of the asymptote	9.15	9.71	10.13	9.96
Olsen P, mg kg^{-1}, giving the above yield	29	25	19	18

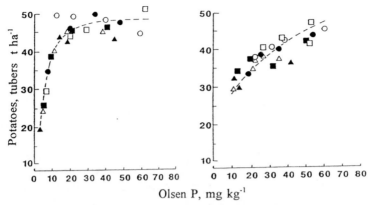

FIGURE 2. Relationship between Olsen P in soil and yields of potatoes grown on soils with either (a) 2.4% or (b) 1.5% soil organic matter, Agdell, Rothamsted.

the ability of the roots to grow through the soil to absorb sufficient P for optimum yield at a lower level of Olsen P. This explanation was supported by data from a glasshouse experiment. A soil sample from each of the plots contributing to the data in Figure 2 was dried, mixed with quartz to make a better rooting medium and cropped with ryegrass in pots. The relationship between yield and Olsen P was not affected by the level of organic matter in the soil and the shape of the response curve was similar to that for crops grown in the field on the better structured soil. Similarly soil depth and stoniness might affect the critical value for P. This type of result poses a serious question. Would it be better to try to improve soil structure or excessively enrich soils with P? For soils vulnerable to P loss by erosion, runoff or leaching, then it must be preferable to try to improve soil structure.

In these experiments every attempt was made to minimise the effects on yield of factors other than Olsen P. Over the wide range of values tested, Olsen P accounted for much of the within year variance in yield. It accounted for 84% for potatoes (Figure 1a), 73% for sugar yields (Figure 1b), 63% for winter wheat given either 80, 120, 160 or 200 kg N ha[-1] (Table 1) and 87% for the wheat yields which varied by 1.5 t ha[-1]. This value for potatoes was higher than that reported by Mattingly (1968) but for sugar yields the value was similar to that of Draycott et al., (1971) who accounted for 60% and 62% of between-site variance by linear regression of sugar beet yield on log soil P in two groups of experiments at different sites.

Thus whilst there might be some reservations about generalising the use of Olsen P, it can be very reliable as a tool for assessing possible causes in yield variation within fields. This is especially so because the soil samples for comparisons will be taken from small, well defined areas within a field in much the same way as small plots are sampled within field experiments.

Critical values for exchangeable K

Johnston and Goulding (1990) reviewed the use of plant and soil analysis to predict the K supplying capacity of soil and concluded that the K status of soils can be classified as well by exch. K as by any other rapid analytical procedure. There are limitations; for example field grown plants in temperate climates rarely grow continuously and, when there is no plant demand for K, non-exch. K can diffuse to exchange sites (see Johnston, 1986b). Also, roots of fully grown crops can explore deeper soil horizons than the surface soils normally sampled for diagnostic purposes (Kuhlmann, 1985).

Experiments made at Rothamsted have sought to relate crop yield to exch. K in soil. Yields of potatoes and field beans (*Vicia faba*) grown on a silty clay loam soil had not reached a plateau yield at 200 mg kg^{-1} exch. K. Figure 3 shows responses of barley grain and sugar from sugar beet also on a silty clay loam. Spring barley yielding about 6 t ha^{-1} grain did not need more than 80 mg kg^{-1} exch. K but sugar yield was still increasing up to 200 mg kg^{-1} exch. K. For both crops the general relationship was clear, but there was some scatter about the mean, more for sugar beet than for barley. This was probably because sugar beet roots explore subsoils more efficiently than do barley roots and subsoil K could have supplied part of the K demand by the beet crop. The amount of subsoil K taken up would vary according to the exch. K content of the subsoils and the amount of root in them. Unless the amount and function of roots in subsoils could be assessed easily there is probably little to be gained from sampling and analysing subsoils for exch. K on a routine basis. However, when investigating yield variability within a field, there could be a justification for subsoil sampling and analysis.

An initial assessment of this limited data set for K suggests that, whilst 100 mg kg^{-1} exch. K would be acceptable for cereals, much more, perhaps in excess of 200 mg kg^{-1}, would be required for potatoes, sugar beet and field beans. This wide range of values can create problems when giving general advice to farmers about the level of exch. K at which to maintain fields. However, it does not invalidate the value of exch. K as a diagnostic tool to decide whether lack of K could be the limiting factor in a poor yielding area within a field.

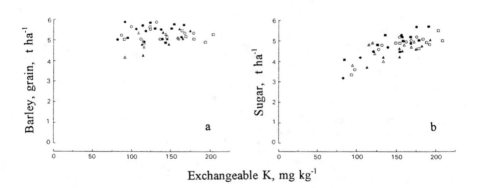

FIGURE 3. Relationship between yields of spring barley (a) and sugar from sugar beet (b) and exchangeable K in soil. Crops grown on soils with 1.5 and 2.4% organic matter (not shown) and manured from 1848-1951 with: NPK, circles; PK, squares; unmanured, triangles; and when cropping was a 4-course rotation: turnips, barley, fallow, wheat, open symbols; turnips, barley, clover, wheat, closed symbols.

CONCLUSIONS

Soil analysis has always been most useful when used to follow changes in soil properties, like nutrient status, within a field as a result of the cropping and fertilising policies adopted by the farmer. This has been amply shown for small plots within experiments. The difficulty has always been to scale up to farm fields because of the problems of getting representative samples from fields inherently lacking uniform soils. Suitable protocols for sampling soil to be analysed for P and K have been discussed elsewhere (Oliver *et al.*, 1997).

In precision farming soil analysis will be an essential management tool. Soil samples will only be taken from small areas shown to differ in yield to assess whether the nutrient status of the soil is the cause of the variation in yield. These areas can be resampled periodically using GPS techniques to record and identify them. If the P and/or K status are identified as factors causing yield differences then more of one or both nutrients can be applied to the areas identified as low in either nutrient. If soils have too high a P or K status then less can be applied. How much more or less and for how long requires that the critical levels of both Olsen P and exch. K are estimated with reasonable accuracy. The data presented here suggest that this can be done. Once Olsen P and exch. K are just above the critical value then soil analysis can be used as a periodic check to ensure that the correct value is being maintained. This approach fits well with the concept of integrated plant nutrient management because, irrespective of the form in which P or K is applied, (fertiliser, organic manure or other amendment) and the availability of the nutrient in the different sources, the aim will be to maintain the appropriate level in the soil.

REFERENCES

Anderson, G., Pidgeon, J.D., Spencer, H.B., Parks, R. (1980) A new hand held recording penetrometer for soil studies. *Journal of Soil Science*, **31**, 279-296.

Boyd, D.A. (1965) The relationship between crop response and the determination of soil phosphorus by chemical methods. *Soil Phosphorus. Technical Bulletin 13*, London, HMSO, Ministry of Agriculture, Fisheries and Food, pp 94-102.

Draycott, A.P., Durrant, M.J., Boyd, D.A. (1971) The relationship between soil phosphorus and response by sugar beet to phosphate fertiliser on mineral soils. *Journal of Agricultural Science, Cambridge*, **77**, 117-121.

Johnston, A.E. (1986a) Soil organic matter, effects on soils and crops. *Soil Use and Management*, **2**, 97-105.

Johnston, A.E. (1986b) Potassium fertilization to maintain a K balance under various farming systems. *Nutrient balances and the need for potassium*, Basel, International Potash Institute, pp 177-204.

Johnston, A.E. (1991) Soil fertility and soil organic matter. *Advances in soil organic matter research: the impact on agriculture and the environment*, W.S. Wilson (Ed.), Cambridge, Royal Society of Chemistry, pp 299-313.

Johnston, A.E., Lane, P.W., Mattingly, G.E.G., Poulton, P.R., Hewitt, M.V. (1986) Effects of soil and fertilizer P on yields of potatoes, sugar beet, barley and winter wheat on a sandy clay loam soil at Saxmundham, Suffolk. *Journal of Agricultural Science, Cambridge*, **106**, 155-167.

Johnston, A.E., Goulding, K.W.T. (1990) The use of plant and soil analyses to predict the potassium supplying capacity of soil. *The Development of K-fertilizer Recommendations*, Basel, International Potash Institute, pp 177-204.

Johnston, A.E., McEwen, J. (1984) The special value for crop production of reserves of nutrients in the subsoil and the use of special methods of deep placement in raising yields. *Nutrient balances and fertilizer needs in temperate agriculture*, Basel, International Potash Institute, pp 157-176.

Kuhlmann, H., Claasen, N., Wehrmann, J. (1985) A method of determining the K uptake from subsoil by plants. *Plant and Soil*, **83**, 449-452.

Mattingly, G.E.G. (1968) Evaluation of phosphate fertilisers. II. Residual value of nitrophosphates, Gafsa rock phosphate, basic slag and potassium metaphosphate for potatoes, barley and swedes grown in rotation with special reference to changes in soil phosphorus status. *Journal of Agricultural Science, Cambridge*, **70**, 139-156.

Mercer, W.B., Hall, A.D. (1911) The experimental error in field trials. *Journal of Agricultural Science, Cambridge*, **4**, 107-132.

MAFF (Ministry of Agriculture, Fisheries and Food) (1994) *Fertiliser Recommendations for Agricultural and Horticultural Crops*, London, HMSO, 112 pp.

Oliver, M.A., Frogbrook, Z., Webster, R., Dawson, C.J. (1997) A rational strategy for determining the number of cores for bulked sampling of soil. *Proceedings of the First European Conference on Precision Agriculture*, J.V. Stafford (Ed.), London, SCI.

Olsen, S.R., Cole, C.V., Watanabee, F.S., Dean, L.A. (1954) Estimation of available phosphorus in soils by extraction with sodium bicarbonate. *United States Department of Agriculture*, Circular No. 939, 19 pp.

RES (1926, 1928, 1930) *Annual Reports for 1926, 1928 and 1930, Rothamsted Experimental Station*, Lawes Agricultural Trust, Harpenden.

Webber, J., Boyd, D.A., Victor, A. (1976) Fertilizers for maincrop potatoes on magnesian limestone soils: experiments in Yorkshire 1966-71. *Experimental Husbandry*, **31**, 80-90.

Weir, A.H., Rayner, J.H., Catt, J.A., Shipley, D.G., Hollies, J.D. (1984) Soil factors affecting the yield of winter wheat: analysis of results from ICI surveys 1979-80. *Journal of Agricultural Science, Cambridge*, **103**, 639-649.

Williams, R.B.J., Cooke, G.W. (1965) Measuring soluble phosphorus in soils, comparisons of methods and interpretation of results. *Soil Phosphorus, Technical Bulletin 13*, London, HMSO, Ministry of Agriculture, Fisheries and Food, pp 84-93.

ON-SITE DIAGNOSIS OF SOIL STRUCTURE FOR SITE SPECIFIC MANAGEMENT

H. DOMSCH

Institut für Agrartechnik Bornim e.V. (ATB), Max-Eyth-Allee 100, D-14469 Potsdam-Bornim, Germany

O. WENDROTH

ZALF, Institut fuer Bodenlandschaftsforschung, Eberswalder Str. 84, D-15374 Muencheberg, Germany

ABSTRACT

The soil structure of a field site has to be considered as an important factor which must be taken into account for precision farming decisions. Measurements of the soil structure are not trivial and require effective methods. Penetration resistance measured with an improved penetrometer in combination with statistical approaches well known from time series analysis was tested in a moraine soil of Brandenburg. The results clearly reflected the physical impact of former management practice.

INTRODUCTION

Precision farming is a tool for increasing the economic efficiency of agricultural production and a way of better accounting for ecological requirements. In order to account for heterogeneity of agricultural field soils, the temporal and spatial variability patterns of their soil properties have to be known.

Biomass production and crop yield are highly influenced by physical, biological and chemical soil properties. However, there are hardly any experimental tools that can be used at the field scale with a high spatial resolution. Grid sampling can be laborious and is often too labor intensive for an intensive sampling of soil parameters across whole field sites.

Basic information for producing cheap maps of selected soil properties is available in Germany from a very intensive soil survey of the whole of Germany (Bodenschätzung), from elevation maps, and from aerial photographs. For soil parameters relevant to soil structural description, such as dry bulk density and permeability, hardly any results exist for intensive field scale measurements, although the effects of these soil structural properties on crop yield have been identified as highly important factors in small scale experiments.

Dry bulk density of soil is hardly measurable at the field scale. Soil core sampling would be too laborious. An alternative to core sampling may be radiometric density probes. However, the required effort remains very high. Measurement of penetration resistance, i.e., the force that soil exerts against a vertical penetrating or horizontally pulled cone, is a rather effective measure of soil structural state (Lüth, 1993). Vertical penetrometers allow for the determination and simultaneous data storing of penetration resistance values down to a depth of about 70 cm with a resolution of about 1 cm. Results can be recorded on a PC where the data are further processed. Horizontal penetrometers measure the cone resistance at a predetermined soil depth. There is still a need for better safety devices and for solving mechanical problems due to worn parts.

The aim of this study was to investigate vertical soil penetration resistance as an effective measure of soil structural state along two transects in an experimental field with a well known history of traffic load and soil tillage regime. Furthermore, field observations should be analyzed with special statistical tools that allow a focus on the spatial covariance structure of on-site observations. The task was to evaluate whether penetration resistance is an adequate tool for identifying the spatial variability pattern of soil structure at the field scale. For this investigation, a vertical penetrometer was used with a pressure transducer directly at the tip of the cone in order to avoid any influence due to shaft friction when driving it into the soil.

MATERIALS AND METHODS

The experimental site is part of the younger glacial drift areas. It is located in an area of valley sand deposit with a depth of 1 to 3.7 m. The soil was classified as a Cambic Podsol/Spodic Arenosol from dry dystrophic sand deposits. For a long time, the field was under agricultural management. In 1990, it was set aside after ploughing in the autumn. The field traffic history is not known in detail. In 1993, traffic lanes were installed with varying loads 3.2 m apart from each other across a band of 80 m. Eight different load treatments were repeated three times with a varying sequence. The traffic lane distance was 1.8 m, and the lane width was 0.5 m.

In summer 1995, part of this systematically loaded area was loosened with a paraplough down to a depth of 50 cm. After twofold tine cultivation (20 cm deep), the whole field was planted to rye. Perpendicular to the load plots and the main tillage direction, two transects, transect 3 and transect 5, with a spatial separation distance of 100 m were selected (Figure 1). Along both transects, soil texture was investigated in 0 - 30, 30 - 50, and 50 - 70 cm soil depths, respectively, each in a distance of 30 m. There were no significant differences between the six sampling locations. Because of this, no further studies were carried out.

Transect 3 leads across the soil loosened with the paraplough whereas transect 5 leads across the soil not loosened. Along both transects, penetration resistance was determined at locations separated by a distance of 15 cm. The penetrometer was equipped with a cone which had a basis area of 1 cm^2. The pressure transducer was mounted directly behind the cone tip. Measurements were registered every 1 cm down to a depth of 70 cm and were downloaded to a computer at the end of each day (leaflet, 1996). In order to increase the productivity of measurements and to provide the force necessary, the penetrometer was mounted on a tractor and driven into the soil hydraulically.

An hydraulic driven penetrometer is still advantageous for studies along lanes, because it makes it possible to work the whole day without getting tired. It also guarantees vertical penetration and constant penetration speed. About 100 measurements per hour

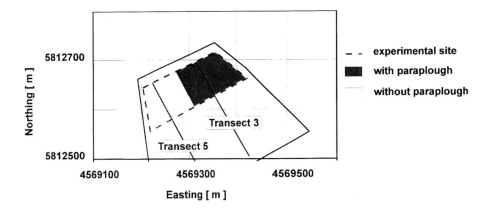

FIGURE 1. Experimental design.

are possible in the average of a day. For this a tractor driver and an operator for the penetrometer are necessary. The operator gives the driver a sign, when the next point shall be targed and when it is reached, using a tape measure. He types the numbers for the measure points (they are identical to the distance to the lane zero point), serves the hydraulic system and takes care that there is no damage done to the penetrometer.

The permanent control of the penetrometer would not be necessary if a more stiff penetrometer with bigger cone basis area would be used. This was not done to keep the measurements comparable to the ones of the hand penetrometer.

If the penetrometer meets a stone or a layer with very high soil strength (> 10.2 N/mm²), the hydraulic pressure is increased until the penetrometer measure limit is reached. The penetrometer stops and the measuring is ended automatically. For those locations with the maximum value, no values could be recorded for greater depths (Figure 2, 40 cm-layer).

The period of time of the penetrometer use is limited by the soil water content. Water tension should not exceed the range of 60 mbar to 200 mbar on sandy soils. These conditions are given in German particularly in some weeks in spring (between the thaw out of the soil and the start of pre summer dryness) respectively after extended rainfalls in the course of the year.

RESULTS AND DISCUSSION

Penetration resistance

The magnitude of the penetration resistance depends on soil texture, soil moisture and bulk density. Penetration resistance values only reflect soil structural differences if soil texture and soil moisture are known or are fairly constant. During the measuring campaign, soil water matric potential was 115 mbar ± 15 mbar in transect 5 and 120 mbar ± 23 mbar in transect 3. For field measurements, this range can be considered as sufficiently constant.

From the data recorded, measurements for soil layers at 20 and 40 cm depth were considered for further analysis. In order to decrease the variability due to measurement variance, values at the respective depth ± 3 cm were averaged.

The shape of penetration resistance profiles and their spatial distribution is coincident with the spatial distribution of traffic load paths (Figure 2). Maximum penetration resistances were found in the 40 cm layer of the soil which was not loosened with the paraplough.

Tine cultivation loosened compacted zones in the 20 cm layer. But even after twofold tine cultivation, highly compacted zones in the wheel tracks maintained a significantly

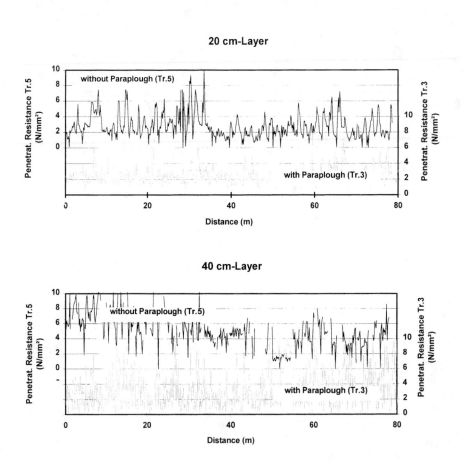

FIGURE 2. Penetration resistance measured at 0.15 m intervals within the 20 cm and 40 cm soil layers along two 80 m long transects (with and without paraplough).

higher density than in the untrafficked soil in between. Therefore, the wheel track pattern is clearly reflected even after tine cultivation.

After deep loosening with the paraplough (transect 3), the standard deviation of penetration resistance increases in the 40 cm layer. Dense aggregates with high penetration resistances remain, although loosening was intensive. However, paraplough loosening in combination with tine cultivation destroyed a large fraction of the aggregates with high density.

Autocorrelation function

As a measure of the spatial correlation of neighboring observations separated by increasing distances, autocorrelation functions for both transects and both soil depths are shown in Figure 3. In general, autocorrelation decreases with increasing lag distance. However, especially in the 20 cm layer of both transects, a sinusoidal shape of the autocorrelation function became apparent, indicating variance components of penetration resistance with repeating or regular patterns. For the 40 cm layer, this wave pattern became less pronounced and fluctuations over shorter distances than in the 20-cm layer became important.

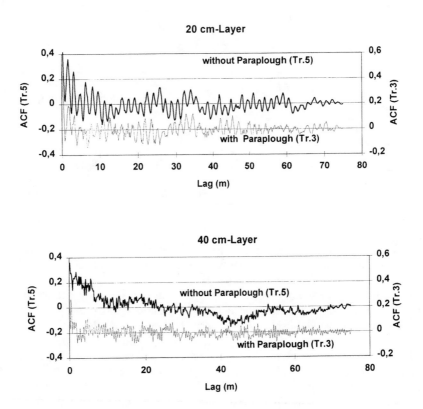

FIGURE 3. Autocorrelation function (ACF) of the penetration resistance data shown in Figure 2.

Spectral density

As an integrative measure of autocorrelation, spectral density is a tool for decomposing variability components with a periodical fluctuation (Nielsen *et al.*, 1983). In the 20 cm layer wavelengths of 1.6 and 3.2 m, respectively, were significant frequency dependent variations (Figure 4). These wavelengths correspond to the width of plots (3.2 m) and the distance between untrafficked zones (1.6 m), i.e., the space between the plots and between the wheel tracks. A peak in the power spectrum at a wavelength of 38.4 m corresponds to the width of 12 plots, i.e., half of the plot area.

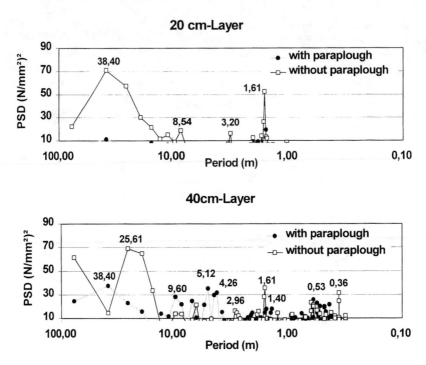

FIGURE 4. Power spectra of the values in Figure 2.

The widening of compacted zones at 40 cm depth in transect 5 hides a clear periodic variance component and a peak at the wavelength corresponding to the plot width. This layer exhibits a clear fluctuation with a period of 1.6 m. A peak at 25.6 m corresponds to the width of eight plots or a third of the plot area. The short wave fluctuation with a 0.36 m period may not result from additional loads but was probably caused by the working width of a plowshare of the formerly used moldboard plow.

Tillage with the paraplough overlays some relations and causes others: e.g., in the 20 cm layer, periods of 1.6 m and 38.4 m are hardly pronounced, the width of the paraplough share (0.5 m) is exhibited in the 40-cm layer as is the total working width of the paraplough (1.4 to 1.5 m). Further peaks at 4.2 m (equal to three times 1.4 m), 9.6 m (three times 3.2 m) and 38.4 m (twelve times 3.2 m) correspond to multiples of known widths and measures of plots and tools.

CONCLUSION

Penetration measured at about field capacity reflected effects of previous loads or tillage operations both qualitatively and quantitatively. The measuring method is very effective to characterise the soil structure along lanes compared to other procedures. The use range is limited by soil humidity so that measurements are particularly carried out in spring. Statistical approaches well known in the discipline of time series analysis provide tools for the identification of periodic variance components which are typical for tilled soils surveyed perpendicular to their regular patterns of traffic paths and tillage operations. This means that sampling in certain zones along transects in a field may already be sufficient to identify critical and/or typical states of soil structure.

REFERENCES

Lüth, H.-G. (1993) Entwicklung des Längs-Penetrographen als Meßverfahren zur Bodenverdichtung. Diss., Agrarwissenschaftliche Fakultät der Christian-Albrechts-Universität zu Kiel.

Nielsen, D.R., Tillotson, P.M., Vieira, S.R. (1983) Analyzing field-measured soil water properties. *Agric. Water Management,* **6**, 93-109.

Leaflet (1996) Penetrometer „sensopress 17" v96e. *Company MediUM-SENSOR,* Berlin.

OPTIMISATION OF LIME APPLICATION TO TAKE ACCOUNT OF WITHIN- FIELD VARIATION IN pH AND SOIL TEXTURE

E.J. EVANS[a], R.S. SHIEL[b], S.B. MOHAMED[a]

Departments of Agriculture[a] and Agricultural and Environmental Science[b], University of Newcastle upon Tyne, NE1 7RU, UK

ABSTRACT

Soil pH varied greatly across two uniformly managed arable fields in Northumberland. The geostatistical range was large, resulting in maps with substantial areas having a uniform pH. These areas crossed field boundaries, demonstrating that current practice is not reducing the extent of pH variation. The large areas of uniform pH would allow differential application of liming materials. A uniform application of lime, however, based on an average pH for the entire field was calculated to be less effective than spatially targeted applications; half the area would either not reach the target pH of 6.5, or would exceed pH 7.0. Relationships between lime requirement, soil texture and soil organic matter were examined, and it is shown that variation in texture and organic matter can be ignored unless they vary greatly across the field.

INTRODUCTION

The regular application, at an appropriate rate, of a liming material - such as chalk or ground limestone - is essential in humid regions if plant growth is not to be restricted (MAFF, 1981). Although the target pH for arable soils in Britain is usually taken as 6.5, the pH below which crop growth is adversely affected varies substantially from species to species (MAFF, 1973). At excessively high pH, plant growth is also damaged and even species which prefer a high pH, such as sugar beet, can be affected by deficiencies, such as manganese, at pH > 6.5 (Farley and Draycott, 1973). The effect of acidity (and of alkalinity) on an individual species also varies with the soil texture and organic matter content. Soils which contain little clay or contain a large amount of organic matter have small reserves of micronutrients, and therefore tend to have an increased risk of micronutrient deficiency (MAFF, 1984); the target pH for liming organic or peaty soils in Britain for arable crops is therefore reduced to 6.1 and 5.8 respectively (MAFF, 1985).

The amount of lime needed to raise pH depends on the current pH, target pH, and the buffer capacity of the soil. The last is affected by texture and organic matter content; and for advisory purposes lime requirement in Britain is now estimated from a two-way table of soil pH and soil texture/organic matter content (Archer, 1988). Current commercial practice involves taking a composite soil sample from, usually, a whole field, but this method ignores the effects of within-field variation of soil pH, soil texture and soil organic matter (SOM) content. Borgelt *et al.* (1994) have shown substantial variation in pH and texture within a field, leading to considerable variation in optimum lime recommendation within the field.

If large numbers of samples are to be taken so as to estimate the extent of within-field variation, then cheap and rapid methods of soil analysis are required. Fortunately, the texture and organic matter content change relatively slowly so that on future occasions lime requirements can be obtained from pH measurements only. The number of samples, and hence the spacing within a field, is affected by the minimum numbers necessary to apply geostatistical procedures effectively (Wollenhaupt et al., 1994). It is essential to ensure that the range of the variation in properties is sufficiently large to allow machinery to apply liming materials at an appropriate, variable rate across the field (Webster and McBratney, 1987). As the machine will operate in parallel lines, it is more convenient if the smooth contoured maps produced by geostatistical procedures are adapted to give a grid pattern of uniform cells. Shiel et al. (1997) have shown that conversion to a grid pattern results in a negligible loss of accuracy. The grid also ensures that the best values for pH, clay and SOM are available for each square, and allows the lime requirement to be calculated easily.

In this paper, the variation in soil pH and texture across two commercial fields has been examined and the lime requirement measured both on a conventional and on a precision farming basis. From the difference we have then calculated the error in application that results from conventional methods of liming, and have estimated the likely impact on future crop production.

MATERIALS AND METHODS

Soil samples were collected in January 1991 from two adjoining 16 ha fields used for arable cropping on a commercial farm in Northumberland, UK (55° 17' N, 1° 3' 30" W; NZ217988). Samples were spaced approximately 20 m apart in parallel lines 40 m apart. Ten subsamples were collected from 0-15 cm at each sampling point, using a cheese auger, and bulked. The 244 samples from North field and 235 from South field were dried and ground to pass a 2 mm sieve. Soil pH was measured in a 1:2.5 soil:water ratio using a glass electrode (MAFF, 1986). The soil texture was assessed by hand texturing, working the moistened soil between thumb and fingers. The hand textured results were calibrated using 10 randomly selected samples from each field which were dispersed with calgon after removal of organic matter and the clay and silt were measured by the pipette method (Avery and Bascomb, 1974). Organic matter content was measured by shaking a 5 g soil sample with 50 ml 0.43 M HNO_3 for 1 hour and filtering through a Whatman No. 42 into a polythene flask. The optical density of the filtrate was measured in a spectrophotometer at 540 nm (Haneklaus and Schnug, 1996). The optical density was correlated against the carbon content measured in a LECO C-S determinator using 36 of the samples.

The data were analysed using UNIMAP (UNIRAS, 1989) and MINITAB and were converted into map form using the 'fault' procedure in UNIMAP. This has been shown to be the most successful of the map realisation methods for modelling datasets from these fields (Mohamed et al., 1997).

RESULTS

The mean pH of both North field and South field was low enough to justify application of lime (Table 1) but, within both, the pH at different sampling points varied from very acid (4.64, North; 5.13, South) to neutral (6.94, North; 7.15, South); the variation in pH, as measured by CV, was relatively small and in neither field was the distribution

TABLE 1. pH, clay and soil organic matter (SOM) content of North and South fields.

	North field				South field			
	Mean	SD	CV(%)	Skewness	Mean	SD	CV(%)	Skewness
pH	5.84	0.43	7.4	0.21	6.10	0.364	6.0	-0.14
Clay (%)	31.5	4.73	15.0	-0.16	29.1	7.677	26.4	-0.77
SOM (%)	4.73	0.88	18.5	2.42	5.19	1.853	35.7	2.60

severely skewed. Clay content, by comparison, was much more variable within each field, and in South field was negatively skewed. The distributions of sand and silt contents were also variable, and that of sand was skewed. There was no clear pattern to the distribution of silt samples (data not shown). The distribution of soil organic matter was also variable and strongly positively skewed in both fields. There were a small number of samples with large SOM contents, particularly in South field.

Although SOM and clay were not significantly correlated in North field, there were significant correlations between pH and both properties (Table 2). In South field, pH was not correlated with clay or SOM, but there was a highly significant negative correlation between SOM and clay. On the basis of the sand, silt and clay contents, all of the samples were allocated to one of the twelve soil textural groups (Hodgson, 1976).

In both fields clay loam was the commonest textural class; it constituted just over half the samples in South field (Table 3), and the textural class found by averaging all the sand, silt and clay contents for the samples was also clay loam in both fields. Using the mean pH and textural group, lime requirements of 6.0 and 5.0 t $CaCO_3$/ha are found for North and South fields respectively (MAFF, 1985).

The distribution of clay, SOM and pH within both fields had a large range (> 140 m), and sills were substantially larger than nugget values (data not shown). This indicates that the variation of all three properties is non-random, and justifies the production of maps by interpolation (Webster and McBratney, 1987). There were substantial areas of low pH in North field, and these extended across the boundary into South field (Figure 1).

TABLE 2. Correlations between pH, clay and SOM in North and South fields.

	North Field		South Field	
	pH	SOM	pH	SOM
SOM	0.210*		-0.088	
Clay	0.390***	0.055	0.028	-0.442***

TABLE 3. Proportion of North and South field samples in each soil texture class.

	North field	South Field
Sandy loam	0	10.6
Sandy clay loam	8.6	19.1
Clay loam	73.0	54.5
Clay	18.4	15.7

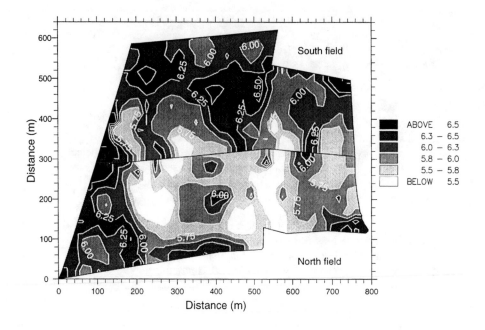

FIGURE 1. pH of North and South fields.

Areas of high pH also crossed the field boundary. The pattern of pH variation was consistent with the large range found by kriging. Clay content was lowest in one corner of South field, although this area sweeps in a large curve along the boundary between the two fields (Figure 2). There are substantial areas of both fields with clay contents in excess of 35%. Two areas in South field have SOM > 6%, and once again there is an extension across the boundary into North field (Figure 3). There is a very large area of low (< 4.25%) SOM in the middle of both fields, and this extends across the boundary.

The maps were converted into rectangular grids of 20 m squares within each of which best estimate values of pH, texture and SOM were calculated using UNIMAP. Using pH only, and the average clay loam texture, the lime requirement of each grid square was found (Table 4). Forty-four % of North field and 31% of South field received a lime application that differed by more than 2 t/ha from the appropriate rate if a uniform rate was applied to the whole field. All of North field comes into the same liming textural group but part of South field is sandy loam (Table 3) and should receive different rates of application. This would only result in 4.3% of the field receiving a different rate of lime from taking an average textural value. Only 1.1% of South field, and none of North field, has a SOM content > 10% and would class as 'organic', although 15.8% of South field has a SOM content greater than 6%.

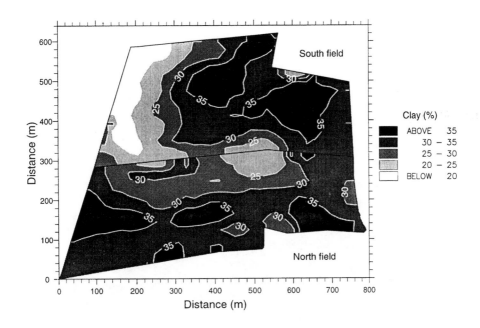

FIGURE 2. Clay content of North and South fields.

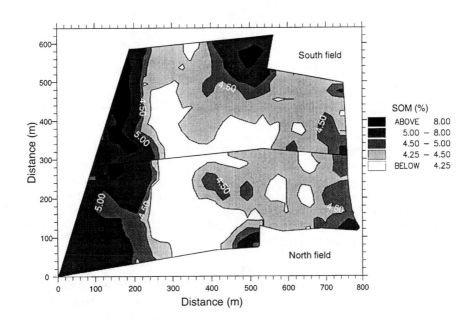

FIGURE 3. Soil organic matter content in North and South fields.

TABLE 4. Lime requirement of areas of North and South fields.

pH	Appropriate requirement t CaCO₃/ha	North field area ha	North field error t/ha	South Field area ha	South Field error t/ha
6.5	0	0.64	6	1.36	5
6.4	2	0.56	4	1.12	3
6.3	3	0.84	3	1.64	2
6.2	4	0.80	2	1.92	1
6.1	5	1.00	1	1.72	0
6.0	6	1.20	0	1.68	1
5.9	6	1.20	0	1.68	1
5.8	7	2.12	1	0.96	2
5.7	8	1.04	2	0.84	3
5.6	9	1.68	3	0.48	4
5.5	10	0.80	4	0.24	5
5.4	10	0.84	4	0.12	5
5.3	11	0.56	5	0.16	6
5.2	12	0.36	6	-	
5.1	13	0.16	7	-	
5	14	0.04	8	-	

DISCUSSION

The use of an average pH for these fields will lead to substantial areas being over or under limed. MAFF (1985) have given a formula for lime requirement (L, t/ha) in terms of target pH (T), present pH (P) and lime factor (F, t/ha) which can be rearranged to give:

$$T = P + L/F \qquad (1)$$

The lime factor for clay loam is 8 t/ha. This suggests that on North field, with a uniform lime application, 44% of the area will not reach pH 6.5 and 14% will exceed pH 7. In South field, 21% will not reach pH 6.5 and 18% will exceed pH 7, ignoring the area of sandy loam. This equation is, however, likely to overestimate the effect of lime at high pH values and underestimate its effect at low pH, as the buffer curve is non linear (Magdoff and Bartlett, 1985). It might be expected that the increased leaching loss from the overlimed areas, and the reduced loss of calcium from underlimed portions of the field (Bolton, 1977), would have reduced the within-field variation, but this does not seem to have occurred. As pH does not respect the field boundary, it appears that the remaining differences are the result of some inherent property of the soil.

The inclusion of texture as a variable has little impact on the resulting pH, as only 8% of one of the fields has a different texture - and there is only a difference in lime application rate between sandy and clay loam for half this area. Ignoring the textural difference, and applying a uniform rate of lime to the whole field will, however, increase the area overlimed, but reduce that underlimed. Edmeades et al. (1985) and Logan and Floate (1985) have suggested relating lime requirement to pH, clay and organic matter content directly, an approach which has the attraction of linking buffer capacity to the change in pH. Unless an accurate map of clay and soil is available, however, such an approach is impractical.

Bailey et al. (1989) have tested the MAFF model against one based on soil incubation with liming material, and found an excellent correlation. They noted, in fact, that SOM was more important as a determinant of lime requirement than was clay content; this may have been because of the correlation ($r = 0.34*$ [$n = 35$]) between clay and SOM in their samples. The samples used here did not exhibit a positive correlation (Table 2) and hence the outcome may be different. As SOM is known to obscure the exchange sites in clay (Evans and Russell, 1959), it may not be surprising that MAFF have included such a wide range of clay content in their 'medium + clay' group (18 - 100%). More precision in measuring SOM may be required if lime applications are to be applied more precisely and lime-induced micronutrient deficiencies are to be avoided on coarser textured soils. Although ground limestone remains a relatively inexpensive material, on a hectare-year basis, the range of pH noted here mitigates against a cavalier attitude to application. Farmers may be less likely to make use of spatially targeted lime applications because the nature of ground limestone makes it more difficult to spread accurately than granulated fertilisers, but the areas receiving different application rates in this study are sufficiently large that, despite these physical limitations, spatial application would be practicable and may be economically worthwhile. pH is cheaper to measure than soil P or K, and can have just as great an impact on crop performance. It will certainly be pointless to improve the accuracy of application of N, P and K if pH continues to vary over the range noted here.

REFERENCES

Archer, J. (1988) *Crop nutrition and fertiliser use.* Ipswich, Farming Press, 265 pp.

Avery, B.W., Bascomb, C.L. (1974) Soil Survey Laboratory Methods. *Technical Monograph,* **6,** Harpenden, Soil Survey.

Bailey, J.S., Stevens, R.J., Kilpatrick, D.J. (1989) A rapid method for predicting the lime requirement of acidic temperate soils with widely varying organic matter content. II Testing the lime requirement model. *Journal of Soil Science,* **40,** 821-829.

Bolton, J. (1977) Changes in soil pH and exchangeable calcium in two liming experiments on contrasting soils over 12 years. *Journal of Agricultural Science, Cambridge,* **89,** 81-86.

Borgelt, S.C., Searcy, S.W., Stout, B.A., Mulla, D.J. (1994) Spatially variable liming rates: a method for determination. *Transactions of the American Society of Agricultural Engineers,* **37,** 1499-1507.

Edmeades, D.C., Wheeler, D.M., Waller, J.E. (1985) Comparison of methods for determining lime requirements of New Zealand soil. *New Zealand Journal of Agricultural Research,* **28,** 93-100.

Evans, L.T., Russell, E.W. (1959) The adsorption of humic and fulvic acids by clays. *Journal of Soil Science,* **10,** 119-132.

Farley, R.F., Draycott, A.P. (1973) Management difficulties of sugar beet in organic soils. *Plant and Soil,* **38,** 235-244.

Haneklaus, S., Schnug, E. (1996) A rapid method for the indirect determination of the organic matter content of soils. *Commun. Soil Sci. Plant Nutr.,* **27,** 1693-1705.

Hodgson, J.M. (1976) Soil Survey Field handbook. *Technical Monograph,* **5,** Harpenden, Soil Survey.

Logan, K.A.B., Floate, M.J.S. (1985) Acidity in upland and hill soils: cation exchange capacity, pH and lime requirement. *Journal of the Science of Food and Agriculture,* **36,** 1084-1092.

MAFF (1973) Fertiliser recommendations for agriculture and horticultural crops. London, GFI, HMSO.

MAFF (1981) Lime and liming. *Reference Book 35,* London, HMSO.

MAFF (1984) Lime and fertiliser recommendations No 1 Arable Crops. *ADAS Booklet 2496,* London, HMSO.

MAFF (1985) Changes in ADAS lime recommendations (GENFER). *Internal Technical Note,* West Midlands, ADAS.

MAFF (1986) The analysis of agricultural materials. *Reference Book 427,* London, HMSO.

Magdoff, F.R., Bartlett, R.J. (1985) Soil pH buffering revisited. *Soil Science Society of America Journal,* **49,** 145-148.

Mohamed, S.B., Evans, E.J., Shiel, R.S. (1997) Mapping techniques and intensity of soil sampling for precision farming. *Precision Agriculture: Proceedings of the third International Conference,* P.C. Robert, R.H. Rust, W.E. Larson (Eds), Madison, ASA, pp 217-226.

Shiel, R.S., Mohamed, S.B., Evans, E.J. (1997) Planning phosphorus and potassium fertilisation of fields with varying nutrient content and yield potential. *Proceedings of the First European Conference on Precision Agriculture,* J.V. Stafford (Ed.), London, SCI.

UNIRAS (1989) *Unimap 2000 Users Manual.* Soborg, Denmark.

Webster, R., McBratney, A.B. (1987) Mapping soil fertility at Broom's Barn by simple kriging. *Journal of the Science of Food and Agriculture,* **38,** 97-115.

Wollenhaupt, N.C., Wolkowski, R.P., Clayton, M.K. (1994) Mapping soil test phosphorus and potassium for variable-rate fertiliser application. *Journal of Production Agriculture,* **7,** 441-448.

PRELIMINARY EXPERIENCE CONCERNING PRECISION AGRICULTURE IN POLAND

M. FOTYMA, A. FABER, M. CZAJKOWSKI, K. KUBSIK

Institute of Soil Science and Plant Cultivation, Department of Soil Fertility and Fertilization, 24-100 Pulawy, Poland

ABSTRACT

The spatial variability of fertility and yields within fields has not been considered previously under Polish conditions. This paper discusses the first attempt to integrate data on spatial variability with site-specific nitrogen and phosphorus fertilization. The studies carried out on a test field of 7.2 ha have shown high levels of variability in soil factor and chlorophyll indices, LAI and yields of winter wheat. Possible relationships between these factors were investigated using WOFOST Crop Simulation Model and statistical analysis. The simulated potential yield was 8.11 t ha^{-1} and the model showed no effect of water shortage on grain yield. Some correlations between yield and chlorophyll index, LAI, soil type, soil complex, Mg and available water holding capacity were found. The analysis has shown non-significant correlations between yield and pH, P and K. The fertilizer rates of N and P were calculated according to P soil test values and extrapolated yields. The differences between calculated and commonly used fertilizer rates were discussed.

INTRODUCTION

In western Poland the medium and large size farms are usually divided into a few uniformly utilized fields. The agricultural landscape in this part of the country is flat and monotonous with a majority of arable lands. The large fields are managed uniformly. The fertilizer rates are blanket applied based on soil quality, average yield and soil test values. The results of soil surveys and soil analyses have shown, however, that the soils in these fields are heterogeneous with respect to soil map type, texture, water status, pH and content of soluble nutrients.

Recently, in some countries a new course of research and a more practical approach to land utilization with regard to soil heterogeneity is being developed (Schnug *et al.*, 1993; Murpy *et al.*, 1994; Stafford *et al.*, 1996). The targeting of inputs made possible by precision agriculture seems to be attractive for the 'developed agriculture'. An important question for countries where the economy is in transition to the open market exists - is this system beneficial for the actual economic situation in these countries? The studies that were carried out sought to address this question.

MATERIALS AND METHODS

In Poland the research on precision agriculture started in 1994 in the specially established experimental station at Baborowko (16.65° E, 52.58° N) in western Poland. The whole area of the station was split into eight fields from 5 to 7 ha each. For these fields soil maps and maps of agrochemical properties have been prepared. The paper refers to just one of those fields.

The field of 7.2 ha was sampled manually within a grid of 25 m. The soil series were identified according to genetic properties of soil profiles, texture and usefulness of soils using the Polish soil description system (Witek and Gorski, 1977). Two general categories of soil series were distinguished: the genetic soil types and the soil usefulness complexes. Under the name complex of agricultural soil usefulness we gather a group of different soils, which have similar agricultural properties and may be utilized in a similar way. Complexes are separated on the qualitative basis of such criteria as: character and properties of the soil itself, terrain relief and moisture conditions. The soil samples were analyzed for pH (1 M KCl), soluble P and K (Egner-Riehm method) and soluble Mg (Schachtschabel methods). The soil units map and macroelement maps were drawn using AUTO-CAD software. The interpolation of pH and nutrient contents in soils were done using an inverse distance technique.

Winter wheat (cv. Ellena) succeeding oilseed rape was grown under rainfeed conditions in 1995/96. The wheat was sown on September 21. The presowing fertilization was 23 kg P ha^{-1} (TSP) and 67 kg K ha^{-1} (KCl). The field was fertilized with 75 kg N ha^{-1} supplied as a water solution of urea and ammonium nitrate. The first 50 kg N was used before the start of spring vegetation and the remaining in GS 30 (Zadoks et al., 1974). The crop was harvested on August 23 using a Massey-Ferguson combine harvester equipped with the GPS and the flow yield measurement device. The map of the yield was prepared by the Massey-Ferguson Datavision System.

The physiological status of the crop was tested by detailed biometric studies. At one to two week intervals measurements of total above-ground biomass, LAI (LAI-2000 Plant Canopy Analyzer, LI-COR Inc.) and Chlorophyll Index (SPAD Chlorophyll Meter, Minolta Corp.) in the upper leaves were done on 0.04 ha plots treated as the entire test field, located on four soil complexes. The LAI and chlorophyll indexes (the relative quantity of chlorophyll by measuring the transmitance of leaf) were used as an integrated measure of growing conditions (Castelli et al., 1996). The LAI dynamics in the spring and summer vegetation were compared with potential LAI (determined by radiation and temperature only) simulated with the WOFOST 6.0 Crop Simulation Model (Supit et al., 1994). The model was calibrated for the winter wheat cultivated in Polish conditions (Faber et al., 1996). From the same model the evapotranspiration and water balance calculations were used to characterize the growing conditions. The simulations were run for soils with available water contents of 140 and 210 mm in soil profile up to 1 m using daily meteorological data. The meteorological data were obtained from an automatic meteorological station (Cambell Sci.) located close to the field.

The soil maps and yield map were used to calculate the optimal nitrogen and phosphorus rates with the computerized recommendation system NAW-2 officially used in Poland. The weighted average rates were compared with rates calculated for the mean conditions of the field.

The statistical package SAS was used to analyze the data. Analysis of variance was performed on the yield data with soil types and complexes as random independent variables. Simple linear correlation and regression were done between yield and such soil characteristics as: pH, P, K, Mg and available water holding capacity in soil profile.

RESULTS

Soil maps

Officially used in Poland, soil maps (1:5000) are of limited value in terms of identifying detailed variation in soil series and physical and chemical soil parameters. The soil studies carried out have shown three genetic soil types and four soil complexes within the tested field (Figure 1). There were the following types: A - podsolic soils, B - brown soils and D - black earths. Under these type four complexes were distinguished (numbers according to classification): 2 - good wheat complex, 4 - very good rye complex (wheat-rye), 5 - good rye complex and 6 - weak rye complex.

The range of values of chemical characteristics measured across the field were: pH 4.6-7.4 and P 3.0-27.1, K 4.6-32.4 and Mg 2.0-11.0 mg 100 g^{-1} soil.

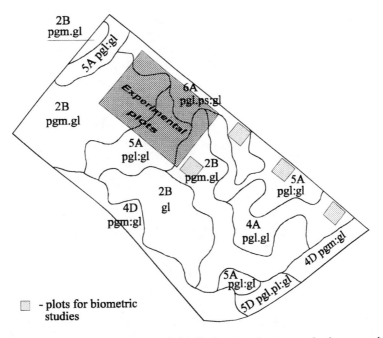

FIGURE 1. Soil map units in the test field (Soil textural groups: pl - loose sands, ps - slightly loamy sands, pgl - light loamy sands, pgm - heavy loamy sands, gl - light loams).

Vegetation conditions

The start of spring growth in 1996 was very late. Freezing and refreezing of the soil at the end of March and beginning of April destroyed plants in the local soil depressions. The average losses of plants were from 5 to 15 %. The losses were irregular and difficult to evaluate specifically. The beginning of spring growth progressed at relatively low temperatures (Table 1). The potential grain yield as determined by crop characteristics, radiation and temperature only was 8.11 t ha^{-1} according to WOFOST simulation.

113

TABLE 1. Weather data for the growth period of 1996.

	Mean daily temperature		Total rainfall	Sum of evapo-transpiration	Mean amount of water available in potential root zone for soils with	
	min. (°C)	max. (°C)	(mm)	(mm)	AWC=140 (mm)	AWC=210 (mm)
March	-3.5	1.9	4	5.0	139	209
April	2.0	14.6	9	14.9	137	207
May	7.7	17.8	94	60.3	135	205
June	10.6	22.9	40	94.0	100	170
July	11.0	20.9	218	73.7	123	193

This is the yield ceiling for intensive production in the absence of distinct water and nutrients stresses. The average grain yield from the tested field was 17 % lower than the potential yield which means that certain factors limited production. Water deficits according to simulation did not limit growth. The maximum water deficit in the growth period reached a value of 63 mm and was not dependent on the water retention capacity of the soils. However, the temperature limited the LAI dynamics in comparison with the potential LAI simulated by the WOFOST. The maximum temperatures of 28 to 33 °C at the beginning of June (GS 43-47) stressed the crop. The effect of this stress was, most probably, different maximum LAI and LAI reductions on the four soil complexes (Figure 2). Regression analysis indicates that maximum LAI which is typically reached shortly before anthesis accounts for some 54 % of the variation in yields in Polish conditions (Nierobca and Faber, 1996).

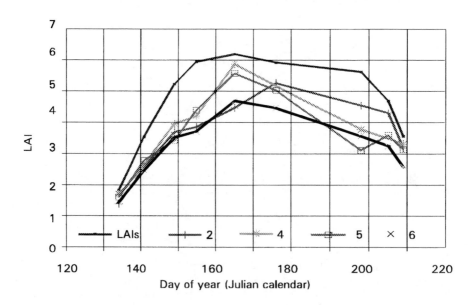

FIGURE 2. LAI dynamics of winter wheat on four soil complexes (LAIs - simulated LAI; 2, 4, 5 and 6 - measured LAI for soil complexes).

Nutrient uptakes seem to be other limiting factors. The integrated measure of growth and trophic conditions were chlorophyll indexes (Figure 3). The average chlorophyll indexes in the growing period reached values of 594, 572, 548 and 527 for soil complexes 2, 4, 5 and 6 respectively. The grain yields from plots for biometric studies located on these complexes were 8.42, 8.06, 6.84 and 5.92 t ha^{-1}. The decrease of yield corresponded with the linear decrease of chlorophyll indexes.

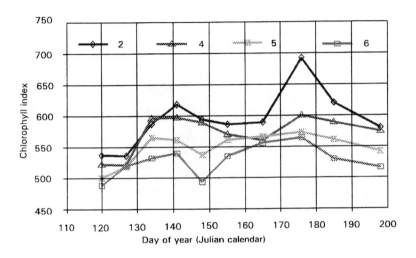

FIGURE 3. Chlorophyll indexes of youngest wheat leaves from four soil complexes.

Yield map

The yield contour map of the winter wheat grain shows a two-fold variation of the yield across the field (Figure 4). The variation of raw data from combine was from 3.15 to 9.15 t ha^{-1}. The frequency distribution of the yield was very close to a normal distribution.

Relationships between soil and yield variability

The yield was interpolated to the soil sample points to investigate the relationship between yield and soil characteristics. The linear correlation between yield and pH, P and K were not statistically significant ($p < 0.05$, not referred here). The relationship between yield and Mg and yield and available water holding capacity were significant but weak (not referred here). The one-way analysis of variance has shown significant difference between soil types and soil complexes treated as random effects (Table 2). The results obtained suggest that deeper analysis of yield affecting factors are needed to explain spatial yield variability more accurately.

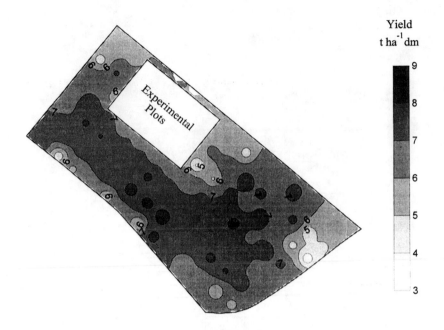

FIGURE 4. Yield contours map (t ha^{-1}, moisture 15 %).

TABLE 2. Mean yield for soil types and complexes at 88 sample sites.

Soil type	Number of sites	Mean yield (t ha^{-1})	Soil complexes	Number of sites	Mean yield (t ha^{-1})
A	32	6.6 a	2	43	7.1 a
B	43	7.1 a	4	21	6.1 ab
D	13	5.7 b	5	19	6.7 ab
			6	5	6.0 b

Harmonic mean of cell size = 12.4, means with the same letter are not significantly different.

Optimization of fertilization rates

The nitrogen rates recommended by the NAW-2 system depend on response curves for the soil complexes. The map of recommended N rates with 40 kg ha^{-1} contour intervals is presented in Figure 5. The optimal nitrogen doses were in the range 10 to 210 kg ha^{-1}. The average rate weighted on area polygons of yields obtained for test field was 140 kg ha^{-1} (interval 10 kg ha^{-1}).

On this field a dose of 75 kg N ha^{-1} was applied. The dose was calculated according to the average yield expected for the whole field. This suggests that the N rate was underestimated and that the full yield potential was not achieved.

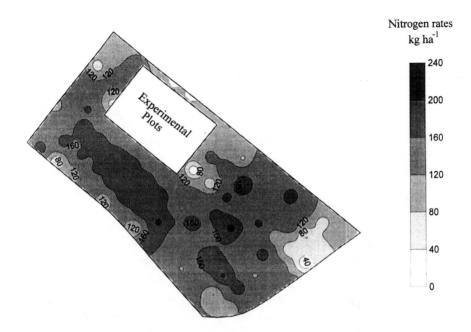

FIGURE 5. Recommended nitrogen rates contour map according to yield and soil complexes.

The phosphorus rates recommended by the NAW-2 system depend on soluble P content of the soil and P uptake by the crop. The doses recommended are equal to the sum of P uptake by the crop and P addition for the increase of soil P content needed when the value of the soil test was evaluated as too low. The soil within the field differs substantially in respect to the content of available P. The optimal P rates were in the range 1 to 45 kg ha^{-1}. The weighted average amounted to 20 kg ha^{-1}, while the applied P rate on the basis of the mean input for the whole field was 23 kg ha^{-1}. This suggests that the P rate was probably close to the needs.

CONCLUSIONS

Based on the factors analyzed, the relationships between yield variability and variability in soil type and chemistry were either non-significant or only weakly significant. A deeper investigation is required to obtain a more complex indication and explanation of yield variability. The most likely factors influencing yield are plant available water and some stresses related to weather, water and nutrition. It seems to be possible to analyse these limitations more accurately by the use of LAI and chlorophyll indices rather than by other approaches. Future work will therefore concentrate on interactions between yields, crop water needs and physiological status of plants. In the economic conditions of Polish agriculture the benefit of optimum site-specific application of fertilizers could need higher N rates. The net improvement in profitability would be achieved by yield increase rather than saving on fertilizer costs.

REFERENCES

Castelii, F., Contiilo, R., Miceli, F. (1996) Non-destructive determination of leaf chlorophyll content in four crop species. *J. Agron. & Crop Sci.*, **177,** 275-283.

Faber, A., Bloch, Z., Nierobca, A., Demidowicz, G., Kaczynski, L. (1996) Growth and yield simulations of winter wheat cultivated in Poland using WOFOST. I. Model calibration. *Fragmenta Agronomica*, **4**, 40-50 (in Polish).

Murphy, D.P., Schnug, E., Haneklaus, S. (1994) Yield mapping - a guide to improved technique and strategies. *Proc. 2nd International Conference of Site-Specific Management for Agricultural Systems.* University of Minneapolis. Soil Sci. Soc. Agr. (in press).

Nierobca, A., Faber, A. (1996) Leaf area index as an indicator of growth and grain yield of winter wheat. *Fragmenta Agronomica*, **3**, 54-66 (in Polish).

Schnug, E., Murphy, D., Evans, E., Haneklaus, S., Lamp, J. (1993) Yield mapping and application of yield maps to computer-aided local resource management. *Soils Specific Crop Management*, P.C. Robert (Ed.), Madison, USA, Am. Soc. Agron., 87-93.

Stafford, J.V., Ambler, B., Lark, R.M., Catt, J. (1996) Mapping and interpreting the yield variation in cereal crops. *Comput. Electron., Agric.*, **14**, 101-119.

Supit, I., Hooijer, A.A., van Diepen, C.A. (Eds.) (1994) System Description of the WOFOST 6.0 crop simulation model implemented in CGMS. Luxembourg: Office for Official Publications of the European Communities. Agriculture series. pp 146.

Witek, T., Gorski, T. (1977) *Evaluation of the Natural Capability of Agricultural Areas in Poland.* Institute of Soil Science and Cultivation of Plants, Pulawy, pp 21.

Zadoks, J.C., Chang, T.T., Konzak, C.F. (1974) A decimal code for the growth stages of cereal. *Weed Res., Oxford,* **14,** 415-421.

SELECTIVE SOIL SAMPLING FOR SITE-SPECIFIC NUTRIENT MANAGEMENT

D.D. FRANCIS, J.S. SCHEPERS

USDA-ARS, 119 Keim Hall, University of Nebraska, Lincoln, NE 68583-0915, USA

ABSTRACT

Most fields have considerable spatial variability related to soil fertility. The intent of site-specific nutrient management is not to remove spatial variability, but rather to manage fertilizer applications to achieve maximum economic yield for each given site. One problem facing many farmers is finding an affordable means for developing accurate variable rate fertilizer application maps. Soil sampling fields on a grid basis may provide the most accurate guide for variable rate nutrient application, but costs may be prohibitive for most farmers. If high density grid sampling is not possible, selecting sampling areas on the basic principles of soil color, texture, depth, slope and erosion phase as used in past soil testing programs is a good starting approach. In this study the non-mobile nutrients phosphorus (P), potassium (K), and zinc (Zn) generally followed the preceding soil characteristics except where livestock and manure operations had taken place in the past. Selective sampling can also make use of other information such as past field history, yield maps and remote sensing as guides to reduce the number of soil sampling sites. All methods which reduced the size of the managed area for nitrogen (N) and P increased the accuracy of the fertilizer recommendation maps.

INTRODUCTION

Soil test results form the basis on which fertilizer needs are determined. In the past, before fertilizer came into common usage, it was relatively uncommon to find big differences in nutrient levels in different parts of a given field, expect where extreme heterogeneity of soil type existed (Melsted and Peck, 1973). Today, large differences in nutrient levels are often found in samples taken from different parts of the same field. The combination of many factors may have contributed to the variation present today; soil formation, cropping history, past field delineation, crop nutrient removal, and crop input applications such as fertilizer, lime, manure, and pesticides (Sawyer, 1994). Variable rate fertilization becomes the logical management approach for fields that have significant spatial variability related to soil fertility. Properly implemented, variable rate technology (VRT) application has the potential to improve input efficiency, field profitability, and environmental stewardship (Sawyer, 1994). Perceived values and benefits, such as environmental stewardship, are difficult to evaluate and quantify, but in the end, may be the most important consideration for adoption of VRT (Sawyer, 1994).

Equipment and technology to vary the rate of fertilizer application and other agricultural inputs is currently available and in limited use. Traditionally, soil sampling designs were chosen to estimate the field-average soil nutrient level. Variable rate fertilizer application requires soil sampling and mapping procedures that accurately map a field's nutrient spatial variability. Accuracy of soil nutrient maps generally improves with increased sampling density. As we move to a farmer's management system, the accuracy of the application map may in turn depend on acceptable costs. From a farmer's perspective, fertilizer application has always been based on economics. Farmers may not tolerate a negative impact on profitability, even if other perceived benefits are strong.

Using grid sampling to map soil test levels has been shown to give the most accurate application maps (Wollenhaupt et al., 1994; Franzen and Peck, 1995). Data collected on a 32 m grid suggested that sample spacing probably should not exceed 90 m if accurate maps are to be produced for most crop production fields (Wollenhaupt et al., 1994). More recent work has documented significant soil variability to the sub-meter scale (Raun et al., 1996). The relative accuracy of maps constructed by using data from wider grid spacing typically show a rapid loss of resolution when compared to maps constructed from narrower sample spacing. Often many features of a soil fertility map are lost whenever grid sample spacing is increased over three times that used in making the original map. This indicates that the perceived maximum soil sample grid spacing allowed to make accurate soil fertility contour maps are most likely an artifact of the narrowest sample density used to make the spacing determination.

The uncertainty in sampling scale needed along with the substantial cost of high density grid sampling and analysis may limit its use by many farmers. The objective of this study is to evaluate other strategies for mapping location specific soil nutrient levels, thus possibly reducing soil sampling density and costs.

MATERIALS AND METHODS

The study site was a 59 ha irrigated corn field in Buffalo County, Nebraska, near the town of Shelton. Soil samples were collected in late March of 1994 based on a 12 by 24 m alternate grid design that resulted in a network of triangles (0.0288 ha/sample). Sample cores were divided into two depths, 0 - 20 cm and 20 - 92 cm. Surface samples were analyzed for organic matter (OM), nitrate-N (NO_3-N), P, K, Zn, and pH. Sub-surface samples were analyzed for NO_3-N. If soil depth to sand was less than 92 cm, this information was also recorded.

RESULTS AND DISCUSSION

The purpose of soil sampling, whether managing nutrients on a whole field or site specific basis, is to gather more information which will aid in making better decisions. The starting point for any farmer wanting to get into site specific management is being able to geo-reference all field derived information. Eventually soil maps, yield maps, and other data will become informational layers in a Geographic Information System (GIS). All this information can then be integrated to aid in making sound management decisions.

Before designing a soil sampling strategy, consideration must be given to the farmer's nutrient management desires, application equipment, and acceptable costs. A land owner or tenant with a long term contract may have completely different needs and desires than a tenant with a short term contract, although both may want to use VRT technology. Land owners may accept high soil sampling costs in a given year because it can be amortized over a number of years, while tenants with medium or short term farming contracts may never be willing to accept high soil sampling and analytical costs. Soil sampling design may also depend on the soil nutrient in question. Certain soil fertility characteristics may follow soil type, past management history, slope phase, and/or erosion phase.

The OM content of a soil is often an indication of its productivity. In addition to potential productivity, mineralization, and effects on bioactivity, persistence and biodegradability of pesticides are some reasons why OM maps are important to VRT technology. Organic matter generally has lower spatial variability than most soil nutrients and one map may be good for 20 years or more. Soil color changes caused by OM allows aerial photographs of bare fields to be used in soil sampling designs for OM. An aerial photograph of the entire field in Buffalo County taken just after planting indicates that a relativity accurate OM map could be generated from the collection of a limited number of soil samples (Figure 1).

In addition to distinguishing soil type, soil surveys often provide information about past management. The photobase for the most recent soil survey of Buffalo County is a series of aerial photographs taken in 1970. Some of the past management units are clearly depicted on the soil survey (Figure 2). The 59 ha field in this study had been farmed as one unit since the center pivot was installed in 1979. However, the soil fertility contour maps generated from the grid sampling for non-mobile nutrients P, K, and Zn indicate some of these past management features are still overriding soil fertility differences associated with soil type, OM levels, topography, and erosion (Figure 3). The old farmstead in the lower left-hand corner, which would have been dominated by buildings and pens dedicated to livestock production, is noted on all three maps. In addition to the farmstead an old sheep feedlot, which was last operated in the 1920s, can be found in the mid-upper left-hand portion of the Zn fertility map.

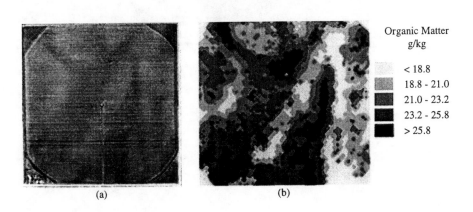

FIGURE 1. Figure (a) is an aerial photograph of the entire field taken just after planting in May, 1994. Figure (b) is the soil organic matter map in g/kg generated from 12 by 24 m grid sampling.

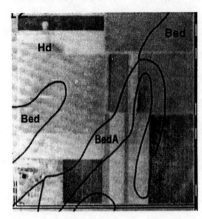

Bed: Blendon loam
     0 to 1 percent slopes

BedA: Blendon loam
     1 to 3 percent slopes

Hd: Hord silt loam
     0 to 1 percent slopes

FIGURE 2. Most recent SCS soil survey map (1974) of field in Buffalo County, NE.

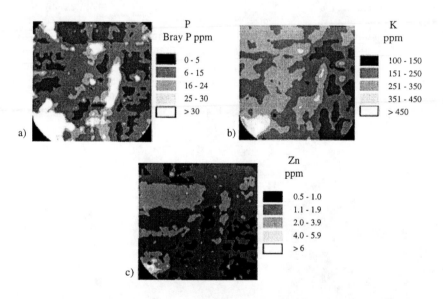

FIGURE 3. Soil fertility contour maps for P (a), K (b), and Zn (c) generated from samples collected on 12 by 24 m alternate grid design.

Phosphorus was the only non-mobile nutrient found to be deficient in this field. If we look at increasing grid sample spacing to reduce sampling costs, it does not take long for our nutrient map to lose many of the features of the finer sampling (Figure 4). If sampling on a fine grid is out of the question because of high cost, then we must decide on how to select sampling areas within a field. It is obvious that a reduced sampling design should match the anticipated VRT management unit.

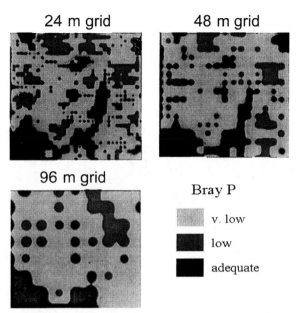

FIGURE 4. Soil Bray 1 P maps constructed from data collected on 12 by 24 m alternate grid design. Average concentrations were used to generate data for the 48 and 96 m grid maps.

Yield maps often do not accurately portray nutrient availability to the crop, while slope phase, erosion phase, and OM levels are generally known to be good indicators of production levels. As taught in all beginning soil courses, soil sampling should always avoid odd areas of a field which would include dead furrows, old livestock lots and lanes, old fence lines, and field depressions. Past historical knowledge on crop growth patterns and management practices should also be used when available. In this field, the farmer knows from past cropping history that the east, southeastern portion of the field is often moisture stressed due to shallow topsoil. Figure 5(a) depicts the anticipated VRT management units and proposed soil sampling areas based on OM, slope and erosion phase, and past crop growth patterns. Using composite samples from these units would have resulted in some changes to the soil fertility maps for P, K, and Zn, but still would have indicated that P was the only deficient non-mobile nutrient (Figure 5(b)). The farmer had been uniformly treating this field and samples collected by a consultant at approximately the same time as the grid sampling was done indicated no deficiency of P for the field although over 75% of the field was deficient.

The final VRT application map for P would include the farmer's fertility goals. This may be a minimum application to only meet the current crop's anticipated needs or it could be a program to build up and maintain the nutrient at a given level. A soil map of non-mobile nutrients may be good for up to 5 years or more. When soil sampling sites are geo-referenced, future soil samples can be used to evaluate the fertility program or to check nutrient levels in other areas of the management unit to further define the field. If soil fertility levels for a non-mobile nutrient are at a point where a farmer just desires a maintenance program, then application maps could be drawn from yield maps to cover harvest removal.

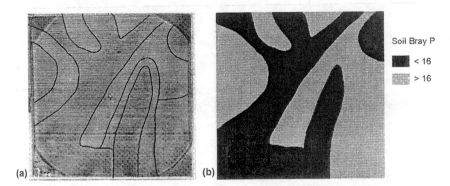

FIGURE 5. Proposed soil sampling areas (a) based on OM, slope and erosion phase, and past crop history. Figure (b) indicates where Bray P deficiencies would exist if composite soil samples were taken from these areas.

In humid regions where substantial leaching may occur during the non-growing season, soil sampling for NO_3-N is usually not necessary. However, in dryer areas such as Nebraska with limited leaching during the non-growing season, residual soil NO_3-N considerations and use, are important to achieving high N use efficiencies and reducing potential for groundwater contamination. Doing annual soil sampling on a fine grid basis for NO_3-N would be economically unfeasible for most farmers. If fine grid sampling is unacceptable, the next logical approach is most likely the old time-honored method of using soil color, texture, depth, slope and erosion phase to determine sampling units. Our OM map could be substituted for soil color. As such, our selected sampling areas for this field would basically be the same as that for the non-mobile nutrients (Figure 5(a)).

Unlike non-mobile nutrients, N management would be similar for most farmers in that they would be applying rates to maximize economic returns. The University of Nebraska's N recommendation for corn incorporates the effect of soil OM and residual soil NO_3-N levels on grain yield response to N fertilizer. The equation used to determine recommended N is:

$$Nrec = 35 + 1.2(EY) - 8(NO_3\text{-}N) - 0.14(EY)(OM) - LC - MC - WC \qquad (1)$$

where Nrec is the recommended N application rate, in pounds per acre; EY is the 5-year historical average grain yield in bushels per acre plus a 5% increase; NO_3-N is the average root-zone NO_3-N concentration in parts per million; OM is the percentage of organic matter in the soil; and LC is legume credit; MC is manure credit; and WC is irrigation water NO_3-N credit.

The above algorithm was originally developed for use on a field basis with inputs representing means for a given field. When applied to a given site or location, EY would have to be estimated if historical information is not available. For lack of better information, we used the field-average EY for all sites in our calculations. Differences in patterns are noted between the N recommendation maps developed from selected sampling versus grid sampling (Figure 6). If the N recommendation map developed from the finest grid sampling is considered the correct recommendation, applications based on using selected sampling would have increased the accuracy of N applied as compared to a uniform application rate (Table 1). Selected sampling would result in an additional 11% of the field falling within 17 kg N/ha of the recommended amount.

Aerial or satellite remote sensing appears to have the potential to monitor in-season crop N status and also to provide an application map for sidedressing if an N deficiency is detected. However, at the present time, the cause of patterns in remote-sensed images is often a matter of speculation. Considerable research is still needed in ground-truthing remote-sensed images. Currently, yield maps and remote-sensed images can be important tools used to facilitate directed sampling. Successful implementation of VRT N application will most likely rely on a combination of methods such as yield maps, remote-sensed images, soil topography, and soil sampling information (Ferguson, 1996).

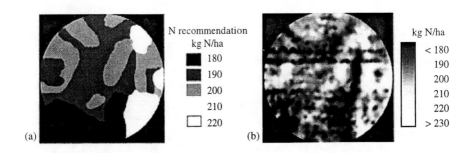

FIGURE 6. In figure (a) the N recommendation map was developed from OM map, selected soil sampling from Figure 5(a), and field average EY. Figure (b) is the N recommendation map constructed using the residual NO_3-N and OM from the staggered 12 by 24 m grid soil sampling data.

TABLE 1. Differences for the same location in the field between N recommendations calculated from 12 by 24 m grid sampling data versus uniform N rate and N recommendations calculated from selective sampling plus OM data.

| Strategies | Percent of N recommendations at deviation intervals from grid sample values (kg N/ha) | | | | | | |
	-(67-50)	-(50-34)	-(34-17)	-(17-0)	+(0-17)	+(17-34)	+(34-50)
Uniform rate	0.5	2.8	13.9	31.0	35.7	14.5	1.6
Selective sampling + OM	0.2	1.5	9.8	35.5	42.0	10.5	0.5

CONCLUSIONS

The amount of variability occurring within a given field is unique for that field. Development of a field's nutrient application map will also be unique and will depend on the farmer's nutrient management goals, what the farmer sees as acceptable costs, and the ability of the farmer's application equipment. Lack of ability to describe all of the soil fertility variability in a field should not preclude the use of VRT. Clearly, all methods that reduce the size of the managed area should increase the accuracy of fertilizer recommendations as compared to a uniform whole field approach. If high density grid sampling is not possible, selecting sampling areas on the basic concepts of soil color, texture, depth, slope and erosion phase as used in past soil testing programs is a good starting approach. When selecting soil sampling areas, avoid all odd or dissimilarly treated areas. Knowing previous history on manure application and livestock management will aid in selecting appropriate sampling areas. If VRT is the best management approach for most fields, then the old adage of keeping it simple while limiting costs and perceived risks will result in the most rapid and widespread implementation of VRT.

REFERENCES

Ferguson, R.B., (1996) Site-specific nitrogen management. *Proceedings of the North Central Extension-Industry Soil Fertility Conference*, **12**, 1-11.

Franzen, D.W., Peck, T.R., (1995) Field soil sampling density for variable rate fertilization. *J. Prod. Agric.*, **8**, 568-574.

Melsted, S.W., Peck, T.R., (1973) The principles of soil testing. *Soil Testing and Plant Analysis*, L.M. Walsh and J.D. Beaton (Eds), Madison, WI, SSSA, pp 13-21.

Raun, W.R., Johnson, G.V., Lees, H.L., Taylor, S.L., Solie, J.B., Stone, M.L., Whitney, R.W. (1996) Potential replacement for traditional soil test calibration - sensor based plant analysis. *Agronomy abstracts*, Madison, WI, ASA, pp 312.

Sawyer, J.E., (1994) Concepts of variable rate technology with considerations for fertilizer application. *J. Prod. Agric.*, **7**, 195-201.

Wollenhaupt, N.C., Wolkowski, R.P., Clayton, M.K., (1994) Mapping soil test phosphorus and potassium for variable-rate fertilizer application. *J. Prod. Agric.*, **7**, 441-448.

VARIATION IN SOIL CONDITIONS AND CROP PERFORMANCE

R.M. LARK

Silsoe Research Institute, Wrest Park, Silsoe, Bedford, MK45 4HS, UK

ABSTRACT

Existing approaches to the description of soil variation are discussed and their possible application to the description and management of variation within fields is considered. Some of the methods are illustrated using data collected from Cashmore field at Silsoe Research Institute.

INTRODUCTION

"Each divers soile hath divers toile"
Thomas Tusser 'Five hundred points of good husbandry'

The spatial variability of soil at many scales is a familiar phenomenon. The spatial variability of crop yields at within-field scales, despite uniform management practices, is also well established (Mercer and Hall, 1911). It may well be that the requirement of a crop for particular inputs also varies spatially at within-field scales. This is not a logically necessary implication of soil and yield variability; but it is a not unreasonable intuition, and the possible financial and environmental benefits of spatially variable application certainly justify further study.

Soil variation is likely to be important in determining the spatial variation of crop yield and of requirements for an input. How might appropriate information on soil conditions be collected and interpreted? Soil variation may be described simply and quantitatively, in terms of the change in values of one or more properties from point to point, by sampling the soil (usually on a regular grid). The resulting data may be stored on computer and manipulated mathematically (Webster and Oliver, 1990). Soil variation may be also be described by classification. Classes may be defined by threshold values of critical properties. Alternatively a soil classification may be constructed by recognising a number of distinct 'types' among soil profiles in the region of study. Any profile can then be allocated to a class representing the 'typical' profile which it most closely resembles. The former type of classification is generally used for specific purposes (e.g., inventory of salt-affected soils). The latter type of classification (e.g., the soil series of England and Wales) is used for general purposes, and, ideally, reflects the processes of soil genesis within the region - grouping together soils formed under similar conditions over similar parent material. Classification may be useful for making generalised statements about soil conditions based on trials and experience. It also provides a framework for sampling soil properties. If soil classes differ significantly with respect to a given property, then the mean value of the property within a given class will be a good predictor of the value of that property at sites where the class occurs. Soil classification also forms the basis for soil mapping. In free survey, the soil surveyor delineates 'map units', regions of the landscape which are as uniform as possible with respect to soil profile classes and which can be recognised by reference to topography, vegetation and other surface clues. Ideally 'simple' map units are delineated, each corresponding primarily to one profile class (after which it is named), but there are inevitably inclusions of other classes - typically over 20-40% of the area of the map unit (Bie and Beckett, 1971).

In this paper classification of the soil and grid sampling are considered as alternative or complementary strategies for collecting, organising and interpreting information on within-field soil variation. Possible approaches are compared, illustrated with data from a study field at Silsoe Research Institute.

METHODS

Cashmore field is a 6 ha field at Silsoe Research Institute, which has been the subject of detailed observations on soil and yield variation (Stafford et al., 1996). Yield mapping has been conducted using a flowmeter on a combine harvester equipped with a GPS positioning system. A soil map of the field, based on soil series, was produced by Professor J.A. Catt of IACR Rothamsted. The map was made by manual interpolation from the observations made on a 20 m interval grid across the field. Seven soil series were recognised in the field. These correspond to three broad divisions of parent material. Drift of variable depth and texture over the Lower Greensand (Lowlands, Hallsworth and Nercwys series), Gault clay with or without superficial loamy drift (Evesham and Bardsey series respectively) and clayey alluvium (Enborne and Fladbury series). The yield maps for three successive harvests of winter barley were overlaid and vectors representing the yield in each season were extracted for locations whose spatial co-ordinates agreed to within 5 m. The corresponding map unit for each vector was then identified. Figure 1 shows mean yields within each map unit for the three seasons. An analysis of variance was conducted to test the significance of the main effect of soil map unit and the map unit by season interaction in accounting for yield variability. Both were highly significant (p < 0.001 in both cases).

Soil sampling was carried out at 100 sites across Cashmore field in the spring of 1995 and analyses were conducted for gravimetric moisture, mineral nitrogen, pH and organic matter on samples from two depths (0-20 cm and 20-80 cm), (representing all but the

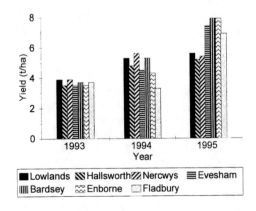

FIGURE 1. Mean yields (Winter barley) within soil map units on Cashmore field.

128

Evesham map unit). Analyses of variance were carried out to test whether the map units differed significantly with respect to each property. Map unit means and the significance of the difference between each map unit are shown in Table 1.

DISCUSSION

Classification

It is generally accepted that free survey, based on a soil classification, is appropriate as a general tool for handling soil information at certain scales. A soil map with simple map units would typically be produced at a scale between 1:25,000 and 1:50,000. At such a scale most fields, at least in the UK, would occur within one - or maybe two - map units. However, it would be expected that 20-40% of randomly located sites within each map unit might be found to correspond to a class other than the eponymous one. In principle, therefore, it may often be possible to account for a good deal of within-field variability of the soil in terms of differences between profile classes. Free survey is unlikely to be appropriate for mapping at within-field scales, at least in Europe. However, the soil class could be identified at sites on a sampling grid using standard criteria (e.g., Clayden and Hollis, 1984), perhaps at the same time as samples are collected for analysis. These point observations could be generalised to a map of classes by indicator kriging (Bierkens and Burrough, 1993) or by manual interpretation by the surveyor using some of the surface clues which would be used in free survey. How might soil classification and mapping be used in the study and management of within-field variation? Four possibilities are considered below.

To interpret past experience of variable performance in the field, as a possible basis for future management. If differences between map units account for a significant proportion of yield variation, then the soil map might provide a basis for generalising from past yield performance. For example, if one map unit consistently gives high yields then it might justify higher levels of certain inputs in future years. If consistently low yields occur in another map unit then this might be explained by field investigation at sites within the map unit, and remedial action may be identified.

TABLE 1. Map unit means for different soil properties.

Property. (Two depths 1: 0-20 cm 2: 20-80 cm)	Map unit mean, and significance of difference between map units (p value)						
	Lowlands	Hallsworth	Nercwys	Bardsey	Enborne	Fladbury	p
Moisture% 1	20.0	23.0	21.8	29.8	24.5	27.7	<0.001
Moisture% 2	17.7	18.4	17.8	22.7	21.4	24.9	<0.001
Org. M.% 1	2.4	3.6	3.3	3.5	3.2	3.3	<0.001
Org. M % 2	1.7	2.7	3.0	1.8	2.0	2.4	<0.001
Mineral N 1	6.2	8.8	7.2	7.7	6.5	6.0	0.003
Mineral N 2	4.0	5.1	4.8	6.2	4.2	4.7	0.06
pH 1	6.6	6.3	6.4	6.7	6.3	6.3	0.003
pH 2	6.6	6.4	6.4	6.8	6.4	6.3	0.025

The results for the ANOVA on yields in Cashmore field by season and map unit, described above and represented in Figure 1, are encouraging in so far as much of the yield variation is related to the differences between soil classes. However, the significant interaction with season suggests that it will be difficult to predict performance with confidence at the stage in crop development when decisions on inputs must be made. In this particular case seasonal weather is likely to be important. Figure 2 shows the potential soil moisture deficit at Silsoe (inferred from local rainfall and class A pan evaporation) from 1 March to harvest in each season. 1995 was clearly a drier year, and this is reflected in the poor relative performance of the coarse-textured soils (Lowlands, Hallsworth and Nercwys map units). However, it is notable that the Bardsey map unit performed relatively well in both 1994 and 1995. This soil comprises loamy drift over Gault clay, and may be reasonably well 'buffered' against seasonal differences in weather. The topsoil will be better drained than a clay soil (Evesham, Enborne and Fladbury map units) giving good conditions for crop establishment, while the sub-soil will have larger reserves of soil moisture than a sandy loam (Lowlands, Hallsworth and Nercwys map units) during a dry summer. This information on the relative risk of poor performance in different map units may be useful in future management.

As a tool for generalising from sampling. The significant differences between map unit means with respect to a number of potentially important soil properties on Cashmore field (Table 1), imply that it is possible to make generalisations about these properties which might be useful for explaining spatial variations in crop performance. For example, the interpretation in the previous section that the Lowlands, Hallsworth and Nercwys map units may be more susceptible to dry weather is supported by the (significantly different) mean gravimetric water contents in Table 1. Decisions on spatially variable inputs might also be aided by such information. For example, the significant pH differences between map units may imply that lime should be applied at different rates, and the significant differences in soil organic matter content may have implications for the efficacy of certain pre-emergence herbicides. Further sampling of the field for these, or related properties, may be conducted more efficiently by sampling within the map units. In some cases it may be possible substantially to reduce the costs of soil analysis by taking a very few bulked samples from each map unit.

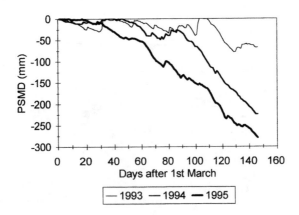

FIGURE 2. Potential soil moisture deficits from 1 March to harvest, 1993-1995.

<u>Use of classification and generalisation in combination with process models</u>. Leaching risk, if spatially variable within fields, may be an important factor to consider when determining local optimum levels of certain inputs. Leaching models may be used for assessing risk, but require detailed information on soil properties. For example, to run the SLIM model of Addiscott and Whitmore (1991) requires, among other inputs, information on the volumetric water content of the soil at three suctions and a permeability parameter, α. Pedotransfer functions may be useful in such circumstances. These are essentially regressions of important, but costly to determine, soil properties on more easily determined properties. For example, Hall *et al.* (1977) present regressions of volumetric water content at several tensions on bulk density, percentage of certain particle size classes and organic matter content. Addiscott and Whitmore (1991) present a regression of α on clay content. Figure 3 shows the simulated profile distribution of a solute applied to the soil surface on 1 October by 31 January the following year (using 1994/95 weather data). The results represent the Evesham and Lowlands series on Cashmore, suggesting a greater risk of leaching from the latter. Water contents were obtained using particle size data from bulk samples taken from each map unit, mean organic matter contents from samples collected in spring 1996 and information on bulk-density from samples previously taken on cognate soils in adjacent fields.

There are potential problems in this approach. Each prediction from a regression has an associated error. Each regression used will contribute to the error of the final result, as will sampling error of each independent variable. Furthermore, the regression for the a parameter on clay content and one of the regressions used for water content each have a quadratic term. If property i is represented by a non-linear function of property j then applying this function to the mean value of j for a region gives a biased estimate of the corresponding mean of i. In this case the regressions used are not markedly non-linear over the relevant range. However, further work is needed on scaling behaviour of pedotransfer functions. Similarly, a response corresponding to the mean soil properties within a map unit is not equal to the mean response for the map unit, if the model is non-linear. Addiscott and Whitmore (1991) suggest that this is not a serious problem for the SLIM model. However, in general it is important to retain information on the variances of soil properties within map units, as well as the mean values, to allow the use of methods, such as Addiscott and Wagenet's (1985) 'sectioning', to overcome non-linearity.

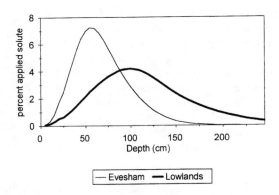

FIGURE 3. Simulated distribution of solute applied 1 October 1994 by 31 January 1995 for two map units.

Special purpose or 'functional' classes. The soil classes identified on Cashmore field are based on profile characteristics related to soil genesis, and as such they are 'general purpose' classes. Where a particular aspect of soil behaviour is concerned, 'functional' or 'special purpose' classes of soils, which behave similarly, may correspond to more than one general purpose class. An interesting use of functional classification in the study of within-field variation is the work of Verhagen et al. (1995). Here four 'functional layers' were recognised in the soil within a field. Any one profile could be simply described as a combination of one or more of these layers. Sampling for hydrological properties could be directed to the estimation of properties for each of the four layers. This information could then serve as input for a hydrological model for a variety of distinct soil profiles.

Point sampling and interpolation

Soil data may be gathered from a grid of sample sites across a field by direct sampling. The usefulness of these data can be substantially enhanced by the procedure of kriging, an interpolation method which generates optimum (in the least-squares sense) and unbiased estimates of *either* the value of the property at unsampled points, *or* the mean value of the property within a specified region (block kriging). Not only are kriged estimates optimal, but the mean squared error of the prediction is also estimated. Kriging is based on the variogram, a model of the spatial dependence of variation in the values of the property. Once the variogram is estimated, an optimal sampling grid can be designed to achieve a specified level of precision for kriged estimates with the minimum sample cost. Webster and Oliver (1990) discuss the kriging procedure. Probably the major obstacle to the adoption of kriging as a working tool for managing within-field variability is the sample effort required to estimate the variogram. Webster and Oliver (1992) showed that a minimum of 100 data are required to obtain an adequate estimate when the variability of the soil property conforms to a simple, isotropic variable. It may be possible to obtain data sets of this size routinely for certain variables (e.g., from simple hand-held sensors). However, most soil properties cannot be sampled this intensely in a commercial context due to the field effort required and the costs of analysis, particularly when there is a requirement for samples from different depths. Assuming that adequate sampling is possible, how might a set of kriged estimates of soil properties be used? Three options are discussed below.

Threshold models. The experience of experts and the results of field trials may be encoded in threshold values for different properties. For example, advice for P and K fertilisers in the UK may be based on index levels defined by threshold values of a standard soil analysis. Figure 4(a) shows a kriged map of soil pH (0-20 cm) for Cashmore field based on 100 samples. This could be converted by a threshold rule to a binary map showing where pH conditions are or are not suitable for growth of a particular crop. However, the pH values on which such a map is based are estimated with error. A more useful map for decision making would be a map of the conditional probability that the true pH at a site falls below the threshold. This can be obtained using the non-linear disjunctive kriging estimator (Yates et al., 1986). Figure 4(b) shows the map of the conditional probability that soil pH falls below a specified threshold for Cashmore field.

Fuzzy sets. A problem with threshold models is that in virtually all circumstances they are artificial. Most biological systems will show a continuous response to variables such

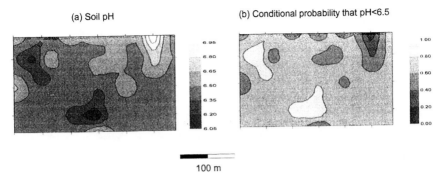

(a) Soil pH

(b) Conditional probability that pH<6.5

100 m

FIGURE 4. Kriged estimates of (a) soil pH and (b) conditional probability that pH<6.5 Cashmore Field.

as soil pH rather than a binary response at a threshold value. The experience of experts and the results of field trials may not be well represented by a threshold value which divides the soil into two mutually exclusive sets of those suitable for a crop and those unsuitable. It may be better represented by a 'membership function' in the 'fuzzy' set of suitable soils which equals 0 at pH values which are unambiguously unsuitable, 1 at pH values which are unambiguously suitable and some value between 0 and 1 at intervening pH values. The form of the membership function may be adapted to reflect the judgement of one or more experts, or even experimental data. The methods of fuzzy set theory (Burrough, 1989) may then be used to manipulate the membership values in several fuzzy sets and generate output, relating perhaps to the requirement for a particular input at each site.

Process models. Sets of kriged estimates of soil properties across a field could be combined to give useful output by running a mechanistic model of the key processes at each location. A model of the fate of applied nitrogen, soil nitrogen and crop nitrogen use could be run repeatedly for each location under different 'what if' management scenarios in order to identify optimum local rates of nitrogen application. This strategy is ideal from a scientific point of view, in that it is based on good understanding of the processes involved. However, progress is still required in the development of appropriate models which are sufficiently robust and validated to be run with confidence at this scale. Serious obstacles to the practical implementation of such a strategy, even with well founded models, are the costs of sampling all the relevant input variables and the magnitude of the error of the model output due to 'propagation' of the estimation errors for each input variable (Heuvelink et al., 1989).

CONCLUSIONS

In conclusion, existing approaches to the study of soil variability may be useful for managing variability within agricultural fields. Soil classification may be a useful framework for interpreting patterns of yield variation, for sampling the soil to study quantitative variation in key properties and for interpreting this variation in terms of possible management responses. Some of the information gathered about soil classes may be used to generate input for models, but care is needed when a model of processes at one scale (the profile) is run with input variables referring to the map unit. There are various ways in which grid sampling of the soil and interpolation might be used to determine management responses to variation within fields, but the major obstacle to the application of this approach is the cost of adequate field sampling.

ACKNOWLEDGMENTS.

This work was supported by funding from the Ministry of Agriculture, Fisheries and Food under project CE0204.

REFERENCES

Addiscott, T.M., Whitmore, A.P. (1991) Simulation of solute leaching in soils of differing permeabilities. *Soil Use and Management,* **7,** 94-102.

Addiscott, T.M., Wagenet, R.J. (1985) A simple method for combining soil properties that show variability. *Soil Science Society of America Journal,* **49,** 1365-1369.

Bie, S., Beckett, P.H.T. (1971) Quality control in soil survey I. The choice of mapping unit. *Journal of Soil Science,* **22,** 32-49.

Bierkens, M.F.P., Burrough, P.A. (1993) The indicator approach to categorical soil data. *Journal of Soil Science,* **44,** 361-368.

Burrough, P.A. (1989) Fuzzy mathematical methods for soil survey and land evaluation. *Journal of Soil Science,* **40,** 477-492.

Clayden, B., Hollis, J.M. (1984) Criteria for differentiating soil series. *Soil Survey of England and Wales,* Technical Monograph 17.

Hall, D.G.M, Reeve, M.J., Thomasson, A.J., Wright, V.F. (1977) Water retention and porosity of field soils. *Soil Survey of England and Wales,* Technical Monograph 9.

Heuvelink, G.B.M., Burrough, P.A., Stein, A. (1989) Propagation of errors in spatial modelling with G.I.S. *International Journal of Geographical Information Systems,* **3,** 302-322.

Mercer, W.B., Hall, A.D. (1911) The experimental error of field trials. *Journal of Agricultural Science,* **4,** 107-132.

Stafford J.V., Ambler, B., Lark, R.M., Catt, J. (1996) Mapping and interpreting the yield variation in cereal crops. *Computers and Electronics in Agriculture,* **14,** 101-119.

Verhagen, A., Booltink, H.W.G., Bouma, J. (1995) Site-specific management: balancing production and environmental requirements at farm level. *Agricultural Systems,* **49,** 369-384.

Webster, R., Oliver, M.A. (1990) *Statistical methods in soil and land resource survey.* Oxford University Press.

Webster, R., Oliver, M.A. (1992) Sample adequately to estimate variograms of soil properties. *Journal of Soil Science,* **43,** 177-192.

Yates, S.R., Warrick, A.W. and Myers, D.E. (1986) A disjunctive kriging program for two dimensions. *Computers and Geosciences,* **12,** 281-313.

AN EVALUATION OF SOIL MINERAL N, LEAF CHLOROPHYLL STATUS AND CROP CANOPY DENSITY ON THE YIELD OF WINTER WHEAT

A.R. LEAKE

Focus on Farming Practice, CWS Agriculture, The White House, Gaulby Lane, Stoughton, Leicester, LE2 2FL, UK

G.A. PAULSON

Norsk Hydro ASA, Bygdøy allé 2, N-0240, Oslo, Norway

ABSTRACT

Considerable differences in soil mineral nitrogen (SMN) levels were recorded within individual fields at a fenland farm. To ascertain how this influenced yield, a 10 ha field was sampled, to a depth of 90 cm, using a pneumatic sampler fitted with GPS, on a 36 m grid. Each core was divided into 30 cm horizons and analysed individually.

Chlorophyll status of the flag leaf was measured around each sampling point using the Hydro 'N tester', and crop canopy density was measured using computer enhanced satellite imagery. The crop was subsequently yield mapped and the maps of all collected data compared. There was a weak correlation between relative chlorophyll content and yield (R2 = 0.2459) and between chlorophyll content and SMN levels (R2 = 0.2614). No statistical analysis of the NDVI values and grain yield was carried out although visual analysis suggests a direct relationship.

INTRODUCTION

Focus on Farming Practice is a experimental project in integrated crop management, seeking to combine traditional techniques of crop rotation and cultural husbandry with modern technology to optimise the use of inputs. The use of mineral fertilisers forms the basis of crop nutrition in conventional agriculture at present and blocks of land are generally treated uniformly. In conjunction with Hydro Agri UK Ltd, Focus on Farming Practice is seeking to develop field specific fertiliser programmes which not only take account of inherent and rotational soil fertility, but also of seasonal variations such as sowing date, plant populations, estimated yield and winter rainfall, all of which may affect nutrient availability and requirement. To provide accurate data on the soil nutrient levels present prior to formulating such specific programmes, soil samples are taken prior to the application of fertilisers in the spring. These data are then used in the programme.

Field observations carried out at CWS Agriculture's Coldham Hall Farm, Wisbech, Cambridgeshire showed considerable within-field yield variation, which could not obviously be attributed to pest, disease or weed infestation. As fertiliser is applied uniformly, the differences were thought to be attributable to soil variation in one or more of the major elements required for crop growth. To establish precisely which elements might be causing the effect, a number of fields were soil sampled using the Hydro Geonor hydraulic sampler. This equipment enabled ninety individual cores to be taken on a 36 m grid from the main rooting zone, 0-90 cm. Full soil analysis of each sample showed only small variations between the samples in phosphate, potassium, magnesium or pH. Estimated available mineral N levels were highly variable. The farm rotation (Table 1) includes both vining peas and brassicas, which have the potential to contribute significant amounts of organic nitrogen to the soil when residues are incorporated (Kuhlmann and Engels, 1989). Consequently, mineral N levels are relatively high compared to rotations without such break crops. High levels of mineral N can be balanced with reduced fertiliser inputs, resulting in optimum applications and minimal loss to the environment (Macdonald *et al.*, 1989). The problem highlighted by the analysis of individual field samples was the within-field variation, which, at its extreme, ranged from 59 kg/ha N to 916 kg/ha N. Variations of this degree are problematical to precision calculation of field fertiliser requirements where each field is treated as a common entity, in line with current system requirements. Where such variations exist, it is unclear whether the variations are attributable directly to the variations in yield. It is, therefore, also unclear whether the high mineral N levels measured are enhancing or reducing yield and in order to ascertain this information, a field scale trial was established.

TABLE 1. Previous cropping during 1986-1995 in field 1187 at Coldham Hall Farm.

Year	Crop	Year	Crop
1986	Winter wheat	1991	Vining peas
1987	Winter wheat	1992	Cauliflower
1988	Calabrese	1993	Winter wheat
1989	Vining peas	1994	Canning potatoes
1990	Winter wheat	1995	Winter wheat

MATERIALS AND METHODS

A 10 ha field of marine alluvial deep stoneless calcareous coarse silty soil of the Wisbech series, with a history of variable within-field yields was selected. Canning potatoes preceded the experiment and were likely to have increased residual N levels across the field. Winter wheat cv. Riband was drilled on 9th October at 168 kg/ha, 310 seeds/m^2. Emergence was even across the field giving an established plant population of 165 plants/m^2. Effective weed control was achieved across the field with 1.883 kg/ha isoproturon applied in early November. Soil sampling was carried out in late February prior to the application of fertilisers. To ensure that sampling points were accurately spaced, a measuring wheel was used to check the distances between points and flexible marking canes inserted adjacent to the soil core holes. A base station was established to provide a fixed reference point. The Hydro Geonor hydraulic sampler was fitted with a Global Positioning System (GPS) antenna and lap top computer which enabled the progress of the sampler to be tracked across the field and the coordinates of each sampling point to be recorded. The regular shape of the field and the tramlining system used in drilling the crop enabled the sampling sites to be verified by ground based observations.

Soil samples were taken in a continuous column to a depth of 90 cm and sub-divided into three sections on extraction from the corer, separating the profiles in depths of 0-30, 30-60 and 60-90 cm. Each horizon sample was then bagged individually and labelled with the satellite coordinate reference and a manual field reference. A total of 72 cores was taken giving a total of 216 individual samples. Additional samples were taken at 2 m distances from several of the sampling points to identify the extent of variation over a limited distance. Samples were transported to the laboratory and frozen prior to mineral N analysis.

The crop received three applications of nitrogen fertiliser. Bulk prilled urea (46%N) was applied at 87 kg/ha on 21st March (40 kg/ha N) and again at 57 kg/ha on 25th April (26 kg/ha N). A final application of Extran (34.5%N) was made on 4th May at 75 kg/ha (26 kg/ha N) giving a total for the crop of 92 kg/ha N.

Since the period from soil sampling to harvest is relatively long, the decision was made to gather some intermediate information which might indicate if the trends measured through soil analysis were being reflected in the growth of the crop. Such measurements would normally consist of tiller numbers/m^2, or biomass/m^2. However, as these in themselves are not necessarily a measure of N availability, a chlorophyll meter was used to ascertain flag leaf 'greenness' which, following extensive training by Hydro Agri (Deutschland), was considered to provide a closer correlation with plant N status (Fox *et al.,* 1994). The flag leaves of thirty individual plants were measured within a 1 m radius of each sampling point. The meter automatically aggregates these readings and provides an average for each position. Each of the 72 sampling points was thus measured taking readings from a total of 2160 plants.

The field was photographed at GS73 in mid July by satellite and this imagery used to obtain the Normalised Difference Vegetation Index (NDVI). This provides a relative comparison of crop canopy density within the field.

The crop was harvested in mid August using a Massey Ferguson 40 Series combine fitted with a global positioning yield mapping system which records yield variations across the field.

RESULTS

Final grain yield varied significantly across the field with a range of 5.5 tonnes/ha to over 10.5 tonnes/ha. Some of this variation can be directly attributed to headland effects and known soil physical variation in the field, however much of the variation present can not be directly attributed to either of these factors.

There was no significant correlation, r2 of -0.0115, between the soil mineral nitrogen level and the final grain yield (Figure 1). The results from the 72 soil mineral nitrogen samples showed considerable variation both down the profile and spatially across the field (Table 2).

Further analysis correlating soil mineral nitrogen levels at the different depths showed a significant relationship (r2 = 0.727) between the 30 to 60 cm sample and those from 60 to 90 cm. There was little correlation between the 0 to 30 cm sample and either the 30 to 60 cm or 60 to 90 cm samples. This also highlights the need to sample to at least a depth of 60 cm and preferably 90 cm to accurately determine the soils nitrogen content at the start of the growing season.

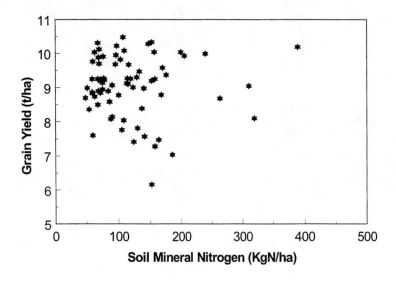

FIGURE 1. Correlation of soil mineral nitrogen and grain yield in field 1187.

TABLE 2. Soil mineral nitrogen from 36 metre grid samples in field 1187.

Profile depth (cm)	Soil Mineral Nitrogen (kgN/ha)			
	0 to 30	30 to 60	60 to 90	0 to 90
Range	10 - 69	5 - 145	19 - 214	48 - 389
Mean	22	37	61	120
S.D.	10.80	26.57	38.84	65.64

The variation in soil mineral nitrogen was further highlighted by intensive sampling at 2.0 m distances from the main sampling point (Table 3). Sampling point 18, part of the 36 m grid across the field, had a soil mineral content of 75 kgN/ha in the 0 to 90 cm sample. Adjacent samples in the 2 m grid had nitrogen contents ranging from 80 to 156 kgN/ha. Similar results were obtained from sampling point 57 where the grid sample soil nitrogen was 107 kgN/ha and the range of the detailed points was 80 to 107 kgN/ha.

The soil mineral nitrogen data was aggregated using a spherical variogram model to provide a total soil mineral nitrogen map (Figure 2). This indicates that the majority of the field had a soil nitrogen supply between 75 and 150 kgN/ha with only isolated areas with higher mineral nitrogen levels.

The measurements of flag-leaf relative chlorophyll content, determined using the Hydro 'N Tester' also showed considerable variation across the field from 495 to 823 with a mean of 691 (Figure 3). The majority of the values were within the 650 to 750 range. There was a weak correlation between the relative chlorophyll content and yield, with an r2 of 0.2459. There was also a weak correlation between the chlorophyll readings and the soil mineral nitrogen levels (r2 of 0.2614).

TABLE 3. Soil Mineral Nitrogen from intensive sampling regime.

Profile depth (cm)	Soil Mineral Nitrogen (kgN/ha)			
	0 to 30	30 to 60	60 to 90	0 to 90
18A	16	18	46	80
18B	13	43	71	127
18C	18	70	68	156
18D	14	37	76	127
Grid Sample 18	22	13	40	75
Range	13 - 22	13 - 70	40 - 76	75 - 156
Mean	17	36	60	113
S.D.	3.40	22.63	16.19	34.59

Profile depth (cm)	0 to 30	30 to 60	60 to 90	0 to 90
54A	25	18	49	92
54B	21	20	51	91
54C	24	36	62	121
54D	28	20	32	80
Grid Sample 54	24	27	55	107
Range	21 - 28	18 - 36	32 - 62	80 - 121
Mean	24	24	50	98
S.D.	2.61	7.43	10.97	15.96

MAP 1. **Total Mineral Nitrogen, 0 - 90 cm, Coldham 1187, Spring 1995**

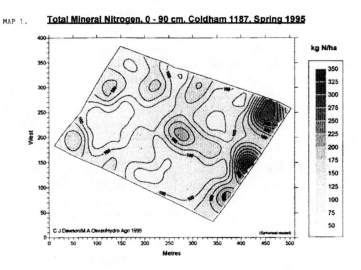

FIGURE 2. Total mineral N status of the soil in spring to 90 cm by interplation using kriging.

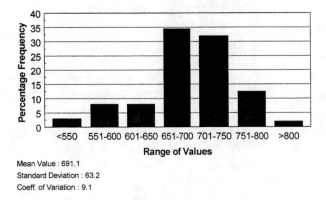

Wheat flag-leaf chlorophyll contents as measured by Hydro 'N Tester'
(adapted from Minolta SPAD-502 meter)

Mean Value : 691.1
Standard Deviation : 63.2
Coeff. of Variation : 9.1

FIGURE 3. Frequency distribution of leaf relative chlorophyll values.

Analysis of NDVI satellite images indicated variation in the measured crop canopy density across the field. There were areas of relatively low density concentrated on the headlands, with much higher densities across the field. No attempt was made in this particular study to statistically analyse the relationship between NDVI values and grain yield. However, visual analysis suggests a direct relationship between crop density and yield with dense areas giving the higher grain yields. There was no visual correlation between NDVI and soil mineral nitrogen levels.

DISCUSSION

The analysis of mineral N from the samples showed less general variation across the field than expected with levels typically between 75 and 150 kg N/ha, the levels above this only occurring in isolated areas. The application of fertiliser at 92 kg N/ha would provide sufficient total N even at the lower end of the range to achieve the targeted yield of 10 tonnes/ha. Where the mineral N level was at the top end of the range excess N above that required for optimal yield would probably have been applied. It is possible that this has masked the effects of variations in mineral N on crop yield.

The weak correlation between the chlorophyll meter readings and the SMN suggest that even though yield was not influenced by SMN the chlorophyll content of the plants was to a small degree. There was also a weak correlation between the chlorophyll meter readings and final yield which will need to be explored further before we can ascertain whether this can be manipulated to alter yield.

The extent to which the inadequately defined data collection methodologies affect the results is unknown. The localised differences measured adjacent to sampling points 18 and 54 suggests that in further trials more than one individual core should be taken at each point.

ACKNOWLEDGEMENTS

The authors would like to thank Massey Ferguson for provision of the GPS equipment and yield mapping combine, C. J. Dawson and Associates for the production of the mineral nitrogen and chlorophyll leaf status maps, Dr Margaret Oliver, Reading University for the provision of the spherical variagram model and Logica UK Limited for the provision of the NDVI map.

REFERENCES

Fox, R.H., Piekielek, W.P., Macneal, K.M. (1994) Using a Chlorophyll Meter to Predict Nitrogen Fertiliser Needs of Winter Wheat. *Commun. Soil Sci. Plant Anal.*, **25 (3 & 4)**, 171-181.

Kuhlmann, H., Engels, T. (1989) Nitrogen Utilisation in relation to N-fertilisation. The Fertiliser Society, London, Proc. No. 287.

Macdonald, A.J., Powlson, D.S., Poulton, P.R., Jenkinson, D.S. (1989) Unused Fertiliser Nitrogen in Arable Soils - its Contribution to Nitrate Leaching. *J. Sci. Food Agric.*, **46**, 407-419.

AN EVALUATION OF INDICATOR PROPERTIES AFFECTING SPATIAL PATTERNS IN N AND P REQUIREMENTS FOR WINTER WHEAT YIELD

D.J. MULLA

Department of Soil, Water, and Climate, University of Minnesota, St. Paul, MN 55108, USA

A.U. BHATTI

Department of Soil Science, Northwest Frontier Province University, Peshawar, Pakistan

ABSTRACT

A methodology is needed to divide fields into management zones which receive different application rates of N and P fertilizer. Since intensive grid sampling is costly, this paper investigates the use of surrogate soil properties that can be estimated from remotely sensed images, or easily measured crop and soil properties to develop management zones for precision agriculture. Data for soil test nutrient levels and yield goals collected at 172 locations in a wheat farm were used to estimate N and P fertilizer requirements for each sampled location. These 'best estimates' for fertilizer requirements were compared with requirements estimated from mean properties in zones delineated by patterns in crop yield and soil surface organic matter content. The criteria of crop yield and organic matter content were both acceptable indicators of spatial patterns in nutrient requirements and soil fertility. The variability in crop yield was a particularly sensitive indicator of spatial patterns in nitrogen fertilizer requirements, but was insensitive to variations in phosphorus requirements. In contrast, the variability in surface organic matter content was moderately sensitive to variations in nitrogen fertilizer requirements, and highly sensitive to variations in phosphorus fertilizer requirements. We conclude that spatial patterns in surface organic matter content could be useful in delineating fertilizer management zones, from which soil samples for nutrient analysis could be collected and composited. These samples could then be the basis for recommendations on fertilizer rates for each management zone.

INTRODUCTION

The Palouse region of Eastern Washington is characterized by steep, rolling hills formed in loess. As a result of high rates of erosion, most farms show light-colored exposed subsoils with low organic matter content along top slope, ridge top, and clay knob positions. Along bottom slope positions, deposition has produced thick layers of topsoil with high organic matter content.

It would be desirable to divide large spatially variable farms into management zones that each receive different rates of nitrogen and phosphorus fertilizer to better match patterns in soil fertility and crop productivity (Mulla *et al.*, 1992; Wollenhaupt *et al.*, 1994). The objective of this study was to use spatially intensive measurements of soil properties and grain yield across a wheat farm to determine whether or not spatial patterns in these properties would be a useful indicator of patterns in nitrogen and phosphorus fertilizer requirements.

MATERIALS AND METHODS

A summer fallow/wheat farm near St. John, Washington (SE 1/4, sec. 14, T 19 N, R 42 E) was selected for study. This site is located in steep, rolling topography with exposed (lighter colored) subsoil or shallow topsoil on eroded hilltops, ridges and clay knobs.

Extensive soil sampling was conducted in September 1987 along four 655 m long east-west oriented parallel transects separated by 122 m. Soils were collected with a hydraulically driven Giddings core sampler to a depth of 1.8 m at intervals of 15.2 m along each transect. Each sample was analyzed for available phosphorus (P), organic matter content, pH, nitrate-nitrogen (NO_3-N), and ammonium-nitrogen (NH_4-N) in the surface 0-30 cm increment (Page, 1982). In addition, total profile available water content and total nitrate-nitrogen were determined for the total depth sampled.

The soil test analyses at each sampling location were used to make nitrogen and phosphorus fertilizer recommendations according to guidelines in Washington State University Extension bulletins (Halvorson et al., 1982). The transects were divided into three fertilizer management zones using a customized geographic information system (GIS) described by Mulla (1991). Along the south side of each transect, fertilizer was applied at a uniform rate of 73 kg N ha^{-1} and 6 kg P ha^{-1}. This represents the grower's typical management practice.

In October 1987 Stephens winter wheat (*Triticum aestivum*) was sown uniformly throughout the entire farm. At harvest in August 1988, a Suzui binder was used to cut wheat strips that measured 7 to 10 m in length and 0.6 m in width. A total of 172 plots were harvested in each of the uniformly fertilized strips at intervals of 15.2 m. Plots in the uniformly fertilized strips were located about 5 m from the site where soil samples were collected the previous September. Bundles were collected and mechanically threshed and weighed in the field. Grain yield for each plot was calculated in kg ha^{-1}.

Statistical analysis of the data on grain yield and soil properties were computed using standard classical methods (Steel and Torrie, 1980). Data were grouped into classes according to one of the following criteria: fertility status, grain yield, and organic matter content. For fertility status, yield, and organic matter, the cutoff levels separated the field into three zones (Table 1). Cutoff levels for yield and organic matter were set at approximately the values of the mean plus or minus half a standard deviation.

TABLE 1. Cutoff values used to divide the study site according to criteria based upon fertility status, grain yield, and organic matter content.

Criteria	---------Management Zone----------		
	1	2	3
Fertility status			
nitrogen (kg ha^{-1})	<45	<45	>45
phosphorus (kg ha^{-1})	>5	<5	<5
Grain yield (kg ha^{-1})	<3360	3360-4704	>4704
Organic matter (%)	<1.5	1.5-2.4	>2.4

Differences in yield and soil properties in each zone for a given criteria were compared using an LSD test for an experimental design with unequal numbers of replications (Steel and Torrie, 1980, p. 146).

RESULTS AND DISCUSSION

A comparison of three different criteria for dividing the study site into zones that are managed with different fertilizer applications is given below. The criteria discussed are fertility status, grain yield, and organic matter content.

Fertility status

The steep, rolling topography at the study site (Figure 1) is typical of much of the dryland Palouse farming region. The sampled transects were divided into three zones using the criterion of fertilizer recommendation rate (Table 2). Nitrogen and phosphorus fertilizer requirements were estimated using standard soil fertility recommendation procedures. For nitrogen these involved estimates for profile nitrate-N, surface ammonium-N, mineralizable N, and a yield goal based on soil profile available water content and expected precipitation. For phosphorus these involved estimates for soil test phosphorus in the surface soil. The fertilizer requirements estimated from detailed soil testing data at each of the 172 sampled locations were considered to be the 'best' recommendations against which fertilizer requirements from all other approaches were compared. In general, this approach for estimating variable rates of fertilizer has proven to be valid when N response curves are generated in each zone, with additional precision being obtained only by directly estimating the unit nitrogen requirement in each zone (Fiez et al., 1994).

TABLE 2. Comparison of mean measured soil properties and wheat grain yields for management zones divided on the basis of fertility status. Also given are the number of locations and the average recommended rates of nitrogen and phosphorus fertilizer in each management zone.

| | ------Fertilizer Management Zone (kg ha^{-1})------ | | |
| | zone 1 | zone 2 | zone 3 |
Measured property	<45 N & > 5 P	<45 N & < 5 P	>45 N & < 5 P
No. of locations in each zone	43	77	52
Nitrogen fertilizer (kg ha^{-1})	22	22	90
Phosphorus fertilizer (kg ha^{-1})	18	0	0
Grain yield (kg ha^{-1})	3407 a	4142 b	4669 c
Available profile water (cm)	14.8 a	15.7 a	20.4 b
Organic matter (%)	1.2 a	2.3 b	2.3 b
Soil pH	6.4 a	6.0 b	5.9 b
Profile NO$_3$-N (kg ha^{-1})	114.4 a	142.4 b	98.4 a
Surface NH$_4$-N (mg kg^{-1})	5.3 a	2.8 b	2.4 b
Available phosphorus (mg kg^{-1})	7.2 a	17.8 b	18.2 b

Means followed by similar letter(s) in each row are not significantly different from one another at a 5% level of significance.

The low applied nitrogen and low applied phosphorus fertilizer zone (zone 2) generally corresponds to regions located in bottom slope positions, while the low applied nitrogen and high applied phosphorus fertilizer zone (zone 1) corresponds to regions located on eroded ridgetops and knobs. The high applied nitrogen and low applied phosphorus fertilizer zone (zone 3) is generally located in regions at mid slope and bottom slope positions of the landscape.

Table 2 shows that measured grain yields along the uniformly fertilized strips were significantly different between each of the fertilizer management zones. Yields were highest in zone 3 (4669 kg ha^{-1}) and lowest in zone 1 (3407 kg ha^{-1}) where erosion has removed topsoil and reduced productivity. No significant differences existed between measured soil properties (except profile available water and profile nitrate-nitrogen) in zones 2 and 3. All soil properties except profile available water content were significantly different in zone 1 than in zones 2 and 3.

Fertilizer application rates were significantly different for each of the management zones. Table 2 shows that the recommended rates for nitrogen in zones 1, 2, and 3 are 22, 22, and 90 kg ha^{-1}, respectively, for an overall area-average of 43 kg ha^{-1}. Similarly, the recommended rates for phosphorus in zones 1, 2, and 3 would be 18, 0, and 0 kg ha^{-1}, respectively, with an overall average of 5 kg ha^{-1}.

These rates are significantly lower than those applied by the grower as part of his typical uniform management strategy, i.e., 73 kg N ha^{-1} and 6 kg P ha^{-1}. Thus, a variable rate fertilizer strategy would lead to significant reductions in nitrogen fertilization. Since the variable strategy reduces nitrogen applications in eroded areas where productivity is low (zone 1), and in bottom slope areas (zone 2) where leaching losses are likely to be highest, this strategy reduces the potential for nitrogen contamination of surface and ground water.

The fertilizer recommendations presented above are based upon intensive soil sampling. This is too time consuming and expensive for farmers and consultants to implement. Even a sampling strategy that is based upon a sampling frequency of at least one core every ha would be impractical for most grain crops. Thus, alternative approaches are needed that estimate spatial patterns in fertilizer requirements with reasonable accuracy, yet do not require extensive soil sampling. The three indicator approaches described below are used to estimate spatial patterns in fertilizer requirements and the results are compared with those obtained above from intensive sampling.

Grain yield

The sampled transects were divided into three zones using the criterion of measured grain yield along the uniformly fertilized strips (Figure 2). The low, medium, and high yielding zones generally correspond to top, middle, and bottom slope landscape positions, respectively. Yields in the low, medium, and high productivity zones averaged 3166, 4018, and 4904 kg ha^{-1}, respectively (Table 3). Measured profile available water content, organic matter content and soil pH were significantly different between each zone. Available phosphorus in the low productivity zone was significantly lower than that in the medium and high productivity zones, while there were no significant differences in profile nitrate-nitrogen between any of the productivity zones.

TABLE 3. Comparison of mean soil properties and wheat grain yields for management zones divided on the basis of grain yield. Also given are the number of locations and the average recommended rates of nitrogen and phosphorus fertilizer in each zone.

	Yield Management Zone (kg ha^{-1})		
	zone 1	zone 2	zone 3
Measured property	<3360	3360-4704	>4704
No. of locations in each zone	40	83	46
Nitrogen fertilizer (kg ha^{-1})	15	33	45
Phosphorus fertilizer (kg ha^{-1})	0	0	0
Grain yield (kg ha^{-1})	3166 a	4018 b	4904 c
Available profile water (cm)	12.4 a	16.8 b	20.7 c
Organic matter (%)	1.4 a	2.1 b	2.4 c
Soil pH	6.5 c	6.0 b	5.8 a
Profile nitrate-nitrogen (kg ha^{-1})	110.0 a	120.8 a	134.4 a
Surface NH_4-N (mg kg^{-1})	3.8 b	2.7 a	3.9 b
Available phosphorus (mg kg^{-1})	11.7 a	16.5 b	16.0 b

Means followed by similar letter(s) in each row are not significantly different from one another at a 5% level of significance.

Recommended rates of nitrogen fertilizer in the low, medium, and high productivity management zones (Table 3) were 15, 33, and 45 kg ha^{-1}, respectively, with an overall average of 31 kg ha^{-1}. No phosphorus would be recommended if the field were divided on the basis of productivity. Using measured grain yield as a criterion for dividing a field into different management zones is an approach that has many merits. One of the advantages is a significant reduction in amounts of applied nitrogen fertilizer (compared to a uniform application of 73 kg N ha^{-1}). A second merit is that in this water deficient cropping region, locations in the field with low grain yield may be fairly persistent in pattern from one growing season to another.

A practical way for dividing the field on the basis of productivity would be to make a measurement of variations in grain yield across a field, and then develop a stratified random soil sampling strategy to assess soil fertility in low, medium, and high yielding regions prior to seeding for the next growing season.

Organic matter content

Patterns in measured surface soil organic matter (Figure 3) were used to divide the study site into separate zones having low (< 1.5%), medium (1.5-2.4%), and high (> 2.4%) organic matter content. The low, medium, and high organic matter zones generally correspond to top, middle, and bottom slope landscape positions, respectively. Measured grain yields in the low, medium, and high organic matter zones averaged 3443, 3933, and 4742 kg ha^{-1}, respectively, (Table 4). Grain yields were significantly different between each of the three management zones.

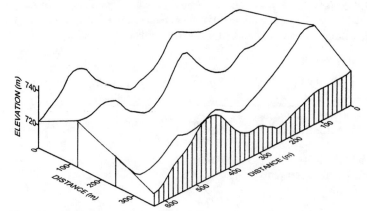

FIGURE 1. Elevation along the four sampled transects at the study site.

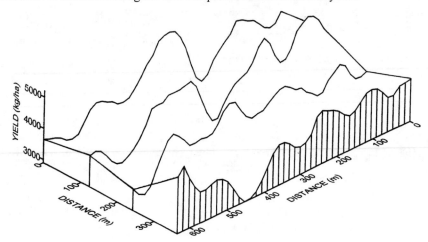

FIGURE 2. Variation in measured grain yield at the study site.

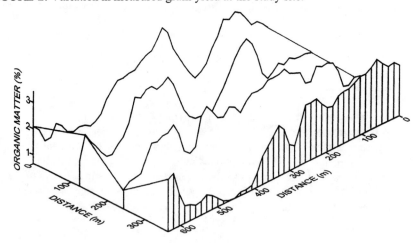

FIGURE 3. Variation in measured organic matter content along sampled transects.

TABLE 4. Comparison of mean measured soil properties and wheat grain yields for management zones divided on the basis of organic matter content. Also given are the number of locations and the average recommended rates of nitrogen and phosphorus fertilizer in each management zone.

| Measured property | ------Organic Matter Management Zone (%)------- | | |
| | zone 1 | zone 2 | zone 3 |
	<1.5	1.5-2.4	>2.4
No. of locations in each zone	56	53	63
Nitrogen fertilizer (kg ha^{-1})	28	45	37
Phosphorus fertilizer (kg ha^{-1})	20	0	0
Grain yield (kg ha^{-1})	3443 a	3933 b	4742 c
Available profile water (cm)	14.1 a	16.4 b	19.8 c
Soil pH	6.3 b	6.2 b	5.8 a
Profile nitrate-nitrogen (kg ha^{-1})	106.0 a	114.8 a	142.4 b
Surface NH$_4$-N (mg kg^{-1})	4.7 c	3.2 b	2.0 a
Available phosphorus (mg kg^{-1})	9.7 a	16.3 b	19.4 c

Means followed by similar letter(s) in each row are not significantly different from one another at a 5% level of significance.

Profile available water content and available phosphorus increased significantly from the low to the medium and from the medium to the high organic matter management zones (Table 4). Thus, the most highly eroded locations, on average, had the lowest profile available water contents and soil test phosphorus levels. Soil pH was significantly lower in the zone having the highest organic matter content than in the other zones, while profile nitrate-nitrogen was significantly higher in the high organic matter zone than in the other zones.

Based upon the average soil properties measured in each organic matter management zone, a set of recommended rates of nitrogen and phosphorus were developed (Table 4). Recommended rates of nitrogen fertilizer in the low, medium, and high organic matter management zones were 28, 45, and 37 kg ha^{-1}, respectively, with an overall average of 37 kg ha^{-1}. Recommended rates of phosphorus fertilizer were 20, 0, and 0 kg P ha^{-1}, respectively, in the low, medium, and high management zones with an average of 7 kg ha^{-1}. An advantage of dividing the field on the basis of organic matter content is that the phosphorus deficiency and low water content in the low organic matter zone are both clearly delineated. This allows over-application of nitrogen to be minimized in this zone while correcting phosphorus deficiencies.

It is commonly observed that eroded soils have shallower topsoil and lower organic matter content than non-eroded soils. In the Palouse region, highly eroded soils are visually distinct from less eroded soils due to the lighter color of the exposed subsoil. Recently, Frazier and Cheng (1989) have shown that it is possible to accurately predict detailed spatial patterns in soil organic carbon content from Landsat Thematic Mapper images of bare soil in the Palouse region. Based upon the results in Table 4 and the existing Landsat technology, it appears that there is a high feasibility of using organic matter as a practical criterion for spatial patterns in crop productivity and soil fertility.

DISCUSSION AND CONCLUSIONS

This study was conducted to compare and evaluate the feasibility of using different criteria for dividing fields into management zones in which rates of fertilizer application can be varied to match patterns in soil fertility and crop productivity. The criteria evaluated included fertility status, organic matter content, and grain yield.

The three criteria were determined to all be acceptable since they allowed the field to be divided into zones that each had significantly different grain yields and recommended rates of fertilizer (Table 5). Generally, the lowest rates of nitrogen were recommended in eroded top slope positions where productivity was low. By applying low rates of nitrogen in low productivity areas, the potential for contaminating surface and ground water is minimized. For each criteria, the field-averaged rates of recommended nitrogen were about 35 kg ha^{-1}, which represents a significant reduction compared to the grower's typical uniform application of about 73 kg N ha^{-1}. The strategies based upon fertilizer management and organic matter showed that 25% and 32%, respectively, of the field required about 20 kg P ha^{-1}.

In terms of practicality, two criteria seem to offer the best feasibility for dividing fields into fertilizer and productivity management zones. These criteria are organic matter content and grain yield, with the former being easily estimated using Landsat Thematic Mapper images of bare soil (Bhatti et al., 1991), and the latter being easily estimated using yield mapping systems (Schnug et al., 1993). Assuming fertilizer costs of $0.31/kg and winter wheat prices of $0.15/kg, the net economic return over the typical fertilizer management strategy was more profitable when the field was divided by organic matter content ($15.36/ha) than when the field was divided into management zones based on grain yield ($4.77/ha). To implement the remote sensing approach, a remotely sensed image could be used to delineate zones having low, medium, and high organic matter contents (Bhatti et al., 1991). Each zone could then be sampled to obtain its soil fertility status and fertilizer recommendations for nitrogen and phosphorus.

TABLE 5. Comparison of recommended nitrogen and phosphorus fertilizer rates for typical management versus variable management zones divided on the basis of fertility status, grain yield, and organic matter content.

Criteria	-------Management Zone-------			Area Average
	1	2	3	
	--------------Nitrogen Rate (kg ha^{-1})--------------			
Typical management	73	73	73	73
Fertility status	22	22	90	43
Grain yield	15	33	45	31
Organic matter content	28	45	37	37
	--------------Phosphorus Rate (kg ha^{-1})--------------			
Typical management	6	6	6	6
Fertility status	18	0	0	5
Grain yield	0	0	0	0
Organic matter content	20	0	0	6

REFERENCES

Bhatti, A.U., Mulla, D.J., Frazier, B.E. (1991) Estimation of soil properties and wheat yields on complex eroded hills using geostatistics and Thematic Mapper images. *Remote Sensing Environ,* **37**, 181-191.

Fiez, T.E., Miller, B.C., Pan, W.L. (1994) Assessment of spatially variable nitrogen fertilizer management in winter wheat. *J. Prod. Agric.,* **7**, 86-93.

Frazier, B.E., Cheng, Y. (1989) Remote sensing of soils in the Eastern Palouse region with Landsat thematic mapper. *Remote Sens. Environ.,* **28**, 317-325.

Halvorson, A.R., Koehler, F.E., Engle, C.F., Morrison, K.J. (1982) Fertilizer guide: Dryland wheat, general recommendations. Cooperative Extension Service FG-19, Washington State University, Pullman, WA.

Mulla, D.J. (1991) Using geostatistics and GIS to manage spatial patterns in soil fertility. *Automated Agriculture for the 21st Century,* G. Kranzler (Ed.), St. Joseph, MI, Am. Soc. Agric. Eng., pp 336-345.

Mulla, D.J., Bhatti, A.U., Hammond, M.W., Benson, J.A. (1992) A comparison of winter wheat yield and quality under uniform versus spatially variable fertilizer management. *Agric. Ecosys. Environ.,* **38**, 301-311.

Page, A.L. (1982) In: *Methods of Soil Analysis Part 2- Chemical and Microbiological Properties,* A.L. Page (Ed.), Agronomy 9, Madison, WI., Agron. Soc. of America.

Schnug, E., Murphy, D., Evans, E., Haneklaus, S., Lamp, J. (1993) Yield mapping and application of yield maps to computer-aided local resource management. *Soil Specific Crop Management,* P.C. Robert, R.H. Rust, and W.E. Larson (Eds), Madison, WI, Agron. Soc. Am., pp 87-94.

Steel, R.G.D., Torrie, J.H. (1980) *Principles and Procedures of Statistics.* 2nd ed., New York, NY, McGraw-Hill.

Wollenhaupt, N.C., Wolkowski, R.P., Clayton, M.K. (1994) Mapping soil test phosphorus and potassium for variable-rate fertilizer application. *J. Prod. Agric.,* **7**, 441-448.

A RATIONAL STRATEGY FOR DETERMINING THE NUMBER OF CORES FOR BULKED SAMPLING OF SOIL

M.A. OLIVER, Z. FROGBROOK

Department of Soil Science, University of Reading, Reading RG6 6DW, UK

R. WEBSTER

Rothamsted Experimental Station, Harpenden, Hertfordshire AL5 2JQ, UK

C.J. DAWSON

Chris Dawson and Associates, Strensall, York YO3 5TD, UK

ABSTRACT

Precision farming requires accurate maps of plant nutrients in the soil within fields. Such maps can be made by kriging from sample data and variograms of the nutrients. Standard technique assumes that data are accurate, but short-range fluctuation in the soil means that measurements on small cores represent the area around them with error. This error is largely a sampling effect, and it can be diminished by bulking cores before analysis. In principle one can make individual measurements as accurate as desired by bulking. The number of cores to be bulked depends on the local variation and the error that can be tolerated.

The number of cores to bulk for a given tolerance can be determined from the variogram. Kriging equations are set up and solved to determine the estimation variances for a range of sample sizes and configurations before sampling. From these one can choose the combination that just meets the tolerance. It is the optimal solution.

We have tested the technique for potassium, phosphorus and magnesium on soil in two physiographic situations in England, one an old flood plain, the other a chalk plateau. The geostatistical analysis and the results are presented. In the absence of strict tolerance limits we recommend that 16 cores be bulked within a few square metres as a 'rule of thumb'.

INTRODUCTION

Patchy growth and yield of arable crops are evident in fields in many parts of the world. Farmers are aware of this, and agronomists have been measuring the variation in uniformity trials throughout this century. Now, with recording combine harvesters operating on commercial farms, we are seeing the almost pervasive nature of the variation and its magnitude, commonly with two-fold variations in yield between patches which may be no more than a few tens of metres across. The consequences where the crop is poor are both economic and environmental; potential yield may be lost by applying too little fertiliser, and nutrients may leach into aquifers if more is applied than the crop can take up. Locally poor performance can sometimes be explained by thin soil. More often it is caused by excess traffic on headlands, for example. When these more obvious causes have been taken into account the question arises as to what extent local deficiencies in nutrients might affect the crop. To answer it we need to relate nutrient concentrations to the crop yield.

The classical approach for investigating such relations has been to measure the concentrations of nutrients in the soil and yields in designed experiments under controlled conditions. This is hardly feasible in ordinary commercial practice, and an alternative approach being explored is to map both yield and nutrient status. Evidently, if we are to attempt to explain yield variation in terms of nutrient supply then the nutrient status of the soil must be determined and mapped.

All properties of the soil vary from place to place. Nutrient concentrations are no exception, but they differ from some properties, such as drainage status, in that their variation within fields appears to be largely or even entirely stochastic. They vary at all scales in an apparently random way, and the relative contributions to the variation from different scales are themselves variable and can be ascertained only by sampling. If the short-range component is large then it can mask longer-range patterns that might be of interest and which could be mapped.

If one has sufficient data from numerous cores of soil then the short-range fluctuation can be smoothed by taking local averages. This is usually too expensive, and the alternative is to determine nutrient concentrations on bulked samples. Bulking followed by thorough mixing may be regarded as a physical averaging. The concentration measured in a bulked sample should equal the arithmetic mean of the individual cores contributing to it, unless some chemical reaction takes place within the sample. We might also think that the variance of the bulked sample would be equal to the variance of the individual cores divided by their number, as a first approximation. However, this disregards the spatial correlation between the cores.

Since we do not know before sampling the short-range variation and the strength of the spatial correlation, we cannot predict the reliability of bulked values or of nutrient maps made from them. We must first determine the magnitude and spatial structure of the short-range variation from individual cores. From these we can calculate the variance associated with any bulking scheme, including the number of cores, their configuration and the size of support, i.e., the area over which to bulk to obtain reliable data for mapping or comparison with local yield. It might then be possible to choose an optimal combination for mapping the large-scale variation.

Using the theory of regionalized variables (Matheron, 1965) and its adaptation by Webster and Burgess (1984) with data from a preliminary survey, we can estimate the variance that we should expect with any bulking design. The procedure includes computing an experimental or sample variogram from the sample data and fitting a suitable model to it. Using the variogram model with the kriging equations, we can determine the variance for different combinations of support (size and shape of area) and configuration of sampling points within the blocks. The estimation variances are calculated for a range of sample sizes and configurations. From these one can choose a combination that meets the tolerance for a given nutrient. The nutrient concentrations measured from the bulked sample provide the data for estimating the values from which to map the nutrient in the whole field. In this way the noise from short-range fluctuation is effectively removed.

THEORY

Consider a small block of land, B, designed to be the support for mapping a variable of interest, Z. We assume that the variable is random, taking values at all places and obeying the intrinsic hypothesis, so that its variation can be expressed by its variogram:

$$\gamma(h) = \frac{1}{2}E[\{Z(\mathbf{x}) - Z(\mathbf{x} + \mathbf{h})\}^2] , \tag{1}$$

where $Z(\mathbf{x})$ and $Z(\mathbf{x} + \mathbf{h})$ are the values of Z at any two places, $\mathbf{x}$ and $\mathbf{x} + \mathbf{h}$, separated by the vector $\mathbf{h}$. Suppose now that we wish to estimate the average of Z, say $\mu(B)$, in B from n known values, $z(\mathbf{x}_i)$, at positions $\mathbf{x}_i$, $i = 1, 2, ..., n$ within it by

$$\hat{\mu}(B) = \frac{1}{n}\sum_{i=1}^{n}\lambda_i z(\mathbf{x}_i) , \tag{2}$$

where the λ_i are the weights associated with the positions $\mathbf{x}_i$. The estimation variance of $\mu(B)$ is

$$\sigma^2(B) = 2\sum_{i=1}^{n}\lambda_i\bar{\gamma}(\mathbf{x}_i, B) - \sum_{i=1}^{n}\sum_{j=1}^{n}\lambda_i\lambda_j\gamma(\mathbf{x}_i, \mathbf{x}_j) - \bar{\gamma}(B, B) , \tag{3}$$

where $\gamma(\mathbf{x}_i, \mathbf{x}_j)$ is the semivariance between points $\mathbf{x}_i$ and $\mathbf{x}_j$, $\bar{\gamma}(\mathbf{x}_i, B)$ is the average semivariance between data point $\mathbf{x}_i$ and the block B, and $\bar{\gamma}(B, B)$ is the within block variance. Formally $\bar{\gamma}(\mathbf{x}_i, B)$ and $\bar{\gamma}(B, B)$ the integrals

$$\bar{\gamma}(\mathbf{x}_i, B) = \frac{1}{B}\int_B \gamma(\mathbf{x}_i, \mathbf{x})d\mathbf{x} , \tag{4}$$

and

$$\bar{\gamma}(B, B) = \frac{1}{|B|^2}\int_B\int_B \gamma(\mathbf{x}, \mathbf{x}')d\mathbf{x}d\mathbf{x}' , \tag{5}$$

where $\mathbf{x}$ and $\mathbf{x}'$ sweep over B independently.

In principle, bulking is equivalent to averaging the values at the positions from which the individual cores of soil are taken in B. Every core is the same size and shape, so the weights λ_i are equal. The number and positions of the cores can be varied by the investigator to achieve a particular precision expressed in terms of the estimation variance. For practical purposes the estimation variance is minimised for a given n when the sampling points are on a centred regular grid - see Webster and Burgess (1984). Provided the variogram is known, the investigator can determine the number of cores to bulk from by solving Equation (3) for a range of n on a square grid, plotting the calculated estimation variance or error against n, and reading from the graph the smallest value of n to satisfy the tolerance.

FIELDS AND SAMPLING

We illustrate the approach using data from two arable fields on contrasting physiography and soil type: field 1 is on old riverine alluvium south of York in northern England, and field 2 is on chalk near Benson (Oxfordshire, south-central England). They were sampled as follows. In each field we chose six nodes 100 m apart, at each of which there were two transects, the mid-points of which coincided with the node. The transects were 14 m long, and the soil was sampled at 1 m intervals. Twenty nine samples were taken at each grid node, giving 174 samples in total. The soil was sampled to a depth of 15 cm. The samples were dried and sieved, and the available potassium (K), phosphorus (P) and magnesium (Mg) were determined in mg l^{-1} by the standard methods of MAFF (1986).

RESULTS

The summary statistics for each field and variable are listed in Table 1. For each field and variable there is an entry for the variance within crosses. In all instances it is substantially less than the variance within the whole field. We include these variances for comparison with the variograms, which are computed only to the limits of the crosses. All of the variates had near-normal distributions, apart from P in field 2. Despite the fairly strong skew of the last, we did not transform it because the procedure averages the values in a way similar to bulking.

Experimental variograms were computed using the usual formula:

$$\hat{\gamma}(h) = \frac{1}{2m(h)} \sum_{i=1}^{m(h)} \{z(x_i) - z(x_i + h)\}^2 , \qquad (6)$$

where $z(x_i)$ and $z(x_i + h)$ are the measured values of Z at x_i and $x_i + h$, and $m(h)$ is the number of paired comparisons separated by the lag h, which was incremented in steps. Models were fitted to the resulting ordered sets of estimates by weighted sum of squares

TABLE 1. Summary statistics from data in mg l^{-1}.

Field	Statistic	Variable		
		K	P	Mg
Alluvium 1	Mean	294.4	21.5	393.8
	Standard deviation	63.3	19.9	131.8
	Variance	4009.8	286.17	17383.0
	Skew	0.24	1.86	-0.53
	Variance within crosses	1366.0	24.13	1420.0
Chalk 2	Mean	126.0	43.9	19.3
	Standard deviation	39.1	10.6	5.14
	Variance	1531.7	111.46	26.44
	Skew	0.45	0.08	0.88
	Variance within crosses	441.7	19.92	11.03

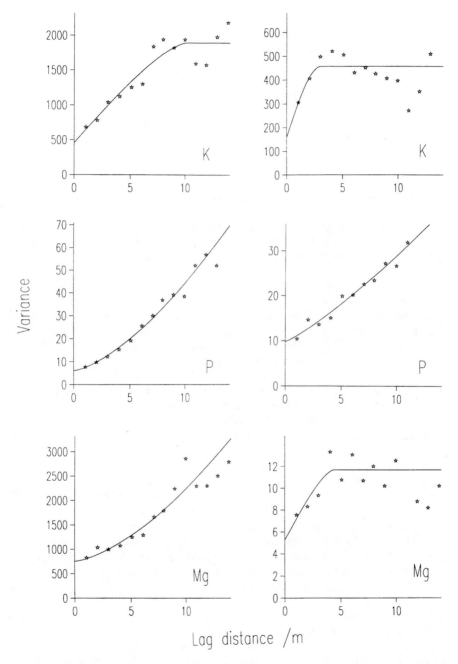

FIGURE 1. Variograms of K, P, and Mg for fields 1 (left) and 2 (right). Experimental values are plotted as points, and the continuous lines show the fitted models.

TABLE 2. Models and coefficients of variograms.

Field	Property	Model	c_0	c	a/m	m	α
					Parameter		
Alluvium 1	K	Circular	456.8	1438.0	10.09		
	P	Power	6.10			1.302	1.46
	Mg	Power	754.5			46.99	1.50
Chalk 2	K	Spherical	159.1	298.0	3.09		
	P	Power	9.82			1.228	1.19
	Mg	Circular	5.24	6.43	4.38		

The parameters are c_0 nugget variance, c the sill of the autocorrelated variance and a the range of the circular and spherical models, and m and α are the gradient and exponent of the power functions.

approximation, and those that minimised the sums of squares were chosen. Figure 1 shows the variograms, and Table 2 lists the models and their parameters.

For each variable and field semivariances were computed from the model and inserted into Equation (3), which was then solved to obtain the estimation variances and their square roots, the estimation errors, for a range of block sizes (of side 2 m, 5 m and 10 m) and sample sizes (n = 4, 9, 16, 25, 36 and 49). The errors were then plotted against n for each block size, and these are displayed in Figure 2. All the blocks were square, and the results are for square sampling configurations. There are no intermediate values of n with these configurations, and the lines are drawn to guide the eye and aid interpretation. The estimation error decreases roughly in inverse proportion to the sample size, and since the larger blocks contain more variance their errors are always somewhat larger than those of the smaller blocks.

To determine the optimal number of cores for bulking from Figure 2 a horizontal line can be drawn across the graph at the specified tolerance expressed as estimation error. The intersections of the line with the graphs of the error indicate the number of cores from which to bulk: it is the next largest square number from the intersection.

There are no well established tolerances that either a farmer or advisers will accept for the precision of estimated values of nutrients. They are also likely to vary according to average concentrations in the soil. For illustration we have chosen some reasonable values to show the consequences for sampling. Let the tolerable error be 7 mg l^{-1} for K. Then in field 1 (top left graph in Figure 2) its value intersects the line for 2 m × 2 m blocks at $n \approx 12$. The next larger square number is 16, and bulking from this number of cores in a centrally placed square configuration will ensure that the tolerable error is not exceeded. The lines for the other two blocks are sufficiently close to that of the smallest that 16 cores should suffice for all three block sizes.

Field 2 is less variable with respect to K, and for a square block of side 2 m only 4 cores would be needed, whereas for a 5 m × 5 m block 9 would be necessary.

An error of 1 mg l^{-1} for P seems reasonable, and Figure 2 shows that a sample of 16 cores is needed in field 2, whereas 9 would be sufficient in field 1.

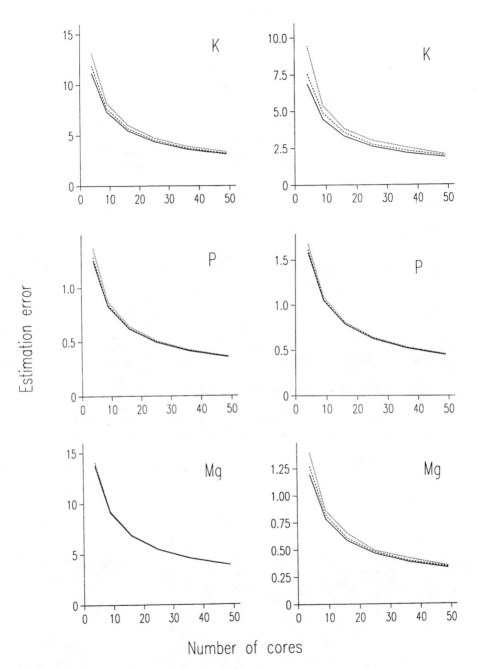

FIGURE 2. Graphs of estimation error against size of sample, n, for K, P, and Mg for fields 1 (left) and 2 (right). The errors were calculated for $n = 4, 9, 16, 25, 36$ and 49, and the lines joining them, solid for 2 m × 2 m blocks, the dashed lines for 5 m × 5 m, and dotted lines for 10 m × 10 m, are to guide the eye only.

The two fields differ enormously in both mean and variances with respect to their magnesium concentrations. It is not evident that any one tolerance should apply to both.

CONCLUSION

The differences in the results for the two fields and the three nutrients shows that there is no single optimal value from which to bulk. To make use of the information in Figure 2 the farmer would need to decide which of the nutrients was critical for the rotation or crop to be grown, and based on this and the estimation variances he would choose the optimal number accordingly. In the absence of prior knowledge, and therefore no variogram from which to compute the estimation variances, a reasonable 'rule of thumb' might be to bulk from 16 cores. It is clear from all of the graphs that the estimation error decreases steeply to this point and thereafter only gently.

ACKNOWLEDGEMENTS

We thank British Sugar Plc and Mr P. Chamberlain for their co-operation and for providing the chemical data on the soil.

REFERENCES

MAFF (1986) *The Analysis of Agricultural Materials*, 3rd edition. Reference Book 427, London, Her Majesty's Stationery Office.

Matheron, G. (1965) *Les variables régionalisées et leur estimation*. Paris, Masson.

Webster, R., Burgess, T.M. (1984) Sampling and bulking strategies for estimating soil properties in small regions. *Journal of Soil Science*, **35**, 127-140.

HOLES IN PRECISION FARMING: SPATIAL VARIABILITY OF ESSENTIAL SOIL PROPERTIES

Ya. PACHEPSKY

Duke University Phytotron, Department of Botany, Duke University, Durham, NC 27708, USA

B. ACOCK

Remote Sensing and Modeling Laboratory, USDA:ARS:BA:NRI:RSML, Beltsville, MD 20705, USA

ABSTRACT

Precision farming prescriptions are often based on grid sampling data. Risk assessment techniques are needed to evaluate uncertainty in the outcome of prescriptions relying on interpolation between sampling points. Our objective has been to develop such a technique. We apply geostatistical simulations to generate equiprobable spatial patterns of essential soil parameters between sampling points. A crop simulator is then used to transform probability distributions of errors in soil parameters into probability distributions of errors in yields. In this work, a genetic algorithm was used to generate random fields of available water capacity. The soybean crop simulator, GLYCIM, was used to find the probability distribution of errors in yield estimates. Typical commercial and research sampling densities were compared. We found that yield uncertainty assessment had to be done for a sequence of years since temporal variability was greater than spatial variability. Back calculation of soil properties from yield maps using crop simulation may be more feasible than interpolation between sampling points, since the results can be less uncertain.

INTRODUCTION

Crop modelling can be a tool in prescription farming. Both crop yield and the impact of farming on the environment are complex functions of weather, soil, cultivar and management. Crop simulators can be run for several input scenarios to find which scenario will be the most appropriate. This scenario will then be a prescription.

Such use of crop models requires soil, weather, and management data at the same scale as the scale of field operations and the scale of yield monitoring. Table 1 presents a comparison of characteristic scales of various parameters related to precision farming. Each estimate of a characteristic length with a standard error is an average over values from 20 randomly chosen publications on precision farming. Yield monitoring and variable rate applications of water and chemicals have fairly fine scales. Soil parameters have different scales, and unfortunately water and nitrogen often have short range variability. Even the soil sampling scale for research is larger than the scales of some essential parameters. Commercial soil sampling may completely miss spatial correlations in soil, and geostatistical interpolation may be very imprecise.

TABLE 1. Characteristic scales related to precision farming and their standard errors.

Data source	Characteristic length, m
Soil sampling grid size	
Research	31 ± 18
Commercial	122 ± 25
Yield sampling	
Grain yield monitors	18 ± 4
Variability range from the spherical variogram	
Soil water content	10 ± 5
Nitrate content	90 ± 60
pH	150 ± 50
K	120 ± 60
Available variable rate applications	
Sprinkler irrigation	10
Chemical application	10

Two possible uses of crop models are feasible when the soil sampling density is low. One use is to run crop simulations to assess the risk of using interpolated soil data for prescriptions. Another use of crop models is to back calculate soil properties from yield maps and other more detailed sources available.

The objective of this work was to develop an approach to the risk assessment of prescriptions. We simulated soil variability, then simulated the corresponding yield variability and inferred how uncertainty in soil data may affect uncertainty in yield predictions.

MATERIALS AND METHODS

To simulate soil variability, a model of the variability has to be selected first, and the most common one is the model of the second order, stationary, spatially correlated, random variable which implies that the mean, variance, and semivariogram of the soil property in question are not dependent upon sampling location. This model has been used in studies that included the interpolation of soil variables with kriging.

The use of a spatial variability model to simulate rather than interpolate spatial distributions is referred to as geostatistical simulation or stochastic imaging (Deutsch and Journel, 1992). To do stochastic imaging, one must assume the mean, the variance, and the semivariogram of the variable in question and select locations in which the values of this variable will be randomly generated. A stochastic image is a set of values of this variable in preselected locations, with the mean, variance, and semivariogram calculated for this set of values being the same as assumed. If the variable was sampled in some of the preselected locations, a stochastic image preserves the sampling results. Because the images are generated randomly, they differ from each other in locations other than sampled ones. Images are equiprobable and give statistical distributions of the variable in question in preselected locations.

164

Sets of stochastic images reflect the full variability of the variogram model as opposed to kriging, hence they capture more of the information available (Journel, 1996). Stochastic imaging was initially developed to correct for the smoothing effect shown on maps produced by kriging, and to eliminate artifacts that kriging may produce (Deutsch and Journel, 1992). Whereas kriging is viewed as an appropriate tool for showing global trends, stochastic imaging allows us to characterize uncertainty about the spatial patterns of a parameter in question. Applications of stochastic imaging in environmental studies have yielded an efficient method of risk and uncertainty assessment (Rossi *et al.*, 1993).

Figure 1 illustrates how we use stochastic imaging to assess uncertainty in soil and yield estimates. Suppose we know the spatial distribution of a soil parameter on a fine grid. We derive estimates of the mean, variance and semivariogram from this distribution. Then we sample this spatial distribution of the soil parameter with a lower density and generate several equiprobable stochastic images of the spatially-distributed soil parameter using the same mean, variance, and semivariogram. Values of the soil parameter at the sampling points are preserved in each image. Then we construct the uncertainty distribution of the error in the soil parameter value from differences between values in the images and true values of the soil parameter at each node of the fine grid.

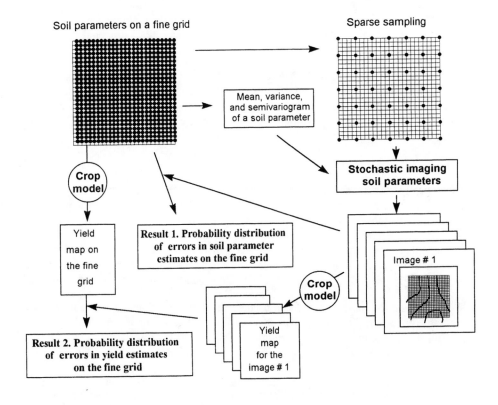

FIGURE 1. Geostatistical simulations to assess uncertainty in yield estimates.

Using a crop model, we calculate the 'true' spatial distribution of yield estimates and spatial distributions of yield estimates for each of the images. Then we construct the uncertainty distribution of the error in yield estimates from differences between the values in the images and the 'true' values at each node of the fine grid.

The differences between the generated and true values of the soil parameter will cause differences between true yield values and yields estimated from generated values. From these, we can construct the uncertainty distribution of the error in the yield estimate.

We can repeat this procedure with other sampling densities. As a result, we will see how the sampling density can affect uncertainty in yield estimations for this soil parameter.

A crop model and a variability simulation technique that we have used to implement this approach to the uncertainty analysis are described below.

Crop model

A comprehensive soybean crop simulator, GLYCIM, has been used in this work (Acock and Trent, 1991). GLYCIM consists of a collection of modules that describe related sets of physical or physiological processes and simulate growth of a plant in a uniform crop that is free of pests and diseases. Balances of materials are kept for individual leaves and petioles on the plant. Balances of materials are kept for other plant organs by type (i.e., stems, flowers, seeds) and for soil by cells. Fluxes of water, heat, nitrate, and oxygen are simulated for the soil while fluxes of carbon, nitrogen, and other structural dry matter are simulated for the plant.

GLYCIM is used on farms in the Mississippi Delta separately for each soil in a field to warn farmers of water stress and to estimate the probable effect of various irrigation scenarios on soybean yield. Our general conclusion after five years of experience is that a mechanistic crop simulator with good crop physiology, calibrated for a crop cultivar, can be used to issue prescriptions.

Simulating soil variability

Theoretically, the mean, the variance, and the semivariogram used to generate a stochastic image should be reproduced in that image. Practically, images generated with the existing techniques may have probability distributions and semivariograms that are substantially different from the assumed distributions and semivariograms. This is considered as an advantage when the 'original statistics are inferred from sparse samples and cannot be deemed exactly representative of the population statistics' (Deutsch and Journel, 1992). The difference depends on the algorithm used for generating the images.

We found that the variability induced by the sequential Gaussian algorithm may be of the same magnitude or larger than the variability that is modelled with the input probability distributions and semivariograms. Other geostatistical simulation techniques like the lower-upper decomposition, and the turning band method give similar results.

A new technique based on using genetic algorithms (Pachepsky and Timlin, 1997) provided sequential realizations of random fields with probability distributions and semivariograms much closer to the input distributions and semivariograms than traditional geostatistical simulation tools. An example of the comparison between the genetic and the sequential Gaussian algorithm is shown in Figure 2.

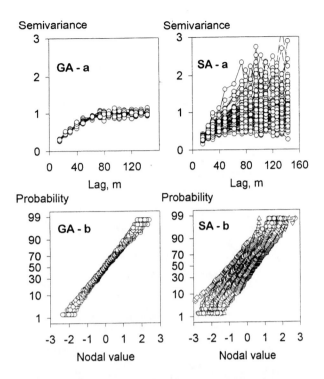

FIGURE 2. Comparison of stochastic images generated with a genetic algorithm (GA) and with the sequential Gaussian algorithm (SA) on the same grid. a - semivariograms of individual images, b - probability distribution functions of the individual images.

RESULTS AND DISCUSSION

We now present an example of the use of geostatistical simulations to assess uncertainty in interpolated soil properties and yields as related to the sampling design. We have compared samplings of 1 per 2 acre (commercial density) and 1 per 0.5 acre (research density) of available water capacity (AWC).

First, AWC has been simulated at the midpoints of 0.125 acre plots (which corresponds to yield monitor sampling density) with a spherical variogram having 100 m range, average 0.06 and variance 0.013. This was considered to be the 'true' distribution.

Then the true distribution was sampled with the commercial and with the research density, and twenty stochastic images were generated at the midpoints of 0.125 acre plots for each sampling density.

The results of the simulation expressed as uncertainty distributions are shown in Figure 3. As expected, the variability of errors in AWC decreases as the sampling density increases. The black bars corresponding to the higher sampling density are shorter at the highest and lowest values of AWC.

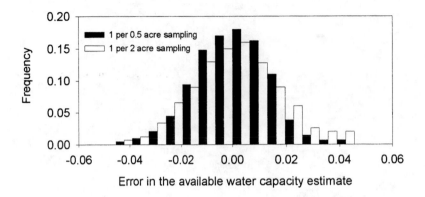

FIGURE 3. Histograms of errors in estimating available water capacity from geostatistical simulations with two sampling densities.

The real value of this improvement in AWC simulations can only be assessed, however, when errors in yield estimates are assessed. GLYCIM was used to calculate yields in the generated fields for 1994, 1995 and 1996. The probability distribution functions of errors in yield estimates are shown in Figure 4. The effect of the variability in available water capacity on the yield variations was markedly different in these three years.

The results shown in Figure 4 can be interpreted from data on weather patterns and simulated yield response to these patterns at a particular available water capacity. These data are shown in Figure 5.

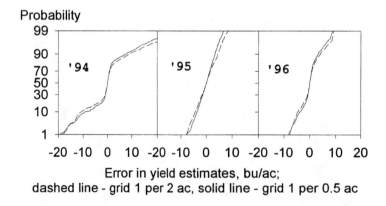

FIGURE 4. Probability distribution functions of the errors in yield estimates from geostatistic simulations of available water capacity.

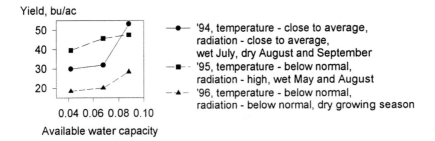

Yield, bu/ac

— '94, temperature - close to average,
   radiation - close to average,
   wet July, dry August and September
—■-- '95, temperature - below normal,
   radiation - high, wet May and August
—▲— '96, temperature - below normal,
   radiation - below normal, dry growing season

FIGURE 5. Simulated soybean yield responses to the available soil water capacity as related to the weather patterns of three consecutive years.

The weather patterns were quite different in 1994, 1995, and 1996. The key month in reproductive development and yield formation is August when the plants are sensitive to the water storage capacity of soil and to the water supply. In 1994 and 1996, high available water capacity increased yields because August was dry in both cases, and yield increased because the soil held water from the July rains. In 1995, August was wet and there was not much advantage in having high available water capacity.

Figure 4 shows that the more sensitive yield is to the soil parameter in a particular year, the wider is the range of error estimates.

The fourfold decrease in soil sampling density compared with the density of yield monitoring made the sampling very inefficient in the example used in this study. The largest errors in yield estimates occurred in 1994 when yield response to soil AWC was greatest and it was most important to know AWC accurately. The relative errors in yield estimates exceeded 30% of the average crop yield in 20% of cases both in 1994 and 1996. The errors in yield estimates were not so large in 1995, but it was the year when the crop did not show a response to AWC, the soil parameter in question. Since the quality of the soil data generated at the 'research' sampling density was so poor, decreasing the density to the 'commercial' sampling level had little further effect.

This study shows why attempts to explain spatial variations in yields from soil sampling data are likely to fail because the low accuracy of the interpolated values becomes most important in years when the crop responds to differences in values of the soil parameter in question.

In the example used in this study, the distribution of errors in AWC itself did not provide an insight on how efficient was the soil sampling. The efficiency of the sampling had to be assessed from the yield estimates derived from these soil parameter estimates. In precision farming projects, such assessment should be done using the model that is being applied to define variable rate applications based on the soil data.

Two additional factors have to be considered to make the risk assessment complete. First, the variogram is never known precisely. Including the uncertainty of the variogram in simulations might <u>increase</u> errors in the yield estimates. Second, geostatistical simulations were done for the midpoints of 25 x 25 m squares. Simulating for points within these squares and averaging over the squares might <u>decrease</u> errors in the yield estimates over the squares. In any case, if the uncertainty appears to be too high, either sampling density has to be increased or an attempt has to be made to back calculate soil properties from yield maps.

CONCLUSIONS

1. Assessment of the uncertainty of soil properties is an important component of risk assessment in precision farming, since soil properties are measured sparsely.

2. Assessment of the uncertainty of soil properties can be done using geostatistical simulations and can be transformed into yield uncertainty by using a crop simulator.

3. Yield uncertainty assessment has to be done for a sequence of years since temporal variability may be greater than spatial variability.

4. Assessment of the yield uncertainty can help to evaluate sampling strategy. Back calculation of soil properties from yield maps and other high density observations using crop simulation may be more feasible than interpolation, because the results can be less uncertain.

REFERENCES

Acock, B., Trent, A., (1991) The soybean crop simulator GLYCIM: Documentation for the modular version. *Miscellaneous Series Bulletin 145,* University of Idaho, Agricultural Experimental Station, Moscow, Idaho.

Deutsch, C.V., Journel, A.G. (1992) *GSLIB: Geostatistical Software Library and User's Guide*, Oxford University Press, New York, 1992.

Journel, A.G. (1996) Modeling uncertainty and spatial dependence: Stochastic imaging. *Int. J. Geographical Information Systems*, **10,** 517-522.

Pachepsky, Ya.A., Timlin, D.J. (1997) Generating spatially correlated fields with a genetic algorithm, *Water Resources Research* (in review).

Rossi, R.E., Borth, P.W., Tollerson, J.J. (1993) Stochastic simulation for characterizing ecological spatial patterns and appraising risk. *Ecol. Applic.*, **3**, 719-735.

PLANNING PHOSPHORUS AND POTASSIUM FERTILISATION OF FIELDS WITH VARYING NUTRIENT CONTENT AND YIELD POTENTIAL

R.S. SHIEL, S.B. MOHAMED, E.J. EVANS

Departments of Agricultural & Environmental Science and Agriculture, University of Newcastle upon Tyne, NE1 7RU, UK

ABSTRACT

Current practice for fertilising winter cereals is to base annual applications of phosphorus and potassium on field-average yield and soil P and K status. The inefficiency of this method, based on results from two commercial fields, is demonstrated. Low intensity soil sampling (4 ha^{-1}) gave a more accurate estimate of differential fertiliser requirements across fields. For both nutrients, 'Fault' was the most efficient method of map realisation. The fertiliser applied is commonly a compound with a fixed P:K ratio, and the result is that only a fraction of the field actually receives the most appropriate amount of nutrients. The result of ignoring within-field yield variation is that high-yielding parts of the field are inadequately fertilised, while elsewhere soil nutrients tend to accumulate in low-yielding areas. This may cause a bias in conventional soil analysis results, leading to even less appropriate overall nutrient recommendations. Crop logging is, however, only of limited value in improving accuracy of application if the yield is normally distributed with a small CV (< 10%). With almost no loss of precision, this system based on smooth contours of yield and nutrient index is rendered into a more practical scheme for fertiliser application by dividing the fields into a 20 x 20 m grid and applying two single-nutrient fertilisers.

INTRODUCTION

After the introduction of fertilisers, one of the major problems was deciding on the most appropriate rate to apply (Hall, 1921). Calculation of nutrients removal by crops would allow the amount lost to be replaced, but this assumes that the content in the soil is non-limiting. The value of soil analysis as a predictor of fertiliser application was demonstrated very clearly over 40 years ago, but there remained the problem that this method was only reliable in a statistical sense, and at an individual field level there was a substantial risk of inappropriate amounts of fertiliser being applied (Cooke, 1967). Because soil properties were known to be variable, the value of taking a composite soil sample had been quickly realised, but the resulting mean value for the field was often representative of only a small proportion of the total area. Furthermore a uniform fertiliser application rate would do nothing to reduce this inherent variability within fields. Yield variability within fields was demonstrated at an even earlier date, making it clear that the offtake of nutrients and the maintenance replenishment, was certainly non-uniform (Mercer and Hall, 1911).

However attractive the possibility of variable application of nutrients within fields may have been, there was no practical opportunity until recently to repeatedly locate areas reliably within fields and there was no machine instrumentation capable of delivering variable fertiliser rates (Searcy, 1995). Variation in nutrient availability within a field may, however, occur over sufficiently short distances to make any variation in application rate impractical. This type of random fluctuation over short distances is conveniently examined as the nugget variance in kriging (Webster and Oliver, 1990); a large range makes variable rate application more practicable and furthermore reduces the intensity and cost of sampling. Mohamed et al. (1997) have developed a method for testing the accuracy of mapping the distribution of a single soil nutrient at different sample spacings using a range of map realisation methods. If both P and K have a large range, then the opportunity will arise of applying both fertiliser materials differentially across the field.

Although available soil nutrient content is measured on a continuous scale, it is common for advisory purposes to convert this to a discrete scale index (MAFF, 1994). In this system, index 0 and 1 indicate likely yield response to applied nutrients, while at 2 and above only a 'maintenance' application, sufficient to replace the amount removed by the crops is necessary. At index 0 the application is sufficiently large to supply maintenance and, over a period of years, raise the index to, at least, 1. To obtain full value from considering such geographic variability, an accurate assessment must also be made of the nutrient offtakes, based on yield logging. This is carried out separately to soil sampling and the two must be harmonised. Also, as the equipment which applies fertilisers tends to operate across a width of, say, 20 m, then if the soil nutrient index and crop offtake vary across this width there may be loss of accuracy in the practical realisation of fertilising, which may reduce the attractiveness of a geographically variable application.

In order to examine the improvements in efficiency in nutrient use that can result from variable rate application, soil P and K contents and crop offtake of P and K have been measured for two fields. Fertiliser requirements based on average, geographically variable and rectangularly gridded systems were then calculated, and these methods compared in terms of the areas of land over- or under-fertilised.

MATERIALS AND METHODS

Two adjoining commercial fields in Northumberland (55°17'N, 1°39'30"W) were soil sampled at 20 m spacings along alternate tramlines 40 m apart and the samples were analysed for P and K content (MAFF, 1986), and clay content (Avery and Bascomb, 1974) in 1991. Based on the P and K results each sample was placed in the appropriate nutrient index (MAFF, 1994). Yields of oilseed and wheat at 20 m (1991) and 12 m (1992) spacings along each 5 m wide cut were measured using a Claydon meter on a Claas combine in both 1991 and 1992. Due to a machine fault, the yield was only recorded for part of South field in 1991. Data was plotted manually onto a 1:2500 Ordance Survey map and digitised. The kriging option within UNIRAS was used to test the variables for randomness and to estimate the sills, nuggets and ranges. Contour maps were prepared using all, two thirds and one third of the soil sample points (Mohamed et al., 1997) with fault, bilinear and kriging routines in UNIRAS (1989). A 20 x 20 m rectangular grid was then superimposed on the data. Values were interpolated for each grid cell, and the area of agreement between this and the smooth contours measured. The measured nutrient index of each sample point was compared with the index attributed by each mapping routine, and the number of incorrectly classified samples was counted. Yield and nutrient data were converted to a regular common spacing of

20 x 20 m squares by interpolation using UNIRAS (1989). The regular grid of soil and yield points were transferred to a spreadsheet and the amount of fertiliser to be applied to each grid square was calculated using the following strategies:

1) **Average method.** The average soil nutrient index and yield were used for each field to calculate nutrient requirement.
2) **Soil index only.** The average field yield was used to calculate the maintenance application of nutrients and, based on MAFF (1994), 25 kg/ha was added if the soil nutrient index of the area was 1 and 50 kg/ha if the index was 0.
3) **Soil index and yield.** The amount of nutrient in each area of the field was based on the soil index, as above, to which was added a maintenance amount based on the yield for that area.

For wheat, the crop was assumed to have removed 7.5 g P_2O_5/kg and 5.6 g K_2O/kg while oilseed rape removed 16 g P_2O_5/kg and 11 g K_2O/kg (MAFF, 1994).

RESULTS

The yields of the crops (Table 1) do not vary greatly within North field. In South field, the CVs are larger and yields are negatively skewed. The negative skew in North field oilseed is due to a single sample with a very low yield. Soil properties were more variable and were more skewed in South field than North. Soil P and K contents are well correlated in both fields (r = 0.43*** in North field and r = 0.51*** in South field). Yield was correlated with clay content in South field (r = 0.68** and 0.47** in 1991 and 1992). Yield correlated negatively with soil P in both fields in 1992. In South field yield correlated negatively in 1991 and positively in 1992 with soil K. None of these relationships explained more than 6% of the total yield variance.

A larger proportion of the lower yielding areas (more than 1 S.D. below mean) in both years and in both fields were associated with larger soil nutrient indices (Table 2). The crop and soil variables had a large range, and the sills were substantially larger than the nuggets (Table 3).

When the areas of each index of P & K, produced using the three smooth contouring routines, were compared with the measured data points that lay within them, 'fault' consistently gave the lowest error rate (Table 4): a similar result was found for gridded data (Table 5).

TABLE 1. Yield and soil parameters of North and South fields.

		Southfield			Northfield		
		Mean	CV	Skewness	Mean	CV	Skewness
Wheat	- t/ha	11.4	15.9	-1.5	9.2	6.3	-0.1
Oilseed	- t/ha	3.6	12.3	-1.3	4.7	9.6	-1.2
P	- mg/kg	16.8	3.05	2.1	11.1	33.4	1.2
K	- mg/kg	146.0	27.6	1.6	109.5	25.3	0.5
Clay	- %	29.8	23.0	-0.8	31.6	12.3	-0.3

TABLE 2. Percentage of samples in various soil P and K indices having yields more than 1 S.D. below mean yield for North and South fields.

North field			South field		
Soil K Index	1991	1992	Soil P Index	1991	1992
0-1	5.9	11.8	1	9.9	11.9
2	23.3	24.1	2-3	11.7	18.9

TABLE 3. Geostatistical parameters of soils and crops in North and South fields.

	North field			South field		
	nugget	sill	range-m	nugget	sill	range-m
Yield 1991	0.4	0.6	210	0.3	0.4	230
Yield 1991	0.8	2.3	250	0.2	0.3	220
P	80	17.0	200	15.0	43.8	175
K	750	1240	250	300	2980	260
Clay	5.0	22.3	150	58.9	58.9	290

TABLE 4. Percentage of sample points attributed by each of the contour mapping routes to a different P and K index from that measured.

	Kriging		Bilinear		Fault		Kriging		Bilinear		Fault	
Spacing	P	K	P	K	P	K	P	K	P	K	P	K
	North field (244 points)						South field (235 points)					
20 x 40	16	13	20	13	9	4	14	10	16	17	9	7
40 x 40	25	18	23	13	16	8	17	15	18	18	14	11
60 x 40	29	15	29	7	24	16	19	19	19	20	16	17

TABLE 5. Percentage of sample points attributed by each of the gridded mapping routines to a different P & K index from that measured.

	Kriging		Bilinear		Fault		Kriging		Bilinear		Fault	
Spacing	P	K	P	K	P	K	P	K	P	K	P	K
	North field (244 points)						South field (235 points)					
20 x 40	17	14	19	11	13	6	15	9	15	17	8	7
40 x 40	22	16	23	10	21	9	15	15	18	17	16	13
60 x 40	31	14	29	16	25	16	7	9	17	20	16	18

There was no consistent increase in the proportion of incorrectly attributed points due to the restriction on boundary position imposed by the rectangular grid, although in the gridded maps each index contained a slightly higher proportion of measured points which were from a different index than did the smooth contoured maps. As the 'fault' procedure consistently misclassified fewest of the measured points, it was used to prepare a table of areas of the two fields having particular combinations of P and K index (Table 6).

TABLE 6. Proportion (%) of North and South fields grid squares having soil in a range of P and K indices.

	P index									
K index	0	1	2	3	Total	0	1	2	3	Total
	North field					South field				
0	0.3	0.5	0	0	0.8	0.7	20.1	7.6	0	28.5
1	35.0	30.2	2.7	0	67.9	0.4	27.8	37.9	2.1	68.1
2	3.0	20.8	7.0	0.5	31.7	0	0	0.7	2.8	3.5
Total	38.3	51.5	9.7	0.5		1.0	47.9	46.2	4.9	

The mean index (1) of P and K for North field corresponds to the most frequent area, but in South field, because of the small area of index 3 soils, the average P index is 2, rather than the more common index 1. In both fields, the **Average method** would result in an inadequate supply of P over a substantial part of the field, while K would be over applied to parts of North field and under applied to parts of South field. A small part of North field would receive excessive P. Only 30% and 38% of North and South fields respectively would have received the most appropriate amount of fertiliser if a single, compound P and K fertiliser were applied. Such a fertiliser, applied to each field, would provide an inadequate amount of one or both minerals in 38% and 55% of North and South fields respectively, and an excess amount in 32% of North field.

If the yield of individual areas is considered, as well as the soil index (**Soil index and yield**) then, compared with **Soil index only,** large deviations (> 15 kg/ha) affect a relatively small part of the fields, except for South field in 1992 (Table 7).

DISCUSSION

Within-field variation of all the properties reported was non-random, and had a large range (Table 3). The use of mapping methods is hence justified, and the number of samples required to provide a satisfactory map is limited more by the minimum needed to produce a map (Wollenhaupt *et al.*, 1994), rather than the maximum justifiable distance between samples. As Mohamed *et al.* (1997) showed for P, the error rate increases relatively slowly as the sample number decreases (Table 4). Conversion from a

TABLE 7. Excess (or shortage) of phosphate fertilizer resulting from using average yield in place of the true yield. Numbers in the table are 20 x 20 m grid squares.

	kg P_2O_5/ha						
	> 35	34 to 25	24 to 15	14 to 5	4 to -5	-6 to -15	< -16
	North field						
1991 wheat	0	1	2	33	300	35	0
1992 oilseed	0	0	10	70	206	76	9
	South field						
1991 oilseed	0	1	9	41	176	58	1
1992 wheat	14	16	16	30	111	147	13

smooth contour map to a grid pattern does not significantly increase the error rate (Table 5), a feature that makes mechanised application on a regular grid practicable. The fault mapping routine consistently resulted in maps which corresponded most accurately to the measured data points. This routine allows for rapid changes to occur ('faults') over short distances, and this appears to give better results for nutrient variation than bilinear or kriging, which attempt to create a more 'smooth' map (UNIRAS, 1989).

A uniform application of fertiliser (**Average method**) only supplies an appropriate amount of nutrients to part of fields, while a compound fertiliser will apply the appropriate amount to an even smaller proportion of the land (Table 6). This confirms the view of Mohamed *et al.* (1997) that substantial areas are currently under-fertilised, while other areas are over-fertilised, and that this is creating environmental and economic costs to the community and farmer respectively. The suggestion must be that non-uniform application should become common practice, and that compound fertilisers should be replaced by those containing only one nutrient. As soil P and K are well correlated and both have a large range, the areas of land with a joint index (0P:0K, 0P:1K etc.) tend to be sufficiently large to make fertiliser application practical (Figure 1).

As crop yield, and hence offtake of nutrients, is not well correlated with soil nutrient index there are many isolated 20 m squares which have to be fertilised differentially to their neighbours (Figure 2). Such a large number of boundaries will lead to inaccuracy in fertiliser application.

Current practice appears to be leading to an increase in nutrients in low yielding areas, and possibly to a depression of soil nutrients in high yielding areas (Table 2). Nutrient accumulation in low yielding areas may bias the result of the field average soil analysis, and possibly lead to an underestimate of nutrient requirement, limiting yield in high yielding areas. This result reverses the common view that high soil nutrient index areas are more productive. It is supported by the large excess of application of nutrients over

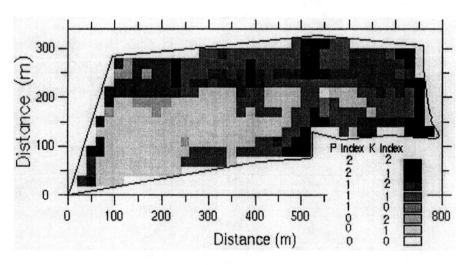

FIGURE 1. Map of P and K index in North field.

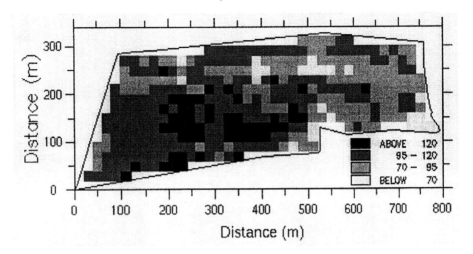

FIGURE 2. Phosphate requirement (kg P_2O_5) of North field, based on soil P index and offtake by the previous crop.

offtake on part of the fields (Table 7), by the strong positive correlation between soil P and K content in both fields and by the weak negative correlations between yield and soil nutrient content. The distribution of soil P and K, which do not respect field boundaries (Mohamed *et al.*, 1997) may therefore be partly due to inherent soil variability and partly to variable offtake within fields.

Although **soil index only** improved the appropriateness of nutrient application substantially in both fields, there was little further improvement obtained from correcting the application rate on the basis of variation in crop offtake (**soil index and yield**) in North field. Even in the more variable crop (1992), only 5% of the field area received an amount which was incorrect by more than 15 kg P_2O_5/ha if average yield was used in place of spot yield. However, in South field in 1992, over 17% of the field would have received an amount of nutrients more than 15 kg P_2O_5/ha from the appropriate amount if spot yield was not considered. CV and skewness (Table 1) are largest for this crop (wheat), and clearly the value of crop logging will be greatest for crops which have a large nutrient offtake per tonne and a large yield variation across the field. The effect of skewness in the 1992 wheat in South field is sufficient for only one third of the area to receive within 5 kg P_2O_5/ha of the appropriate amount.

CONCLUSION

Low intensity (<4 ha^{-1}) soil sampling was sufficient to produce accurate maps of soil P and K content and, with negligible loss of accuracy, these could be converted into rectangularly gridded maps which are practicable for mechanised application. These alone are sufficient to amend P and K application if the yield is normally distributed with a small CV. Fertilisers containing a fixed ratio of P and K increase the extent of nutrient misapplication and should be replaced with single nutrient fertilisers spread independently. Nutrient accumulation is occurring in low yielding areas and, with uniform application of nutrients based on a field average soil analysis, this may result in potentially high yielding areas being inadequately fertilised. The risk of this is greatest where there is a large CV of yield and the yield distribution is skewed. Variable offtake may now be increasing the extent of within field variation in soil nutrient content.

REFERENCES

Avery, B.W., Bascomb, C.L. (1974) *Soil Survey Laboratory Methods.* Soil Survey technical Monograph 6, Harpenden.

Cooke, G.W. (1967) *The control of Soil Fertility.* London, Crosby Lockwood.

Hall, A.D. (1921) *Fertilisers and Manures.* London, John Murray.

MAFF (1986) *The Analysis of Agricultural Materials.* Reference Book 427, London, HMSO.

MAFF (1994) *Fertiliser Recommendations.* Reference Book 209, London, HMSO.

Mercer, W.B., Hall, A.D. (1911) The experimental error of field trials. *Journal of Agricultural Science, Cambridge,* **4**, 107-127.

Mohamed, S.B., Evans, E.J., Shiel, R.S. (1997) Mapping techniques and intensity of soil sampling for precision farming. *Precision Agriculture: Proceedings of the third International Conference,* P.C. Robert, R.H. Rust, W.E. Larson (Eds), Madison, ASA, pp 217-226.

Searcy, S.W. (1995) Engineering systems for site-specific management: opportunities and limitations. *Site specific Management for Agricultural Systems: Proceedings of the second International Conference,* P.C. Robert, R.H. Rust, W.E. Larson (Eds), Madison, ASA, pp 603-612.

UNIRAS (1989) *Unimap 2000 Users Manual.* Soborg, Denmark.

Webster, R., Oliver, M.A. (1990) *Statistical Methods in Soil and Land Resources Survey.* Oxford, Oxford University Press.

Wollenhaupt, N.C., Wolkowski, R.P., Clayton, M.K. (1994) Mapping soil test phosphorus and potassium for variable-rate fertiliser application. *Journal of Production Agriculture,* **7**, 441-448.

THE POSSIBILITIES OF PRECISION FERTILIZATION WITH N, P AND K BASED ON PLANT AND SOIL PARAMETERS

S.E. SIMMELSGAARD, J. DJURHUUS

Danish Institute of Agricultural Science (DIAS), Department of Soil Science, Research Centre Foulum, P.O. Box 23, DK-8830 Tjele, Denmark

ABSTRACT

The variability of yield, plant content of N, P and K, soil mineral N and plant available P and K in soil were studied with the prospects of varying the application rate of these plant nutrients by precision farming. The study was carried out over three years on two fields (10 ha each) grown with winter wheat and winter barley, respectively. Yield mapping was done using a yieldmeter and DGPS positioning system. Soil variables were measured at about 300 grid points in each field, and plant parameters at 35 to 36 grid points. The data were analyzed by linear regression and geostatistical methods. For pedocells of 100 m^2 and as the mean of three years, plant uptake of P ranged from 18 to 24 kg/ha and 13 to 20 kg/ha for the two fields, respectively, and K from 100 to 160 kg/ha and 60 to 108 kg/ha, respectively. Plant uptake of N, P and K depends on both yield level and availability of soil nutrients.

INTRODUCTION

The prerequisite for using site specific management of N, P and K is that the yield response function is known at the required spatial accuracy. As N applied as mineral nitrogen is easily lost by leaching or denitrification, or immobilized, the N rate has to be optimized for the actual year, while applications of P and K can be based on the principle of substitution of harvested P and K because soil available P and K change only slowly. In this paper the possibilities of precision fertilization with N, P and K based on intensive measurements of soil and plant parameters on two fields in Denmark will be discussed.

MATERIALS AND METHODS

Sites and management

The experiments were conducted on two fields, Vindum and Risø, each of about 10 ha. Both fields are of glacial origin and slightly hilly. At Vindum the elevation above sea level is between 54 m and 67 m, and the natural drainage is good except for the southern part of the field and a peat basin (*c.* 1 ha), which has been sub-surface drained. The mineral soil is classified (USDA) as sandy loam. At Risø the elevation is between 1 m and 13 m and the field is sub-surface drained. According to the USDA classification the soil varies from sandy loam to sandy clay loam.

The experimental period covered three years from 1993-95. At Vindum, winter wheat was grown every year and, at Risø, winter barley. The fields were uniformly fertilized with mineral fertilizer. At Vindum the N application was 150 kg N/ha in 1993 and 125 kg N/ha in 1994 and 1995. At Risø the application was 150, 160 and 160 kg N/ha for each year, respectively. At Vindum the application of P and K was 25 kg P/ha and 63 kg K/ha in every year. At Risø the application of P and K was 26, 18 and 23 kg P/ha for 1993 to 1995 and 69, 30 and 74 kg K/ha for the same years, respectively.

Soil sampling

A grid of 20 m by 20 m was established at both experimental fields, and to facilitate geostatistical analyses the grid was supplemented with 20 points in between the ordinary grid points. The grid at Vindum is shown in Figure 1, the grid at Risø showed a similar pattern. In October 1992 samples were taken at all grid points at the 0-25 and 25-75 cm depths for analyses of available P and K. In March 1993, 1994 and 1995 soil samples for analyses of mineral N (N_{min}) were taken at 0-25 cm at each grid point and at 25-75 cm in a 40 m by 40 m grid. Each soil sample was a composite of nine cores, one taken at the center of the grid point and eight taken at a radius of 2.5 m. The available P was found by extraction by a 0.5 molar $NaHCO_3$ solution at a pH of 8.5, and the K content by extraction by a 0.5 molar ammonium acetic solution (Anon., 1995).

Yield and uptake of N, P and K

The fields were harvested by a Dronningborg Industries (Massey Ferguson) combine harvester, equipped with a Differential Global Position System (DGPS) receiver and a yieldmeter. Yield maps were based on position and yield measurements every 2-3 seconds. Measurements were excluded during machine start-ups, near headlands, or when less than a full swath width remained to be harvested. At each field 35 to 36 grid points

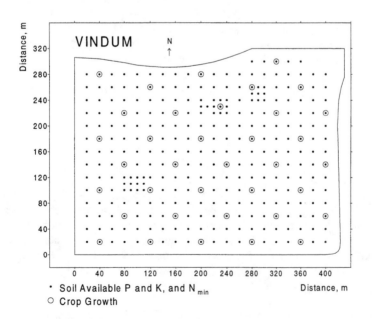

FIGURE 1. The experimental fields at Vindum.

evenly distributed over the fields (Figure 1) were chosen for crop sampling at harvest. At each point four plots, of 0.5 m^2 each, were sampled. The samples were divided into grain and straw, and the contents of N, P and K were analyzed.

Statistical methods

To describe the variability in the fields, semivariograms were estimated for all parameters. The peat area of c. 1 ha at Vindum was excluded from calculations of the semivariograms. For the remaining part of the field at Vindum and the field at Risø intrinsic stationarity was assumed, i.e., no trends in the fields were allowed. Initially, the semivariances were judged for anisotropic variation, and if anisotropic, the data were transformed by rotation and scaling as described in this volume by Kristensen and Olesen (1997), in which the definition of anisotropy angle and anisotropy ratio can also be found. Exponential (1) or linear (2) models were fitted to the semivariances using weighted nonlinear regression analysis:

$$\gamma(h) = C_0 + C_1(l - e^{-h/r})$$ (1)

$$\gamma(h) = C_0 + C_1 h$$ (2)

where $\gamma(h)$ is the semivariance at distance h, r is the range, C_0 is the nugget effect and C_1 is the partial sill (exponential model) or slope (linear model).

Averages of the measured values were predicted for pedocells of 10 m by 10 m using ordinary kriging (Cressie, 1991) using the previously mentioned semivariograms. When kriging the field at Vindum all data from the field were used. Some initial estimations showed that this gave approximately the same results as kriging the two areas separately, except that the method used gave a smoother transition between the two areas. The semivariograms estimated from the non-peat area were used for the whole field. Based on the data of yield of P and K from the 0.5 m^2 plots at harvest, linear regressions for grain and straw, respectively, and available soil P and K at the plots were used to estimate yield of P and K for pedocells of 10 m by 10 m based on corresponding kriged values of soil available P and K.

RESULTS

The range (r) of N_{min} varied between 30 and 122 m, and generally r was greater for the Risø field than for the field at Vindum (Table 1). At both locations N_{min} was lowest in 1995 (Table 1, Figures 2 and 3). The correlations of N_{min} at 0-75 cm were generally low except at Vindum between 1993 and 1995 (Table 2). Except for the area with high organic matter, the range of kriged N_{min} at Vindum was less than 16 kg N/ha at 0-75 cm (Figure 2), ranging mainly between 22 and 38 kg N/ha. At Risø, kriged N_{min} at 0-75 cm (Figure 3) varied mainly between 30 and 60 kg N/ha. As shown in Figures 2 and 3, the correlations between N_{min} and grain yield were low. Thus, the correlation coefficients between grain yield and N_{min} at 0-75 cm were 0.11, -0.18 and 0.04 (n.s.) at Vindum and 0.08, 0.34 and -0.27 at Risø for 1993 to 1995. Soil available P and K were spatially correlated (Table 3), and at both locations the main parts of soil available P and K were found in specific areas (Figures 4 and 5). There was a significant effect of soil available P and K on the

TABLE 1. Nugget effect (C_0), sill/slope (C_1) and range (r) for semivariograms of N_{min}.

Site	Depth cm	Year	Mean mg/kg	C_0 $(mg/kg)^2$	C_1 $(mg/kg)^2$	r m
Vindum	0-25	1993	4.38	0	2.23	33
		1994	3.71	0.33	0.00232	-
		1995	3.04	0	0.63 ·	31
	25-75	1993	3.00	0	0.83	63
		1994	3.04	0.27	0.00041	-
		1995	1.95	0	0.14	32
Risø	0-25	1993	5.34	0.77	1.99	95
		1994	5.93	0	9.67	48
		1995	4.35	0	1.99	45
	25-75	1993	3.28	0	1.65	30
		1994	3.58	1.52	2.20	122
		1995	2.44	0	0.98	36

TABLE 2. Correlation coefficients between years of N_{min} at 0-75 cm.

	Vindum		Risø	
	1993	1994	1993	1994
1994	0.32		-0.41	
1995	0.65	-0.05	0.26	0.17

TABLE 3. Nugget effect (C_0), sill/slope (C_1) and range (r), anisotropy angle (φ) and anisotropy ratio (λ) for semivariograms of soil available P and K.

Site	Depth cm	Mean mg/100 g	C_0 $(mg/kg)^2$	C_1 $(mg/kg)^2$	r m	φ	λ
Vindum							
P, mg/100 g	0-25	3.4	0	0.45	148[1]	-13	2.8
	25-75	2.4	0	1.80	53[1]	-24	2.0
K, mg/100 g	0-25	10.7	0	3.77	25	-	
	25-75	6.5	0	1.75	29	-	
Risø							
P, mg/100 g	0-25	2.6	0	1.1	60	-	-
	25-75	1.5	0	2.9	100	-	-
K, mg/100 g	0-25	13.9	0	38.9	129	-	-
	25-75	6.7	0	10.9	70	-	-

[1] Range depends on direction.

concentration of harvested P and K (Table 4). The average yield of P and K in grain and straw was 17 (11-23) kg P/ha and 84 (53-115) kg K/ha at Vindum and 20 (8-26) kg P/ha and 125 (42-192) kg K/ha at Risø. At Risø the highest yield of P and K was generally found in areas with high values of available P and K in the soil (Figure 5). However, there was no correlation between grain yield and concentrations of P and K in grain.

TABLE 4. Linear regression of concentration of harvested P and K (g/g×100) on soil available P and K (mg/100 g).

Site		Year	Grain			Straw		
			Intercept	Slope	R^2	Intercept	Slope	R^2
P	Vindum	1993	0.19	0.012	0.25	0.067	0	n.s.
		1994	0.27	0.013	0.30	0.021	0.040	0.20
		1995	0.34	0	n.s.	0.029	0.008	0.48
		1993-95[1]	0.31	0.010	0.85	0.036	0.006	0.43
	Risø	1993	0.19	0.026	0.56	0.020	0.0055	0.37
		1994	0.34	0.013	0.44	0.046	0.0118	0.27
		1995	0.37	0.012	0.25	0.078	0.0069	0.18
		1993-95[1]	0.35	0.017	0.91	0.075	0.0081	0.77
K	Vindum	1993	0.45	0	n.s.	0.63	0.020	0.26
		1994	0.38	0.006	0.26	0.93	0	n.s.
		1995	0.55	0	n.s.	1.19	0.027	0.43
		1993-95[1]	0.52	0.0029	0.79	1.31	0.016	0.90
	Risø	1993	0.39	0.0013	0.10	0.82	0.070	0.84
		1994	0.51	0.0037	0.32	1.21	0.075	0.78
		1995	0.52	0.0021	0.24	1.68	0.026	0.43
		1993-95[1]	0.52	0.0024	0.88	1.21	0.057	0.73

[1] Including year as class variable.

DISCUSSION

The spatial variation in N_{min} in spring and between years, especially at Risø, shows that it is difficult to estimate N_{min} based on relatively few measurements. The variation is probably caused by the climate, which affects the N turnover processes differently both in space and time. If precision fertilization were to be based partly on N_{min}, it would therefore be necessary to use modeling or some kind of indirect measurements, e.g., reflectance measurements of winter crops in the spring. However, the low correlations between N_{min} and yield show that precision fertilization cannot be based on N_{min} alone. Thus, the effect of dynamic yield factors such as water (e.g., Thomsen et al., 1997) and N mineralization should also be implemented. Furthermore, other factors which cause low yield potential should be identified and if possible improved. For Vindum preliminary calculations of the effects of nitrogen and water availability have been carried out. Thus, using data from six N fertilizer experiments at Vindum in 1993 and 1994, Simmelsgaard and Andersen (1995) found the following equation:

$$Y = (0.958 N_{av} - 0.00174 N_{av}^2) \cdot f_w \cdot f_{year}; \ hkg \, / \, ha \qquad (3)$$

where Y is grain yield (hkg/ha), N_{av} (kg N/ha) is the total of N_{min} in spring, mineralization of N from spring to harvest (Debosz and Kristensen, 1995) and N fertilization, and f_w and f_{year} are reduction factors ($0 \leq f \leq 1$) for water and year, respectively. For $f_w = f_{year} = 1$ the optimum is at $N_{av} = 275$ kg N/ha and grain yield (Y) is 132 hkg/ha. In 1993 R^2 between yield estimated by eqn (3) and measured yield at the 35 points (Figure 1) and by yieldmeter, respectively, was 0.85 and 0.29. By differentiation of eqn (3) and assuming a

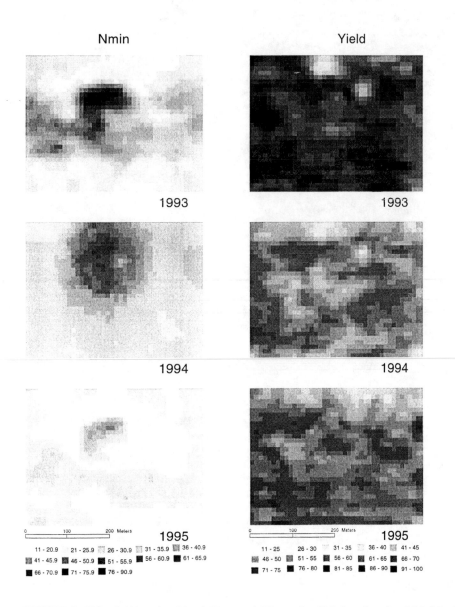

Nmin

Yield

1993

1993

1994

1994

1995

1995

11 - 20.9   21 - 25.9   26 - 30.9   31 - 35.9   36 - 40.9
41 - 45.9   46 - 50.9   51 - 55.9   56 - 60.9   61 - 65.9
66 - 70.9   71 - 75.9   76 - 90.9

11 - 25   26 - 30   31 - 35   36 - 40   41 - 45
46 - 50   51 - 55   56 - 60   61 - 65   66 - 70
71 - 75   76 - 80   81 - 85   86 - 90   91 - 100

FIGURE 2. Kriged (10 m by 10 m) N_{min} at 0-75 cm (kg N/ha) and grain yield (hkg/ha, 85% dry matter) at Vindum.

grain price of 100 Dkr/hkg and an N price of 4 Dkr/kg, the estimated optimum rate of N fertilization ranged from 130 to 220 kg N/ha in both 1993 and 1994. For 1993 a mean N application of 125 kg N/ha should be applied at rates of between 20 and 140 kg N/ha to maximize the yield. However, using this approach the increased yield was only between 1 and 2 hkg/ha.

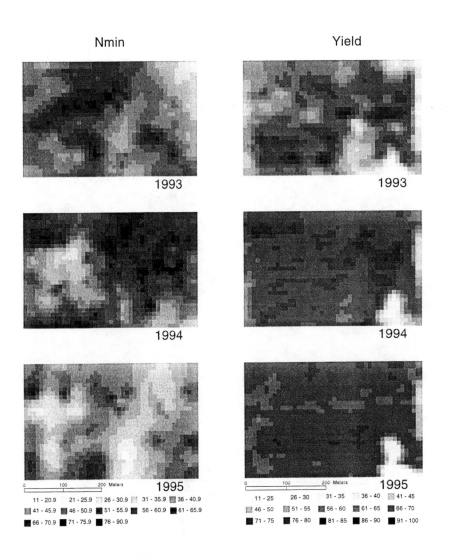

Nmin   Yield

1993   1993

1994   1994

1995   1995

11 - 20.9	21 - 25.9	26 - 30.9	31 - 35.9	36 - 40.9
41 - 45.9	46 - 50.9	51 - 55.9	56 - 60.9	61 - 65.9
66 - 70.9	71 - 75.9	76 - 90.9

11 - 25	26 - 30	31 - 35	36 - 40	41 - 45
46 - 50	51 - 55	56 - 60	61 - 65	66 - 70
71 - 75	76 - 80	81 - 85	86 - 90	91 - 100

FIGURE 3. Kriged (10 m by 10 m) N_{min} at 0-75 cm (kg N/ha) and grain yield (hkg/ha, 85% dry matter) at Risø.

Soil available P or K was not limited at either the Vindum or Risø fields. Thus, the increased concentration in the crop by increasing soil available P and K was a luxury uptake. Therefore soil available P and K above a specific level is not profitable, and in Denmark the generally recommended levels are 3 mg P/100 g and 8 mg K/100 g. However, for many soils these levels may be even lower as for P at Risø. At Vindum and at Risø the average difference between applied and removed P was 8 and 2 kg P/ha, respectively. If soil available P had been at the recommended level for the entire field, harvested P would have been slightly less at Vindum than it actually was and nearly the

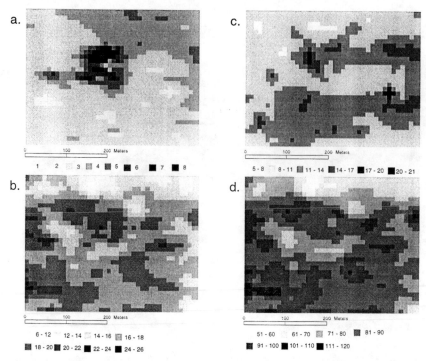

FIGURE 4. Kriged (10 m by 10 m) values at Vindum of a) soil available P (mg/100 g) at 0-25 cm, b) harvested P (kg/ha), c) soil available K (mg/100 g) at 0-25 cm and d) harvested K (kg/ha). Harvested P and K are means of 1993-95.

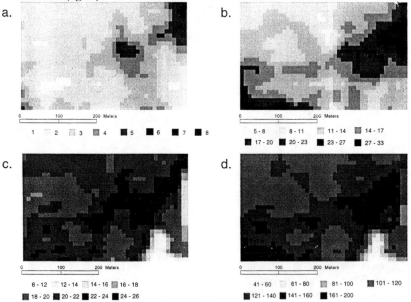

FIGURE 5. Kriged (10 m by 10 m) values at Risø of a) soil available P (mg/100 g) at 0-25 cm, b) harvested P (kg/ha), c) soil available K (mg/100 g) at 0-25 cm and d) harvested K (kg/ha). Harvested P and K are means of 1993-95.

186

same at Risø. At both locations harvested K was greater than that applied, with the main part removed in straw. However, if soil available K had been at the recommended level, less K would have been harvested, especially at Risø. In the Danish Agricultural Advisory Service, fertilization with P and K is based on compensating estimated harvested P and K corrected for soil available P and K. If this principle was used for every pedocell at Vindum and at Risø, the fertilization would be more optimal and reduce the risk of losses of P and K to the environment.

REFERENCES

Anon. (1995) Methods of authorized soil analysis in Denmark. *Ministry of Agriculture and Fishery*. (In Danish).

Cressie, N.A.C. (1991) *Statistics for spatial data*. John Wiley & Sons Inc., 900 pp.

Debosz, K., Kristensen, K. (1995) Spatial covariability of N mineralization and textural fractions in two agricultural fields. *Danish Institute of Plant and Soil Science*, SP report no.26, pp 174-180.

Kristensen, K., Olesen, S.E. (1997) Mapping root zone capacity by co-kriging aerial photographs of available soil moisture. *Proceedings of 1st European Conference on Precision Agriculture*, J.V. Stafford (Ed.), London, SCI.

Simmelsgaard, S.E., Andersen, M.A. (1995) The influence of nitrogen and water availability on crop yield variation. *Danish Institute of Plant and Soil Science*, SP report no.26, pp 99-109.

Thomsen, A., Schelde, K., Heidmann, T., Hougaard, H. (1997) Mapping of field variability in crop development and water balance within a field with highly variable soil conditions. *Proceedings of 1st European Conference on Precision Agriculture*, J.V. Stafford (Ed.), London, SCI.

MAPPING OF FIELD VARIABILITY IN CROP DEVELOPMENT AND WATER BALANCE WITHIN A FIELD WITH HIGHLY VARIABLE SOIL CONDITIONS

A. THOMSEN, K. SCHELDE, T. HEIDMANN, H. HOUGAARD

Danish Institute of Plant and Soil Science, Research Centre Foulum, 8830 Tjele, Denmark

ABSTRACT

Digital processing of small format aerial photography has been used for the detailed mapping of field scale differences in canopy reflection related to crop development and for the selection of sites for intensive monitoring of crop development and water balance.

During three seasons (1992 (field peas), 1993 and 1994 (winter wheat)) with water-limited growth conditions, recurring patterns in crop development were observed. During the 1995 (winter wheat) growing season, the crop development was nearly uniform as a result of sufficient rainfall. The crop development during dry years was found to approximately mirror differences in soil conditions (water holding capacity) mapped from intensive soil sampling.

During 1993 and 1994 crop development and water balance were monitored using spectral measurements and TDR (Time Domain Reflectometry) at six sites representing the full range in soil conditions and expected yields. The detailed measurements and water balance modelling showed a close correlation between i) canopy spectral reflectance and green biomass (potential yield) or leaf area index, ii) aerial photography and ground measured reflectance, and iii) crop development and root zone water holding capacity. Consequently, aerial photography in combination with detailed ground measurements and water balance modelling can be used for high resolution mapping of i) biomass and ii) root zone water holding capacity.

INTRODUCTION

It has been shown by several investigators that measurements of canopy spectral reflectance are well correlated with the amount of green biomass and canopy structure (green leaf area index). In particular, the ratio of near infrared and red reflectance (Ratio Vegetation Index; RVI) has been found to correlate closely with both biomass and leaf area index development (Hinzman *et al.*, 1986; Kleman and Fagerlund, 1987). Further, because of a close correlation between canopy conductance and RVI (Sellers, 1985), RVI measurements can be substituted for laborious measurements of leaf area index in water balance modelling. Canopy spectral reflectance can be mapped in great detail using aerial photography or data from imaging radiometers, or measured from the ground using portable equipment allowing very detailed studies at the plot and field scale of variation in crop development and the related water balance.

Time domain reflectometry (TDR) has been developed into an operational tool for measuring root zone water content. Routine measurements are facilitated by the development of a probe system including measuring probes, cabling and connectors for rapid measurements without disturbance to soil or canopy structure (Thomsen, 1994). As with spectral measurements, TDR measurements are well suited for detailed field studies because of the rapidity of measurements. A water balance model, ET_MODEL, has been developed to use either RVI, measurements of green leaf area index (GLAI), or GLAI calculated from calibrated RVI measurements, to calculate actual daily evaporation from reference (potential) evaporation. The model is developed for use with sparse parameter sets and can be applied to different areas without the need for extensive data collection. Besides reference evaporation, only measurements of canopy RVI values and root zone water content are necessary for model application and validation.

FIELD AND LABORATORY MEASUREMENTS

False colour aerial photographs were taken each year beginning in 1992 using a Hasselblad camera mounted in a light aircraft and Kodak Infrared Aerographic Film 2448. The positive transparencies were colour separated and digitized using a flatbed scanner (ESKOFOT Eskoscan 2540) and analysed using specially developed image processing software. An example of a digitally processed aerial photograph (Figure 1) uses grey tones to show the distribution of relative red reflectance within the research field at the Vindum Estate. Dark areas correspond to areas with high concentrations of active biomass due to light absorption by chlorophyll. Correspondingly, light areas represent low biomass areas with little chlorophyll absorption. Based on July 1992 imagery of the research field, six sites were selected for detailed water balance studies in 1993 and 1994. The sites,

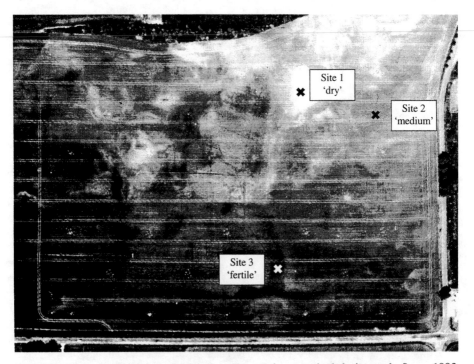

FIGURE 1. Research field at the Vindum Estate photographed during early June, 1993. Visible red band of digitally processed colour infrared aerial photography. Dark areas represent areas with relatively high biomass concentrations. Pixel resolution 0.5 m.

measuring a few square meters each, were located in areas that were locally uniform and covered the range of biomass development in 1992. Three of the sites ('fertile', 'medium', and 'dry') are indicated in Figure 1. The intensive sites were instrumented for TDR measurements of root zone water content. Two probe systems with a total of twelve measuring probes including the lengths 0.2, 0.5, 0.75, and 1 m were installed at each site. Disturbance to soil and canopy structure was minimized by using a probe and cabling design allowing measurements to be made from outside the intensive sites.

During the growing season from April to July, frequent measurements of canopy spectral reflectance were made using a portable dual band (650 and 800 nm) radiometer. The radiometer used was developed using radiation sensors and an A/D converter produced by Skye Instruments Ltd, UK. Spectral measurements were made at the intensive sites and near intersections of a 20 m square grid covering the entire area shown in Figure 1. To establish the relationships between spectral measurements and canopy biomass, leaf area index, etc., destructive sampling for laboratory analysis was performed at selected grid points every 3 to 4 weeks during the entire growing season.

WATER BALANCE MODELLING

A water balance model, ET_MODEL (Thomsen and Hougaard, 1995), developed at Research Centre Foulum, was used to simulate the water balance of the intensive sites during 1993 and 1994. The ET_MODEL calculates daily actual evaporation from Penman calculated reference (potential) evaporation. The model requires daily values of leaf area index or RVI interpolated from measurements and the extractable water content of the rooting zone obtained from soil retention curves or TDR measurements.

Water balance modelling was included in order to investigate the relationship between root zone water availability and canopy development.

RESULTS AND DISCUSSION

At the Vindum research field, frequent measurements of spectral reflectance were used to follow i) the development in leaf area index, ii) biomass, and iii) N-content at a large number of gridpoints. The 0.25 m² areas harvested for laboratory analysis are visible as circular patterns of light areas in Figure 1. In Figure 2 the development of destructively measured green leaf area index, GLAI, is shown as a function of spectral ratio, RVI. In Figures 3 and 4 dry biomass and canopy N-content are shown in the same manner as functions of RVI. A single regression equation could be fitted to the GLAI data and used to calculate GLAI from RVI measurements throughout the season for the highly variable field. If biomass (canopy dry matter content) or total crop area index including green, yellow and dead plant material is to be estimated from RVI measurements, then the fact that the relationship changes over time with increasing amounts of dead material, as shown in Figure 3 for biomass dry matter content, needs to be considered. The same is true for the N-content data shown in Figure 4. The regression relation between GLAI and RVI shown in Figure 2 was found to vary significantly between years which could possibly be due to differences in variety. In other studies including both winter wheat and spring barley a single regression equation was found to represent both crop types well for a three year period.

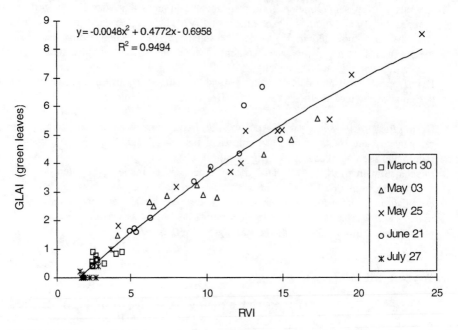

FIGURE 2. Relationship between the spectral ratio vegetation index (RVI) and green leaf area index (GLAI) in 1993.

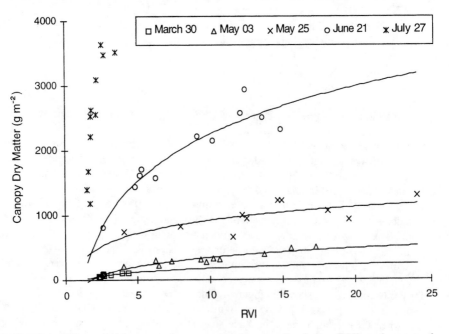

FIGURE 3. Relationship between the spectral ratio vegetation index (RVI) and canopy dry matter content in 1993.

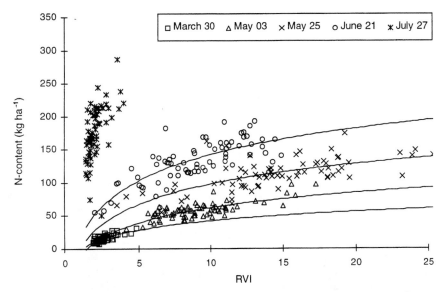

FIGURE 4. Relationship between the spectral ratio vegetation index (RVI) and canopy N-content in 1993.

The relationships between biomass and RVI and between N-content and RVI shown in Figures 3 and 4 could potentially be used for the field scale mapping of biomass and N-content if aerial photography and field measurements were available for approximately the same dates.

At the intensively monitored sites, root zone water content and spectral reflectance were measured frequently during 1993 and 1994. In Figures 5 and 6 measurements made at the driest and the most fertile sites are shown for the two years. The relationship between water availability and RVI development is evident. At the 'fertile' site the crop development was not restricted by water availability whereas the crop wilted completely at the 'dry' site during May 1993.

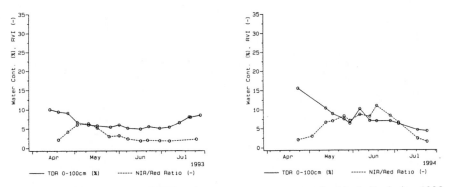

FIGURE 5. Measurements of RVI and root zone water content for 'dry' site during 1993 and 1994.

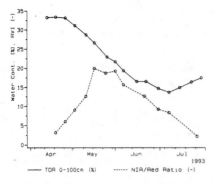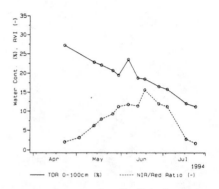

FIGURE 6. Measurements of RVI and root zone water content for 'fertile' site during 1993 and 1994.

The relationship between water availability and crop development was investigated for 1993 and 1994 using data from the intensively monitored sites and the ET_MODEL. Based on site measured RVI development, local precipitation and regional reference evaporation, daily water balance calculations were made for both years without changing the model parameters.

For all sites, the modelled root zone water content closely followed the TDR measurements, supporting the hypothesis that crop development during the dry years 1993 and 1994 was largely controlled by water availability. In Figures 7 and 8, modelled and measured water contents for the most 'fertile' and a 'medium' site are shown for 1993 and 1994. The 'medium' site is situated on an east facing slope receiving less solar radiation than a horizontal surface. Consequently, reference evaporation had to be reduced by 25% to fit modelled and measured water content.

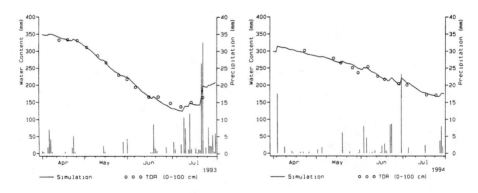

FIGURE 7. Measured and simulated root zone water content for 'fertile' site during 1993 and 1994.

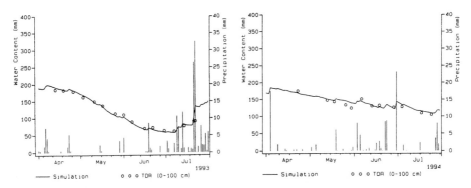

FIGURE 8. Measured and simulated root zone water content for 'medium' site during 1993 and 1994.

CONCLUSIONS

The combined use of field scale aerial imaging and ground truth radiometry and crop sampling was found to be very promising for detailed studies of variability in crop development. For the particular field investigated, crop development in 'dry' years was found (using water balance modelling) to closely follow the root zone water holding capacity.

Further research into robust combinations of

- Yield potential partly based on historical imagery
- Present status of crop development based on tractor mounted instrumentation
- Present status of root zone water availability based on water balance modelling

to a system for e.g. distributed fertiliser application appears highly relevant.

REFERENCES

Hinzman, L.D., Bauer, M.E., Daughtry, S.T. (1986) Effect of nitrogen fertilization on growth and reflectance characteristics of winter wheat. *Remote Sensing Environ.*, **19**, 47-61.

Kleman, J.K., Fagerlund, E. (1987) Influence of different nitrogen and irrigation treatments on the spectral reflectance of barley. *Remote Sensing Environ.*, **21**, 1-14.

Sellers, P.J. (1985) Canopy reflectance, photosynthesis and transpiration. *International Journal of Remote Sensing,* **6**, 1335-1372.

Thomsen, A. (1994) Program AUTOTDR for making automated TDR measurements of soil water content. *SP-report No. 38 (Report series of the Danish Institute of Plant and Soil Science).*

Thomsen, A., Hougaard, H. (1995) Field variability in crop development and water balance. In: (Ed. S.E. Olesen) Proceedings of the Seminar on Site Specific Farming, Aarhus, Denmark, March 20-21, 1995. *SP-report No. 26,* p. 90-98.

MODELING SOIL AND CROP RESPONSES IN A SPATIALLY VARIABLE FIELD

J. VERHAGEN

Department of Soil Science and Geology, Wagenigen Agricultural University, Box 37, 6700 AA Wagenigen, The Netherlands

ABSTRACT

A fertiliser recommendation is defined using a relationship describing declining nitrogen concentration with increasing plant mass. Using simulated, water limited, yield the fertiliser dose is calculated for 65 points over 10 years. By interpolation fertiliser maps are generated and three pattern types are identified.

Definition of a site specific fertiliser level is, however, not possible because the temporal variation overrules the spatial variation. A single application is unsuited for site specific N fertiliser management and introducing more management control points by using split application is needed to be able to adjust N fertiliser management during the growing season. Weather forecasts in combination with simulation models can be used to predict crop growth for periods of several weeks. Based on the simulated expected growth N fertiliser management can be adjusted using the defined fertiliser recommendation.

INTRODUCTION

Site specific management aims at optimising inputs on field and farm level. The approach should benefit the farmer in terms of net return and benefit the environment through lower emission levels. The aims set for site specific farming require a process oriented approach. Based on quantified knowledge of processes farm management can interact with the soil-crop system and adjust where and when necessary.

Guidelines for N fertiliser, in the Netherlands, are based on long term experiments and have been used successfully for a large number of years. The environmental impact, however, is not included in the fertiliser recommendation. For potato, a single application is advised based on the mineral N content in the top 60 cm in early spring. The requirement is defined on a regional level based on the parent material. Spatial variation for site specific N fertiliser management would, using the guidelines, be based on the spatial distribution of mineral N only.

The requirements of the crop for water and nutrients are directly related to the expected growth (MacKerron and Lewis, 1994). Differences in expected production level should be reflected in different N fertiliser levels. Other systems used in, for example, France and Germany calculate the required nitrogen also based on the expected production. This approach is taken here using water limited production as the expected production level in non irrigated conditions. The water limited production is calculated over 10 years for 65 locations in a farm field using a dynamic simulation model. The production levels are translated into N input using a growth related equation (Greenwood *et al.*, 1985; 1990).

Spatial patterns are described in patterns as identified for this field by Van Uffelen *et al.* (1997). The aim is to define site specific N fertiliser levels based on the spatial and temporal variation of the water limited production levels.

The environmental effect of the defined N fertiliser dose can be evaluated on basis of the N remaining in the soil after the growing season. Verhagen and Bouma (1997) showed that N_{min} at 15 September is a good measure from which to determine leaching risk during the wet season. The consequences of a fertiliser application in terms of leaching risks can then be quantified. The results on the environmental impact of this ongoing study will be reported at a later stage.

MATERIALS AND METHODS

Soil survey

The study area is a 2.2 ha field. The soils in the area are strongly layered, texture lenses vary in thickness from 1 mm to 5 cm. Soils were classified as fine-loamy, calcareous, mesic Typic Udifluvents. Pedogenetic soil horizons are lumped into soil physical building blocks or functional layers (Wösten *et al.*, 1986). Four functional layers are identified for this area (Verhagen *et al.*, 1995). Soils are described as a sequence of functional layers of varying thicknesses. Soil data were collected using a regular sampling grid supported by some short distance information (see Figure 1), a cross section is presented in Figure 2 to illustrate the texture gradient from sand to clay.

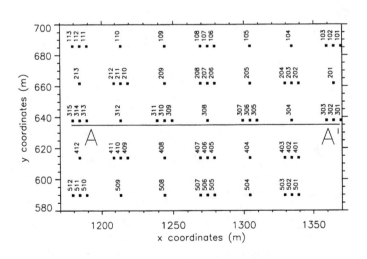

FIGURE 1. Layout of soil survey for the experimental field. Numbers indicate soil augerings.

198

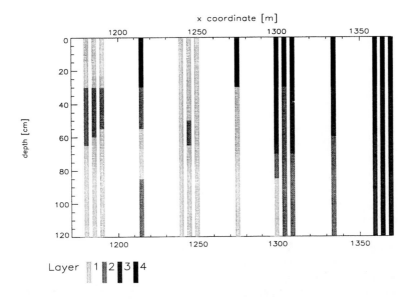

FIGURE 2. Cross-section as indicated in Figure 1. Layer 1 represents the more sandy layers, layer 3 represents the more clayey layers, layer 2 is an intermediate layer and layer 4 is a loamy topsoil.

Simulation model

Simulation of the water limited yields was done using the dynamic simulation model WAVE (Vanclooster *et al.*, 1994). WAVE integrates several existing models, including dynamic simulation of water flow based on the SWATRER model (Dierckx *et al.*, 1986) and the SUCROS crop growth model (Spitters *et al.*, 1989). The total water uptake is calculated as the integral over the root zone. Water stress is calculated according to Feddes *et al.* (1978), in which the maximum water uptake is defined by a sink term as a function of depth. Water uptake is reduced at characteristic high and low pressure head values. The model is used to calculate water limited production of potato (spp. saturna). The optimal water uptake is defined between pressure head values of -10 cm H_2O and -1000 cm H_2O, for the high pressure head a value of -10 cm H_2O was taken and the wilting point was set at -8000 cm H_2O. Maximum rooting depth was set at 60 cm. The growing season started on 15 May (day of emergence) and ended at harvest on 15 September.

N fertiliser recommendation

The requirements of the crop for nitrogen are directly related to the growth that will be made and so are related indirectly, to the expected length of the growing season (MacKerron and Lewis, 1994). In this study we take water limited production as the foundation for the N fertiliser recommendation. The uptake of N (N_u) is defined by the product of biomass production (W) and the N concentration (N_c) in the produced biomass:

$$N_u = W \times N_c \qquad (1)$$

Greenwood *et al.* (1985; 1990) showed an exponential decrease in N concentration with increasing plant mass during the growing season for C3 and C4 plants. The relation used for potato is defined as:

$$N_{cr} = 1.35 + 4.05 \, e^{-0.26W} \tag{2}$$

in which N_{cr} is the critical nitrogen concentration (%) and W the total weight (ton dry matter ha^{-1}). Uptake of more N than defined by Equation 2 does not result in an increase of biomass; it only affects the quality of the product. The minimum N concentration needed to support growth is set at 0.5%. Vos and Marshall (1994) point out that the results of fertiliser dose-response experiments cannot be described by Equation 2, because for N limiting conditions N dilution becomes an important factor, i.e., it is possible to have crops with lower N_{cr} and lower W than described by Equation 2. The optimal N_{cr} is defined as 75% of the range given by Equation 2 and the minimum N concentration (Figure 3).

The supply of nitrogen needed (N_n) to satisfy the required uptake is determined by the efficiency of the plant in exploiting the available nitrogen. The efficiency of the crop in utilising the available nitrogen links the yield-uptake curve to the yield-rate curve and is not constant for all fertiliser rates. In general the efficiency decreases at higher fertiliser rates but results are not conclusive. Smit and Van der Werf (1992) report recoveries of between 30 and 80 % for rates of 120 kg N ha^{-1} and between 20 and 60% for rates of 300 kg N ha^{-1}. Vos (1994) found a recovery of 55% and Neeteson, Greenwood and Draycott (1987) report recoveries varying between 40 and 70%. The dose recovery relation is, however, not quantified and therefore is kept constant at 50 %. The efficiency is fixed at 0.5, indicating that of each unit of N fertiliser 50 % was taken up by the plant. The efficiency of N uptake for N from other sources is also set at 0.5. The nitrogen supply needed (N_n) is:

$$N_n = \frac{N_u}{0.5} \tag{3}$$

The demand side is satisfied by the applied fertiliser, the amount of N in the soil over the rooting depth (60 cm) at the beginning of the growing season (N_s) and the N supply from

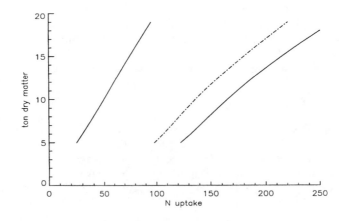

FIGURE 3. Minimum, critical and optimal (dotted line) N uptake rates.

mineralised organic matter (N_m). The average mineralisation rate is set at 0.7 kg ha^{-1} day^{-1} (Neeteson *et al.*, 1987). From emergence on 15 May to harvest on 15 September an estimated 90 kg N is mineralised.

The fertiliser recommendation can be defined as follows:

$$N_f = N_n - N_s - 90 \tag{4}$$

In classical farm management N_s is a field average ignoring spatial variation within the field. The contribution of soil mineral N at the beginning of the growing season is not included in this analysis but can simply be subtracted.

RESULTS

Evaluation of the model

The performance of the model was evaluated for water, nitrogen and end yield for the 1994 growing season. Measured and simulated water content agreed well ($R^2 = 65\%$). Simulated and measured end yield also revealed a good correlation ($R^2 = 65\%$). It was concluded that the model performance was satisfactory allowing use of the model in simulation studies.

Fertiliser scenarios

As can be seen from Figure 4 the temporal variation is much larger than the spatial variation. Spatial patterns are stable as was shown by Van Uffelen *et al.* (1997), the fertiliser dose, however, is not. For the 10 years three main reactions can be distinguished which are displayed in Figure 5. The first pattern is the most common over the simulated period (60%). The second and third patterns in which the lower rates are found on the more clayey areas occur only in 3 years.

Results of the environmental impact will be reported on at a later stage.

DISCUSSION

A fertiliser recommendation based on soil and crop response is suggested and evaluated for a farm field in the Netherlands. The approach used can easily be adjusted for other

FIGURE 4. Fertiliser patterns. Contours at 10 kg N ha^{-1} intervals, lighter colours indicate low values and darker colours indicate high values.

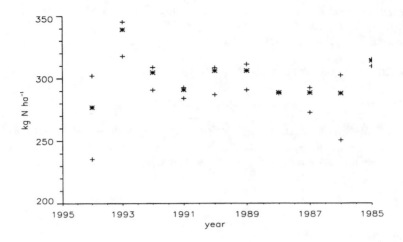

FIGURE 5. Calculated fertiliser rates for the period 1994 -1985, disregarding the early spring N_s. + indicates the minimum and maximum values and * indicates the average value for each year (10 year average = 300 kg ha^{-1}).

crops, Equation 2 can be replaced by a more general equation for C3 or C4 plants. The efficiency factor is crop dependent. The mineralisation rate can be calculated using dynamic simulation models or taken from field observations. The soil mineral N content is preferably measured on a spatial grid.

Spatial N fertiliser patterns are stable over time, however, the fertiliser dose is not. Spatial variation is overruled by temporal variation making it difficult for farm management to optimise inputs based on expected production levels. Long term weather predictions are still not accurate enough to be used in farm management. Booltink *et al.* (1996) used a weather generator in combination with a simulation model to optimise farm management. They found that when approaching the end yield the prediction of end yield increased in accuracy. For generated weather periods of 40 days or longer the variance stabilises at its maximum level.

The fertiliser recommendation is based on a single dose application. Since temporal variation overrules the spatial variation a single dose application cannot be adjusted according to expected production levels. A first application, base dressing, followed by one or more applications adding management control points to adjust fertiliser management should become standard in site specific management. From Equation 2 it follows that half of the required nitrogen has already been taken up when only a quarter of the final crop mass has been produced. The time frame in which 25% of the final crop mass is produced is roughly 30 days. This time frame is less than the 40 day time frame found by Booltink *et al.* (1996) implying that the combination of weather forecasts and simulation models can be used to improve management decisions.

CONCLUSIONS

1. N fertiliser recommendations should be based on expected production levels.
2. Temporal variability is much larger than spatial variability making split application of N fertiliser a prerequisite for site specific N management.
3. Weather predictions in combination with simulation models are essential tools in decision support systems for farm management.

ACKNOWLEDGMENTS

The author acknowledges the contributions of J. Vos, P.M. Driessen and H.W.G. Booltink in defining the fertiliser application rates. This work was partly funded by the EUR-AIR 'IN-SPACE' project 92-1204: 'Reduced fertiliser input by an integrated location specific monitoring and application system', co-ordinated by Dr. D. Goense, Wageningen, The Netherlands.

REFERENCES

Booltink, H.W.G, Thornton, P.K., Verhagen, J., Bouma, J. (1996) Application of simulation models and weather generators to optimize farm management strategies. *Proceedings of the 3rd conference on Precision Agriculture,* P.C. Robert, R.H. Rust, and W. Larson (Eds), Minneapolis, USA, SSSA-spec. publ.

Dierckx, J., Belmans, C., Pauwels, P. (1986) SWATRER, a computer package for modelling the field water balance. Reference manual, Catholic University Leuven.

Feddes, R.A., Kowalik, P.J., Zarandny, H. (1978) Simulation of field water use and crop yield. *Simulation monographs*, Wageningen, PUDOC.

Greenwood, D.J., Lemair, G., Gosse, G., Cruz, P., Draycott, A., Neeteson, J.J. (1990) Decline in percentage N of C3 and C4 crops with increasing plant mass. *Annals of Botany,* **66**, 425-436.

Greenwood, D.J., Neeteson, J.J., Draycott, A. (1985) Response of potatoes to N fertilizer: quantitative relations for components of growth. *Plant and Soil,* **85**, 163-183.

MacKerron, D.K.L., Lewis, G.L. (1994) Modelling to optimize the use of both water and nitrogen by the potato crop. *Potato ecology and modelling of crops under conditions limiting growth. Proceedings of the second international Potato modelling conference,* A.J. Haverkort and D.K.L. MacKerron (Eds), Wageingen, Kluwer Academic Publisher.

Neeteson, J.J., Greenwood, D.J., Draycott, A. (1987) A dynamic model to predict yield and optimum nitrogen fertilizer application rate for potatoes. *Proceedings 262,* London, The fertiliser society. Paper read before The Feriliser Society of London on 10 December, 1987.

Smit, A.L., Van der Werf, A. (1992) Fysiologie van stikstofopname en -benutting: gewas- en beortellings-karakteristieken. *Stikstofstromen in agro-ecosystemen,* H.G. Van der Meer and J.H.J. Spiertz, (Eds), Agrobiologische Thema's 6.

Spitters, C.T.J., van Keulen, H., van Kraailingen, D.W.G. (1989) A simple but universal crop growth simulation model, SUCROS87. *Simulation and systems management in crop protection,* R. Rabbinge, H. van laar and S. Ward (Eds), Wageningen, PUDOC, pp 87-98.

Van Uffelen, Verhagen, J., Bouma, J. (1997) Comparison of simulated crop reaction patterns as a basis for site specific management. *Agricultural Systems,* **54,** 207-222.

Vanclooster, M., Viane, P., Diels, J., Christiaens, K. (1994) Wave a Mathematical Model for Simulating Water and Agrochemicals in the Soil and Vadose Environment - Reference & User's Manual (Release 2.0), Leuven, Institute for Land and Water Management

Verhagen, A., Booltink, H.W.G., Bouma, J. (1995) Site specific management: balancing production and environmental requirements at farm level. *Agricultural Systems,* **49,** 369-384.

Verhagen, J., Bouma, J. (1997) Defining threshold values: a matter of space and time. *Geoderma* (in press).

Vos, J. (1994) Effects of Dicyandiamide on Potato Preformance. *J. Agronomy and Crop Science,* **173,** 93-99.

Vos, J., Marshall, B. (1994) Nitrogen and potato production: strategies to reduce nitrate leaching. *EAPR 93. Proceedings of the 12th triennial conference of the European Association for potato research. Paris-France. 18-23 Juillet 1993.* pages 101-110.

Wösten, J.H.M., Bannink, M.H., de Gruijter, Bouma, J. (1986) A procedure to identify different groups of hydraulic-conductivity and moisture retention curves for soil horizons. *J. of Hydrol.,* **86,** 133-145.

GPS BASED SOIL SAMPLING WITH AN AUGER AND FIELD ANALYSIS OF NITRATE

K. WILD, G. RÖDEL, M. SCHURIG

Bayerische Landesanstalt für Landtechnik, 85350 Freising-Weihenstephan, Germany

ABSTRACT

The determination of soil nutrient content is an essential prerequisite for environmentally friendly and economic fertilization. Current soil sampling and analysis systems do not meet all of the major requirements of farmers and scientists. Major disadvantages are the large sample quantity, delays of up to several weeks between sample extraction and the receipt of soil laboratory analysis results, and the limited consideration of the local variability of soil nitrogen. For these reasons a new soil sampling system and analysis method has been developed. The sampling device, based on an auger, is set up in such a way that there is no shift in soil layers when a sample is taken. A variety of augers can be used for handling different soil types and producing different sample sizes. Samples can be taken down to 60 cm, and a special stripper design allows one auger to sample different layers simultaneously. GPS is used with the appropriate software to determine location, boundaries, field areas, and to guide the operator to selected sampling locations. The third part of the system is a portable analysis kit which allows the samples to be analyzed in the field, immediately after they were taken. Soil nutrient maps can then be created in the field, so that the results are available for further actions without any time delay. The components are mounted on a vehicle, like a car, tractor, or all-terrain vehicle. The system is operated by one person and is designed for fast sampling.

INTRODUCTION AND OBJECTIVES

An economical and environmentally friendly nitrogen fertilization requires the knowledge of the plant available nutrients in the soil at the time of fertilization. Therefore soil sampling is an important task in crop production. The current standard procedure is to extract samples and send them to a soil laboratory for analysis. Due to its design, the auger- and core-based extracting instruments currently in use produce mixed samples that contain about 1 to 2 kg each, which is about ten times the amount required by the laboratory. Because of the lack of appropriate methods it is not possible to homogenize this large soil mixture in the field and extract a representative sample. Therefore a lot of excess soil must be transported to the laboratory, stored and handled. These costs together with disposal increases the costs of soil testing. In addition, the time difference between the sample extraction and the receipt of the test results is often not acceptable. The farmer is often still waiting for the fertilizing recommendation from the laboratory when the fertilizer is due. Also, the inclusion of local variability of soil nitrogen as a basis of local fertilization is very limited with the current procedure. A mixed sample is taken from a large area, leading to a mean value and disregarding the variability within this area. The locations of the sampling sites are not recorded specifically. This makes it difficult to assign the measured nutrient contents to individual sites in a field.

Because of these disadvantages of the current procedure, a new system for sampling soil nitrogen was developed at the Landtechnik Weihenstephan. The goal was to develop the appropriate instrument to remove a small sample quantity (10-20 g), in conjunction with a positioning and navigation system for recording the sample location, and analysis of the soil sample in the field.

SAMPLING INSTRUMENT

The heart of the new soil sampling system, named 'BPEG 60' for 'Bodenprobeentnahmegerät' (soil sample extracting instrument), is a 10 mm diameter, modified concrete auger, made out of steel and propelled by a 12 volt drive motor (Figure 1).

The motor sits on a carriage which is attached to the mounting frame. When the auger is rotated downward, the gripping element at the base of the mounting frame links into the auger's thread spacing. The auger is then drilled into the soil. The feed is consequently no longer dependent on the density or viscosity of the soil, but instead only on the pitch and rotational speed of the auger. Problems like churning or compressing the soil that occur with other systems, especially in loose or very hard soils, do not occur with this system. As soon as the auger is completely screwed in, the gripping element is automatically withdrawn. With the help of a hand crank attached to a cog rail, the auger is pulled out of the soil. The sample material is caught in the space between the threads

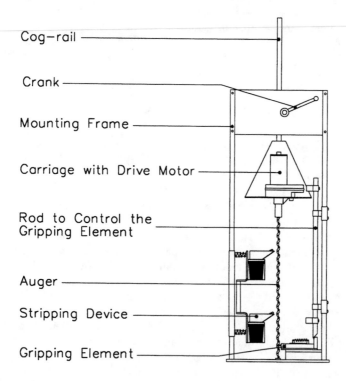

FIGURE 1. System diagram of the soil sampling device BPEG 60.

of the auger. To strip off the sample from the auger, strippers are inserted, which guide the soil to a collection container. By using a vertical arrangement of two strippers, the auger can remove soil samples from two soil layers with one drilling and soil removal process. For example, the higher layer sample from 0-30 cm and the lower from 30-60 cm can each be stripped from the auger and placed in separate containers. By choosing a specific auger diameter the sample mass can be varied, so that quantities of down to 10 g may be taken in one sample.

The soil sampling device is attached to a vehicle with the mounting frame (Figure 2). With a corresponding adapter, it can be mounted on a range of vehicles. The mounting is always configured so that, in a conventional car, the device can be operated by the driver from the driver's seat.

DETERMINATION OF POSITION AND NAVIGATION TO THE SAMPLE LOCATION

An essential component of the new procedure is a positional and navigational system. With the help of the satellite navigation system 'Global Positioning System (GPS)', the position of the sample can be determined. GPS may be used in a differential setting ('DGPS') in which a higher positional accuracy may be reached. For the investigations a mobile GPS-receiver with six channels, plugged into a heavy-duty personal computer, was used. The differential signals were supplied by a 12-channel reference station via radio. The detected positions, a background map and the sample locations are visualized graphically in the vehicle on a computer screen. Before sampling, the field boundaries are determined using GPS and the area is calculated from the boundary's positional data. Then the program divides the area into a number of parcels of equal size, and one sample (or a mixed sample) is taken from each parcel (Figure 3).

FIGURE 2. Soil sampling device and navigation equipment, attached to a vehicle.

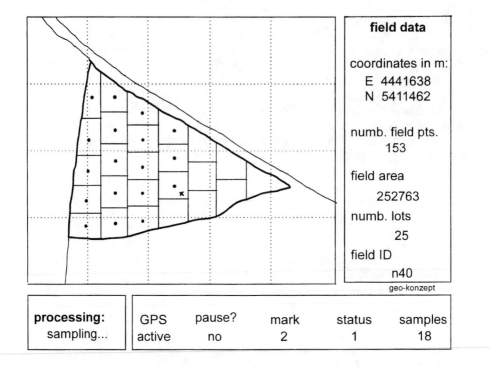

FIGURE 3. Example of computer screen during soil sampling.

The navigation part of the program helps the driver to find the sample locations in their respective parcels. On the screen the position of the vehicle is shown as a cross within the parcels. The coordinates of each soil sample location are saved so that later samples can be taken from the same position. For gaining a higher position accuracy for a sampling location, a mean value is calculated over ten coordinate pairs. On the screen the sample locations are marked with a point.

FIELD ANALYSIS OF NITRATE

By moving the test preparation and nitrogen determination from the laboratory to the sample location, transportation and labor costs are saved. The analysis is conducted with the help of a portable analysis kit which is integrated in a small suitcase. The central instrument is a colorimeter developed by industry. For analysis, the soil is filtered in a calcium chloride solution. Test rods are submerged in the filtrate, and the change in color of these rods is then determined using the colorimeter, from which the nitrogen content of the soil can be calculated. Since only calcium chloride is added to the soil samples, they may be emptied on the field without problems. The test rods are not hazardous waste, and may be discarded with household waste.

RESULTS AND OUTLOOK

So far, the new soil sampling system has been tested on all major soil types in Bavaria (southern part of Germany). On all sites, samples could be obtained in the desired low quantity. Problems occur on very dry sandy soils: like other systems the material does not remain in the spaces between the threads of the auger. On stony soils, little stones are pushed aside by the auger. If the auger hits a larger rock, the mounting frame is lifted up and damage to the extraction instrument is prevented. The sampling device can be moved a few centimeters for a new trial.

Due to the field analysis, the nutrient content is available immediately. The sampling and testing can be carried out by one person, but for faster results a second person can analyze the soil material, while the first takes the sample on the field. Comparison measurements were made for evaluating the accuracy of the portable analysis kit. Samples have been removed with the BPEG 60 and analyzed with the colorimeter. Afterwards the filtrates went to a laboratory and were tested with the current standard method. The resulting nitrate contents are shown in Figure 4.

The average difference between the values of the two methods is -0.8 kg/ha or -2.5 %, the standard deviation is 2.9 kg/ha or 5.0 %. The range of the absolute figures extends from -6 to 8 kg/ha and from -12.5 to 9.3 % for the relative figures. Compared to the accuracies of soil testing results, which were gained by investigating different soil laboratories (Wörle, 1996), the accuracy of the field analysis is relatively high.

Position errors were determined in other investigations, because identical DGPS equipment was also used for other trials (Auernhammer et al., 1994). They were less than 5 m, which is accurate enough for soil sampling (Schön et al., 1994). Together with the sample locations determined by DGPS, soil nutrient maps can be created in the field to determine the necessary local fertilization application. The farmer does not have to face any time delay and can spread the needed fertilizer as soon as required.

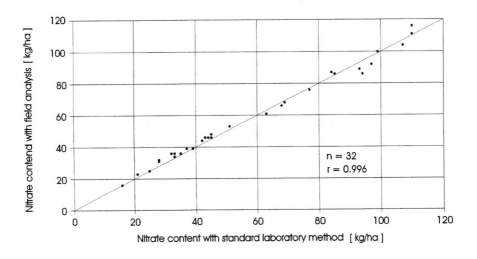

FIGURE 4. Nitrate content with standard laboratory method and field analysis.

ACKNOWLEDGEMENT

The authors thank the Ministry of Agriculture, Food and Forestry of the State of Bavaria for the funding of the project.

REFERENCES

Auernhammer, H., Demmel, M., Muhr, T., Rottmeier, J., Wild, K. (1994) GPS for yield monitoring on combines. *Computers and Electronics in Agriculture*, **11(1)**, 53-68.

Schön, H., Auernhammer, H., Muhr, T., Demmel, M., Stanzel, H. (1994) Positionsbestimmung landwirtschaftlicher Arbeitsmaschinen für die Entwicklung ökologisch optimierter Anbauverfahren. *Gelbes Heft 53*, München, Bayerisches Staatsministerium für Ernährung, Landwirtschaft und Forsten.

Wörle, J. (1996) Schlamperei oder Überlastung? *dlz*, **47(10)**, 40-43.

SPATIAL VARIATION OF PESTICIDE DOSES ADJUSTED TO VARYING CANOPY DENSITY IN CEREALS

S. CHRISTENSEN, T. HEISEL, B.J.M. SECHER

Danish Institute of Plant and Soil Science, Research Centre Flakkebjerg, DK-4200, Slagelse, Denmark

A. JENSEN, V. HAAHR

Risø National Laboratory, Frederiksborgvej 399, P.O. Box 49, DK-4000 Roskilde, Denmark

ABSTRACT

Spatial variation in canopy density has a significant influence on herbicide performance because dense crops have higher competitiveness than sparse crops. Further, canopy density may influence the need for disease control because dense canopies increase disease development. Therefore, a target and optimal pesticide application can be obtained by integrating the spatial variation in a decision algorithm for spatial application of pesticides.

In this paper, spatial variation in canopy density of a winter barley and a winter wheat crop were studied in two fields using spectral reflectance measurements. Incoming and reflected near infrared and red radiation were measured with two sensors mounted 6 m apart on a boom attached to a platform vehicle. The geographical position of all measurements was recorded with a DGPS-system. The ratio between infrared and red reflectance (RVI) was used to calculate the spatial variation in the fraction of photosynthetically active radiation light interception (fPAR) in the fields. The fPAR was mapped with GIS using the geostatistical procedure kriging with a distance weighting as determined from the semi-variogram. The results showed significant spatial and temporal variation in fPAR. Treatment maps with density adjusted herbicide and fungicide doses are shown using simple decision algorithms. The paper discusses how spectral reflectance measurements may be integrated in a real time system for spatially variable pesticide application.

INTRODUCTION

Patch spraying or spatial adjustment of the herbicide dose has several positive environmental and economic effects when there is a great within-field variance of weed occurrence and density (Nordbo *et al.,* 1995) or disease development (Secher, 1997). The main positive effect is that pesticide is saved, reducing the costs and diminishing the total loading of pesticide to the environment.

Several field studies have shown that most weed species are aggregated and often occur as patches within the arable fields (e.g., Marshall, 1988; Heisel *et al.,* 1996). So far, only a few studies on the spatial dispersion of diseases have been carried out (Secher, 1997) but it is known that some epidemic developments emerge and disperse from small areas of the field. Dense canopies are known to expose higher levels of mildew (Jørgensen *et al.,*

1995) and it is also well known that dense canopies increase the competitive ability of the crop and thus reduce the demand for high herbicide doses. However, few studies on spatial variation in fields have been carried out (Christensen et al., 1995).

Two concepts of patch spraying have been suggested (Nordbo et al., 1995). These are the mapping concept where weed and disease monitoring and decision making precede the spraying, and the real-time concept where weed and disease monitoring, decision making and spraying are carried out simultaneously. The mapping system provides the possibility of making decisions on pesticide choice because information about the whole field is necessary to find one or a few suitable pesticides. So far, real time systems have only been developed for weed control in stubble land using simple sensors detecting green vegetation and an intermittent sprayer (Felton et al., 1991).

Automatic crop-weed identification has not yet been implemented for practical purposes. This is mainly because weed species characteristics vary significantly and because of mutual shading among seedlings. Diseases are rather easy to assess in the late epidemic stages, but then it is too late to control them with low doses (Hansen et al., 1992). Benlloch et al. (1996) suggest a system that combines manual and automatic weed assessments in a decision algorithm for patch spraying. A similar concept is followed in this paper. The overall idea is to use data from previous years to make strategic manual surveying and then use highly accurate automatic surveying to adjust the pesticide dose in a real-time spraying system.

The objective of this paper is to show the results of using spectral reflectance measurements to describe spatial variation in canopy density. The measurements are used in a desk exercise with a simple model for density adjusted herbicide and fungicide doses.

MATERIALS AND METHODS

Spectral reflectance was measured in two fields at Risø National Laboratory in 1996. The fields had varying soil parameters and topography. Harvest monitoring had shown significant grain yield variation within the fields. Field I was 9.8 ha with winter barley. The field had a low lying area near Roskilde Fjord in the north-west part of the field. Field II was 2.6 ha with winter wheat (cv Ritmo).

Spectral reflectance measurements were taken on several dates from a vehicle platform manoeuvred in the sprayer tramlines. Only the results from two dates are presented in this article. In field I the results from 11 April were selected to show the spatial variation in crop density at the optimal time of weed control in spring. Weeds were controlled in autumn, i.e., the current measurements were only used as a desk study to show the potential of adjusting the herbicide dose according to spatial variation in crop density. Crop-weed discrimination requires advanced image analysis (Benlloch et al., 1996) that, so far, has not been applied for this purpose. If the density varies significantly it can be included in the decision algorithm for patch spraying (Christensen et al., 1996) using the relative weed cover model (Kropff and Spitters, 1991). In field II, the results from 22 June were used to describe the crop density variation on a date when visual surveying had shown a significant need for disease control. Weeds were controlled in autumn and did not affect the measurements. Visual measurements on this date showed a severe attack of meldew with 100% incidence all over the field.

The distance between the tramlines was 12 m. Two sensors were mounted 6 m apart on a boom attached to a vehicle platform. The sensors (SKYE SKR 1800 with a 25° field-of-view device) consist of a pair of silicon photodiodes and narrow interference filters transmitting the wavelength intervals 640-660 nm (red radiation) and 790-810 nm (infrared radiation). The analogue output from the transducers was amplified and converted into digital signals. The height of the sensors was 2.5 m above the canopy giving a 1 m² circular spot area. Spectral incoming radiation was measured simultaneously with the reflectance measurements using two cosine corrected sensors.

The measurement frequency (1 spot sec^{-1}), the vehicle speed (1-2 m sec^{-1}) and the spot area (1 m²) resulted in a continuous 'string' of spots. A differential global positioning system (DGPS) provided the position of all spots in WGS 84 co-ordinates.

The ratio vegetation index (RVI) was calculated with equation (1):

$$RVI = \frac{NIR(ref) / NIR(in)}{R(ref) / R(in)} \qquad (1)$$

where NIR(ref) and NIR(in) are the reflected and incoming infrared radiation, and R(ref) and R(in) are the reflected and incoming red radiation. RVI is an arbitrary index that depends on the sensor device. Therefore, the mean RVI of the simultaneous measurements on each side of the vehicle was used to estimate the fraction of photosynthetically active radiation (fPAR) intercepted by the canopy as described by Christensen and Goudriaan (1993). The semivariance was computed and the spatial autocorrelation of fPAR was studied by plotting the semivariance as a function of distance classes h (semivariograms). The fPAR was mapped with the geographical mapping software Surfer (Surfer® for Windows, Golden Software, Inc.) using the geostatistical procedure kriging and a distance weighting derived from the semivariogram.

Assuming that the range of fPAR shows the variation in crop competitiveness, the range of herbicide dose Hz around a reference dose (1.0) at a given position was calculated with the linear model:

$$Hz = a - b*fPAR \qquad (2)$$

where:

$$b = \frac{\max(Hz) - \min(Hz)}{\max(fPAR) - \min(fPAR)} \qquad (3)$$

and:

$$a = max(Hz) + b*min(fPAR) \qquad (4)$$

In this paper the minimum and maximum herbicide doses were specified to be 0.8 and 1.2 of the reference dose, e.g., the recommended dose or the dose suggested by PC-Plant Protection (Rydahl and Thonke, 1993). A similar approach was used to generate a fungicide dose map in field II. Assuming that a crop with a low fPAR requires a lower fungicide dose than a crop with a high fPAR, the linear model:

$$Fz = c + d(fPAR) \qquad (5)$$

where:

$$c = \frac{\max(Fz) - \min(Fz)}{\max(fPAR) - \min(fPAR)} \qquad (6)$$

and:

$$d = c*min(fPAR) - min(Fz) \qquad (7)$$

was used to calculate the fungicide dose Fz at a given position. The minimum and maximum fungicide doses in this paper were specified to be 0.85 and 1.15 of the reference, aiming for a constant concentration of fungicide per unit of photosynthetically active leaf (Secher, 1997).

The herbicide and fungicide dose maps were generated using regular 12 m by 12 m blocks and the estimation method block-kriging.

RESULTS

The semivariograms showed a strong autocorrelation of light interception at distances below 100 m. The ratios between nugget and sill were 27% and 28% for fields I and II respectively. The results indicate that the spectral reflectance measurements and the size of the spot area were suitable for describing the spatial patterns of light interception within the field.

Figures 1 and 2 show the maps of the kriged fPAR in the two fields. Light interception in field I ranged between 10% and 40% on 11 April. A major part of field I (7.6 ha) intercepted less than 25% of the incoming PAR on this date. In the low lying area, light interception was 10 - 15%, whereas light interception was higher than 25% in patches in the south and the middle part of the field. In field II, light interception ranged between 79% and 95% on 22 June. Most of field II (1.7 ha) intercepted less than 90% of the incoming PAR on this date.

The dose map in Figure 3 shows the linear adjustment of the herbicide dose to the fPAR in field I. With 12 m by 12 m blocks, the herbicide dose could be reduced by 10 - 20% in 2.2 ha and 0 - 10% in 5.4 ha of the field. The herbicide dose should be increased by 0 - 10% in 2.1 ha and in a small area in the low lying area (0.1 ha) the herbicide dose should be increased by 10 - 20%.

The dose map in Figure 4 shows the linear adjustment of the fungicide dose to the fPAR in field II. The main part of field II (1.6 ha) required a reference dose plus or minus 5%. In 0.9 ha the fungicide dose should be increased by 5 - 15% whereas in a small hill area (0.1 ha) the fungicide dose should be reduced by 5 - 15%.

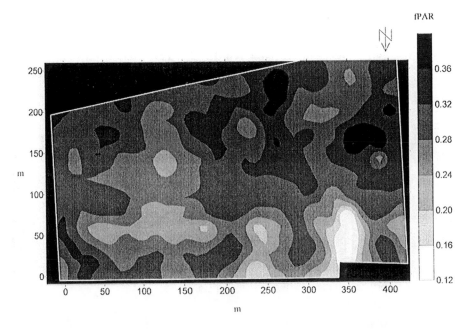

FIGURE 1. Fraction of intercepted radiation (fPAR) on 11 April 1996 in field I.

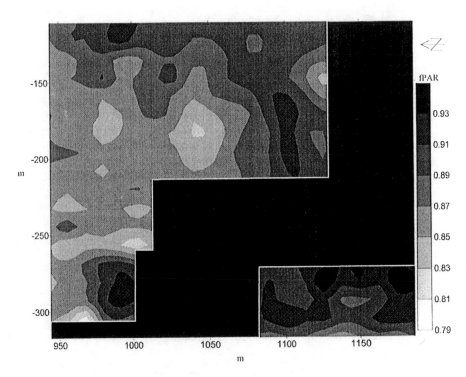

FIGURE 2. Fraction of intercepted radiation (fPAR) on 22 June 1996 in field II.

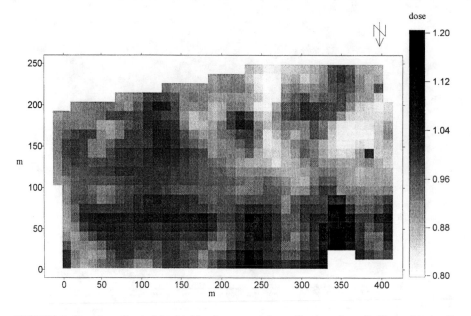

FIGURE 3. Density adjusted herbicide dose around a reference dose (1.0) on 11 April 1996 in field I.

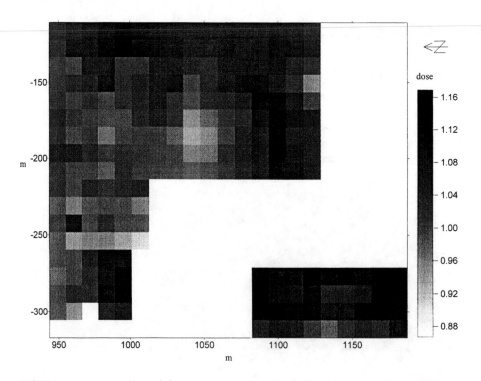

FIGURE 4. Density adjusted fungicide dose around a reference dose (1.0) on 22 June 1996 in field II.

DISCUSSION

The adjustment of the pesticide dose was based on the principle that the geometric mean of fPAR, which represents the mean canopy density of the field, requires a reference dose determined, for example, by using the decision support system PC-Plant Protection (Rydahl and Thonke, 1993; Secher et al., 1995). A simple linear algorithm with specified upper and lower limits was used to adjust the doses around the geometric mean of fPAR. This simple method can be implemented in a real time spraying system when the current range of fPAR is known and the pesticide doses can be specified for maximum and minimum fPAR. The range of fPAR has to be measured prior to a field run, e.g., by measuring fPAR in two reference plots that showed maximum and minimum fPAR in previous years. The reference dose could be the dose recommended by the company. In this case minimum fPAR requires the recommended herbicide dose and the maximum fPAR the recommended fungicide dose. This approach would not change the spatial pattern of the treatment maps with a linear model.

The simple linear algorithm is valid as long as the range of pesticide dose does not cover the non-linear parts of the dose-response curves. The relationships between pesticide doses and weed or disease response are sigmoid, but at high doses the relationship is almost linear and approaches 100% effectiveness beyond the recommended dose. In cereals, 50 - 70% efficacy is often sufficient to avoid yield loss due to weed competition. In this case a non-linear algorithm is required. The spatial pattern of the treatment map may change if a non-linear algorithm is used.

The adjustment of the fungicide dose is based on the goal of applying a constant concentration of active ingredient per unit of fPAR. An improved goal could be to apply a constant rate of active ingredient per unit LAI. Knowing the extinction of PAR per LAI, a non-linear algorithm could have been used to adjust the fungicide dose as fPAR increases approximately exponentially with LAI.

Several studies have shown that it is possible to discriminate weeds from the crop during the early growth stages (Petry and Kühbauch, 1989; Woebbecke et al., 1995; Zang and Chaisattapagon, 1995; Benlloch et al., 1996). A vision system, e.g., a CCD black and white camera, is required when weed seedlings are to be identified. Such an approach also enables weed counting or weed coverage measurements that can be integrated into a decision algorithm for patch spraying (DAPS) (Christensen et al., 1996).

REFERENCES

Benlloch, J.V., Sánchez, A., Christensen, S., Walter, A.M. (1996) Weed mapping in cereal crops using image analysis techniques. *AgEng Madrid* 1996, Paper 96G-0.47, 9 pp.

Christensen, S., Goudriaan, J. (1993) Deriving light interception and biomass from spectral reflectance ratio. *Remote Sensing of Environment,* **43**, 87-95.

Christensen, S., Christensen, K., Jensen, A. (1995) Spatial variability of spectral reflectance and crop production. *Proceedings of the Seminar on Site Specific Farming.* Danish Institute of Plant and Soil Science, SP-report no. 26, pp 56-67.

Christensen, S., Heisel, T., Walter, A.M. (1996) Patch spraying in cereals. *Proceedings of the Second International Weed Control Congress, Copenhagen*, pp 963-968.

Felton, W.L., Doss, A.F., Nash, P.G., McCloy, K.R. (1991) A microprocessor controlled technology to selectively spot spray weeds. *Proceedings ASEA Symposium 1991: Automated agriculture for the 21st century*, pp 427-432.

Hansen, J.G., Jørgensen, L.N., Simonsen, J. (1992) Multispectral radiometry, Source of additional data in field fungicide experiments. *Proceedings from the Workshop on Remote Sensing, Danish Journal of Plant and Soil Science*, S **2207**, 39-45.

Heisel, T., Andreasen, A., Ersbøll, A.K. (1996) Annual weed distributions can be mapped with kriging. *Weed Research*, **36**, 325-339.

Jørgensen, L., Houmøller, M., Secher, B.J.M., Christensen, K. (1995) Mildew in barley and wheat, the national plant disease in Denmark. *12th Danish Plant Protection Conference, Pest and Diseases*, SP-report no 3, pp 57-84.

Kropff, M.J., Spitters, C.T.J. (1991) A simple model of crop loss by weed competition from early observations on relative leaf area of weeds. *Weed Research*, **31**, 97-105.

Marshall E.J.P. (1988) Distribution patterns of plants associated with arable field edges. *Journal of Applied Ecology,* **26,** 247-257.

Nordbo, E., Christensen, S., Kristensen, K. (1995) Weed Patch Management. *Zeitschrift für Pflanzenkrankheiten und Pflanzenschutz,* **102**, 75-85.

Petry, W., Kühbauch, W. (1989) Automatisierte Unterscheidung von Unkrautarten nach Formparametern mit Hilfe der quantitativen Bildanalyse. *Journal of Agronomy and Crop Science,* **163**, 345-351.

Rydahl, P., Thonke, K.E., (1993) PC-Plant Protection: optimizing chemical weed control. *OEPP/EPPO Bulletin,* **23**, 589-594.

Secher, B.J.M. (1997) Site specific control of diseases in winter wheat. *Aspects of Applied Biology. Optimising pesticide applications,* **48,** 57-65.

Secher, B.J.M., Murali, N.S., Gadegård, K.E. (1995) Field validation of a decision support system for the control of pest and diseases in cereals in Denmark. *Pesticide Science*, **45**, 195-199.

Woebbecke, D.M., Meyer, G.E., Von Bergen, K., Mortensen, D. (1995) Shape features for identifying young weeds using image analysis. *Transactions of ASAE,* **38(1)**, 271-281.

Zang, N., Chaisattapagon, C. (1995) Effective criteria for weed identification in wheat fields using machine vision. *Transactions of ASAE,* **38(1)**, 965-974.

QUANTIFYING THE VARIABILITY OF SOIL AND PLANT NITROGEN DYNAMICS WITHIN ARABLE FIELDS GROWING COMBINABLE CROPS

P.M.R. DAMPNEY

ADAS Boxworth, Boxworth, Cambridge, CB3 8NN, UK

G. GOODLASS

ADAS High Mowthorpe, Duggleby, Malton, North Yorkshire YO17 8BP, UK

ABSTRACT

As part of a large collaborative project in England, selected soil and crop nitrogen parameters were measured in 21 fixed, small monitored areas within each of six arable fields growing winter wheat or oilseed rape in both 1995 and 1996. Parameters included soil mineral nitrogen (SMN), wheat grain protein content, nitrogen balance and apparent recovery of nitrogen at harvest. The overall mean Coefficients of Variation were:- 25.7% for SMN; 5.6% for wheat grain protein and 21.7% for wheat nitrogen balance. There was a 2- to 4-fold variation in SMN within individual fields. Apparent nitrogen recovery ranged by up to 2-fold within fields, with mean field values ranging from 61 to 92%. There was a mean range of 2.2% grain protein (at 86% dry matter) within fields. The implications of within field variation in soil and crop nitrogen dynamics are discussed.

INTRODUCTION

Use of optimum rates of fertiliser nitrogen (N) to field crops is of critical importance both to maximise crop yield and quality but also to minimise the risk of leaching of nitrates into watercourses. In Britain, cereal yields are commonly increased 3-fold through use of optimum N rates compared to nil use, and nitrogen has a major effect on the quality of produce harvested from many crops - for instance protein content in wheat grain (Dampney *et al.*, 1995). EC legislation limiting the nitrate content of drinking water has led MAFF and the agricultural industry to strongly promote good nitrogen practices by farmers. Key guidelines for use of nitrogen are provided in Fertiliser Recommendations (MAFF, 1994).

Application rates of fertiliser N are normally determined for individual fields based on soil nitrogen analysis or an assessment of field and crop factors. Nitrogen recommendations issued by the British Ministry of Agriculture (MAFF, 1994) are based on soil type, crop type, crop quality target, and for some crops (wheat, oilseed rape), the expected crop yield. Soil nitrogen supply, assessed from previous cropping and manuring, is classified according to a Nitrogen Index system. Nitrogen supply from applied organic manures is taken into account. More complex computerised systems use broadly similar factors, but incorporate more refined procedures for quantifying soil nitrogen supply and crop demand.

Although imperfect and costly, analysis for soil mineral nitrogen (SMN) is currently regarded as the most precise method for assessing the supply of nitrogen from the soil.

Shepherd (1993) has shown that SMN data is a useful guide to help farm decisions on optimum N use, especially where high quantities of SMN are expected due to previous cropping and manuring. Samples are normally taken from 0-30 and 30-60 cm depth (sometimes 60-90 cm in addition) with analysis carried out for nitrate-N and ammonium-N (extracted by 2M potassium chloride). Potentially mineralisable nitrogen can be assessed by analysis of the 0-30 cm sample for total nitrogen content or by anaerobic incubation.

With the introduction of new technologies, notably GPS (Global Positioning System) and fertiliser spreaders capable of real time variable rate application, it is now possible to consider varying fertiliser applications within fields. However, insufficient is known concerning the cost effectiveness of variable rate N applications, or the opportunity or basis on which more precise N rate decisions might be taken. Care is needed to ensure that improvements in the technologies of spatial location and fertiliser application are matched by an improved understanding of spatial and temporal patterns of nitrogen dynamics and fertiliser response.

This paper reports data from an ongoing collaborative research project funded under the MAFF LINK 'Technologies for sustainable agriculture' programme. The project is seeking to quantify and explain the within-field variation in growth and production of combinable crops in England, mainly winter wheat. It includes a study of the spatial and temporal variation in soil and plant nitrogen dynamics. The implications of this variation are discussed.

METHODS

A wide range of soil, crop growth and yield parameters have been measured in six individual arable fields (two fields on each of three separate farms) in harvest years 1995 and 1996 (Table 1). Field size varied from 7 to 23ha. Yield variation in each field has also been mapped using the Massey Ferguson Yield Mapping system. The data is being examined to identify the causes of variations in yield.

In each field, uniform crop management inputs were applied according to local practice. Data was collected from 21 small, fixed location monitored areas, each measuring 10 metres by 10 metres. The areas were randomly located in each field at the start of the project, with equal numbers positioned in low, medium and high yielding field zones based on past yield maps. There were no monitored areas on the headlands. Data relevant to this paper were collected from each monitored area in each field according to the following procedures.

Soil mineral nitrogen (SMN)

Soil samples were taken by hand gouge corer (Yokefleet) or a machine operated, 1 metre long gouge corer (Boxworth and Bridgets) in depth increments of 0-30, 30-60 and 60-90 cm (Table 1). Each composite sample was combined from six sub-samples, randomly located within each monitored area. Samples were analysed for nitrate-N and ammonium-N using standard laboratory procedures (MAFF, 1986). Values of SMN as kg/ha were calculated assuming a soil bulk density of 1.3 g/cc.

TABLE 1. Field, cropping and sampling details.

Site		1994/95	1995/96
ADAS Boxworth, Cambridgeshire	Soil type	Clay loam over chalky boulder clay, Hanslope soil series	
(2 fields:- Boxworth 1, 9ha Boxworth 2, 7ha)	Crop	Winter wheat (cv. Hereward)	Winter wheat (cv. Hereward, Boxworth 1) (cv. Rialto, Boxworth 2)
	SMN sampling		1 December 1995
	Fertiliser N (kg/ha)	199 (Boxworth 1) 220 (Boxworth 2)	190 (Boxworth 1) 187 (Boxworth 2)
ADAS Bridgets, Hampshire	Soil type	Silty clay loam over shallow chalk, Panholes/Andover soil series	
(2 fields:- Bridgets 3, 10ha	Crop	Winter oilseed rape (cv. Apex)	Winter wheat (cv. Hunter)
Bridgets 4, 7ha)	SMN sampling	6-11 January 1995	28-29 February 1996
	Fertiliser N (kg/ha)	156	200
Yokefleet Farms Ltd,	Soil type	Silty clay over alluvium, Blacktoft soil series	
Humberside (2 fields:- Yokefleet 5, 17ha	Crop	Winter oilseed rape (cv. Rocket)	Winter wheat (cv. Ritmo, Yokefleet 5) (cv. Riband, Yokefleet 6)
Yokefleet 6, 23ha)	SMN sampling	16-23 January 1995	6-13 December 1995
	Fertiliser N (kg/ha)	219	200

Grain yield, protein concentration and nitrogen offtake

At Boxworth and Bridgets, and at Yokefleet 6 in 1996, grain/seed yields were assessed in each monitored area by hand harvesting and hand threshing of an 8 square metre crop area. Grain samples were analysed for nitrogen using near infra-red spectroscopy. Nitrogen offtake in grain was calculated from grain yield and nitrogen data. For wheat, grain protein concentration at 86% DM was calculated by multiplying grain N concentration by 4.9. For both fields at Yokefleet in 1995, and at Yokefleet 5 in 1996, yields were assessed from combine generated yield maps. Except for Yokefleet 5 in 1996, samples of seed/grain were taken from each monitored area and analysed for total N.

Apparent recovery of nitrogen

The apparent recovery of nitrogen by winter wheat was calculated as the percentage of available nitrogen (SMN plus fertiliser nitrogen applied) contained in the grain and straw at harvest in each monitored area in each field. Nitrogen recovery was not calculated where oilseed rape was grown.

Statistical analysis

The distribution and variance of selected data was examined using simple statistical techniques giving values for standard deviation (sd) and Coefficient of Variation (CV). Since only 21 datapoints were available in each field, no attempt was made to investigate the spatial inter-dependency of the data.

RESULTS

Soil mineral nitrogen

Variation in SMN within each field in each season is shown in Table 2. Apart from Boxworth 1 in both years, and Yokefleet in 1996, all SMN levels were low and typical of a previous cereal crop (50-70 kg/ha SMN). In 1996, SMN was higher at Yokefleet reflecting the higher residues from the previous oilseed rape crop. There was a wide range in SMN values within individual fields, from a 2-fold to a 4-fold variation; CVs varied from 18.7 to 38.0% with an overall mean value of 25.7%. There was little difference in the mean CV values for the individual soil types - 24.4% (Boxworth, chalky boulder clay soil); 24.9% (Bridgets, shallow chalk soil) and 27.8% (Yokefleet, alluvial silty clay soil). The mean difference between the minimum and maximum values for individual fields was 69 kg/ha SMN. There were no significant linear relationships between SMN in spring 1995 and SMN in spring 1996 in any field ($p < 0.05$).

Grain protein concentration

Table 3 shows that the variation in grain protein concentration was much lower (overall mean CV = 5.6%) than for other measured parameters. The mean difference between the minimum and maximum grain protein values within individual fields was 2.2%, with a maximum difference of 3.5% (Bridgets 3 in 1996).

TABLE 2. Mean, standard deviation (sd) and Coefficient of Variation (CV%) of soil mineral nitrogen (kg/ha N) sampled at 21 points in six individual fields in spring 1995 and spring 1996.

Site	1995				1996			
	Mean kg/ha	Range kg/ha	sd	CV %	Mean kg/ha	Range kg/ha	sd	CV %
Boxworth 1	94	55-141	19.7	20.9	96	59-158	24.3	25.2
Boxworth 2	61	33-110	17.6	28.7	65	44-100	14.8	22.6
Bridgets 3	51	30-82	11.0	21.8	39	27-58	8.1	20.9
Bridgets 4	36	26-49	6.8	18.7	63	34-137	23.9	38.0
Yokefleet 5	41	24-68	12.7	31.1	109	79-173	26.7	24.5
Yokefleet 6	53	29-102	19.0	35.8	118	86-172	23.5	19.9
Mean	56			26.2	82			25.2

TABLE 3. Mean, standard deviation (sd) and Coefficient of Variation (CV%) of grain protein concentration (% at 86% DM) at 21 points in five individual fields in 1995 and 1996.

Site	1995				1996			
	Mean kg/ha	Range kg/ha	sd	CV %	Mean kg/ha	Range kg/ha	sd	CV %
Boxworth 1	12.6	11.9-13.3	0.44	3.5	10.6	9.6-11.9	0.51	4.8
Boxworth 2	11.4	10.5-12.7	0.65	5.7	9.0	8.6-9.7	0.30	3.3
Bridgets 3*	-	-	-	-	8.8	7.4-10.9	0.84	9.5
Bridgets 4*	-	-	-	-	10.3	9.1-12.0	0.70	6.8
Yokefleet 6*	-	-	-	-	8.8	8.0-9.8	0.52	5.9
Mean	12.0			4.6	9.7			6.1

* oilseed rape grown in 1995.

Nitrogen balance

The nitrogen balance was calculated as the difference between the removal of nitrogen in grain and straw and the supply of nitrogen from the soil and fertiliser (Table 4). Soil analysis for SMN was used as the measure of soil nitrogen supply; nitrogen contained in crop material at the time of soil sampling, and net mineralisation of soil nitrogen were assumed to be constant. There was a 2-3 fold variation in nitrogen balance within individual fields, and an overall mean CV of 21.7% for wheat. The mean difference between minimum and maximum values within individual fields was 97 kg/ha N for wheat, and 93 kg/ha N for oilseed rape.

Apparent recovery of nitrogen

The range of apparent recovery by winter wheat was greatest at Bridgets (Figure 1) and least at Yokefleet and Boxworth in 1996. This may be a soil type effect with the greatest variation occuring on the shallow chalk soil. The overall mean value was 74%, with a range from 61% at Boxworth 1 in 1996 to 92% at Bridgets 4. Looking at Boxworth data over the two years, it appears that the range of variation was greater in 1995 (42%) than in 1996 (28%).

TABLE 4. Mean, standard deviation (sd) and Coefficient of Variation (CV%) of the nitrogen balance (kg/ha N) at 21 points in four individual fields in 1995 and 1996.

Site	1995				1996			
	Mean kg/ha	Range kg/ha	sd	CV %	Mean kg/ha	Range kg/ha	sd	CV %
Boxworth 1	153	91-197	27.0	17.6	145	103-201	25.6	17.7
Boxworth 2	106	74-161	23.6	22.2	118	86-160	18.8	15.9
Bridgets 3	119*	63-173*	24.1*	20.3*	100	42-147	21.0	21.0
Bridgets 4	90*	50-126*	17.1*	19.0*	87	32-152	30.9	35.5
Mean	117			19.9	113			22.5

* data for winter oilseed rape (excluded from annual mean values).

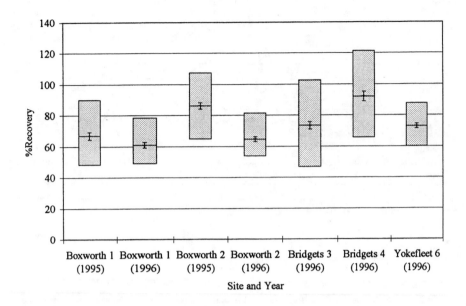

FIGURE 1. Range of nitrogen recovery by winter wheat (grain+straw) expressed as a percentage of soil plus applied N fertiliser. Shaded bars show the range; horizontal bars represent the mean value and associated standard error.

DISCUSSION

All data reported shows that there is significant variability within the studied fields in several important aspects of both soil and crop nitrogen dynamics.

The overall mean CV of 25.7% for the variation of SMN (0-90 cm), agrees well with values of 32 to 41% reported by Lord and Shepherd (1993), 10 to 28% for SMN to 100 cm depth (Knittel and Fischbeck, 1979) and 31.5% for nitrate nitrogen to 100 cm depth (Van Meirvanne and Hofman, 1989). It also agrees well with the general conclusion of Beckett and Webster (1971) that many soil properties, including topsoil total nitrogen concentration, have a CV of 25 to 30% within fields. However, higher CV values are likely following mixed arable or vegetable cropping or where organic manures are used.

Assuming a normal distribution of SMN with a mean of 70 kg/ha and a typical within field CV of 30% (therefore sd = 21), 66% of the field (1 sd either side of the mean) would theoretically have an SMN value of between 79 and 121 kg/ha, and 95% of the field (2 sd either side of the mean) would have an SMN of between 58 and 142 kg/ha. Since the quantity of spring SMN in a soil is regarded as equivalent to fertiliser nitrogen, these data suggest that optimum nitrogen fertiliser rates may vary by over 80 kg/ha within arable fields with this mean quantity of SMN (70 kg/ha of spring SMN is typical of medium and clay textured soils in combinable crop rotations).

The concept of nitrogen balancing, allowing for both N use and the yield/N removal of the previous crop, has been advocated by Sylvester-Bradley (1996) as a means to aid decisions on nitrogen fertiliser use to subsequent crops. Within-field variation in this parameter has been shown to be up to 3-fold, though it is possible that this may be reduced if precision of nitrogen fertiliser use is improved through variable rate application.

Grain protein concentration is a very important wheat quality criterion, and 11.0% protein (at 86% DM) is a common market requirement necessary to sell wheat at a financial premium for breadmaking. The mean range of 2.2% grain protein within the studied fields is highly significant when considered in the context that variations of 0.1% protein below target may lead to financial 'dockage', and 0.5% below target to rejection.

The apparent recovery of nitrogen is often used to indicate the potential risk of nitrate leaching with inadequate recovery of excessive rates of applied fertiliser leading to high residues in the soil (Bloom et al., 1988). Wilson et al. (1996) reported values of 43-80% for recoveries of total available nitrogen (SMN + fertiliser N). These were measured in standard farm crops representing a range of previous cropping and soil types. The latter was observed to have most influence on recovery with 73% on calcareous clay loams and 50-62% on other soil types. Data reported in this paper show a similar range of between field variation of 61-92%. However it is apparent that within field variation is just as great, if not greater, minimum 54-81% and maximum 47-103%. This has implications for those concerned with the reduction of nitrate leaching risk, since at this stage it is not possible to identify whether the within field variation is associated with differences in yield or soil nitrogen supply. Although SMN levels were not measured, Destain et al. (1989) using ^{15}N indicated that on average 74% of nitrogen in the grain was derived from fertiliser nitrogen and that it was not affected by applied N level or previous crop.

This paper provides evidence of significant within field variation in nitrogen dynamics. It is essential that the occurrence of spatial variation is recognised, and integrated with other nitrogen research focused on modelling the effects of specific soil, crop, weather and fertiliser/manure processes. Maximum benefit from an understanding of these processes is unlikely to be achieved unless the effect of natural and man-made spatial variation within fields is more fully understood and, where cost-effective, is managed by farmers. If management of nitrogen variability is shown to be worthwhile, then methods to reliably identify field 'zones' for uniform treatment must be established. The size of the zones must reflect the ability of current or improved spreading equipment to accurately spread fertiliser. There must also be cost-effective means to quantify nitrogen requirement in different zones. Current techniques of SMN analysis are too costly, and there is an urgent need for simpler methods for assessing the soil and/or crop to aid decisions on N use.

ACKNOWLEDGEMENTS

Financial support from the Ministry of Agriculture, Fisheries and Food (MAFF), Crowmarsh Battle Farms Ltd., Massey Ferguson Ltd. and Yokefleet Farms Ltd. is gratefully acknowledged. The authors would also like to thank many people in ADAS and at Silsoe Research Institute (science partner in the project) for their willing help.

REFERENCES

Beckett, P.H.T., Webster, P. (1971) Soil variability - a review. *Soils and Fertilisers*, **34(1)**, 1-15.

Bloom, T.M., Sylvester-Bradley, R., Vaidynathan, L.V., Murray, A.W.A. (1988) Apparent recovery of fertilizer nitrogen by winter wheat. *Nitrogen Efficiency in Agricultural Soils* D.S. Jenkinson and K.A. Smith (Eds), London, Elsevier Applied Science, pp 27-37.

Dampney, P.M.R., Salmon, S.E., Greenwell, P., Pritchard, P.E. (1995) Management of breadmaking wheat: Effects of extra nitrogen on yield, grain and flour quality. Home Grown Cereals Authority, Project Report 109, London.

Destain, J.P., Guiot, J., François, E. (1989) Fate of split applied N Fertilizer to winter wheat. Effect of N level and of preceding crop. A two year experiment with ^{15}N in the Belgian loam region. *Fertilization and the environment*, R Merckx, H Vereecken and K Vlassak (Eds), Leuven University Press, pp 182-188.

Knittel, G., Fischbeck, G. (1979) Die heterogenitat des nitratgehaltes in den profilschichten einer ackerbraunderde zu beginn des fruhjahrs. *Z. Pflanzenernaehr. Bodenkd*, **142**, 689-695.

Lord, E.I., Shepherd, M.A. (1993) Developments in the use of porous ceramic cups for measuring nitrate leaching. *Journal of Soil Science*, **44**, 435-449.

MAFF (1986) The analysis of agricultural materials. *Reference Book 427*, HMSO, London.

MAFF (1994) Fertiliser recommendations for agricultural and horticultural crops. *Reference Book 209*, HMSO, London.

Shepherd, M.A. (1993) Measurement of soil mineral nitrogen to predict the response of winter wheat to fertilizer nitrogen after applications of organic manures or after ploughed out grass. *Journal of Agricultural Science*, **121**, 223-231.

Sylvester-Bradley, R. (1996) Progress in accounting for nitrogen left by the last crop. *Aspects of Applied Biology - Rotations and Cropping Systems*, **47**, 67-76.

Van Meirvanne, M., Hofman, G. (1989) Spatial variability of soil nitrate nitrogen after potatoes and its change during winter. *Plant and Soil*, **120**, 103-110.

Wilson, W.S., Moore, K.L., Rochford, A.D., Vaidynathan, L.V. (1996) Fertilizer nitrogen to winter wheat crops in England: comparison of farm practices with recommendations allowing for soil nitrogen supply. *Journal of Agricultural Science*, **127**, 11-22.

SPATIAL CHARACTERIZATION OF ADULT EMERGENCE PATTERNS AND OVIPOSITION FOR CORN ROOTWORM POPULATIONS IN CONTINUOUS AND ROTATED CORN

M.M. ELLSBURY, W.D. WOODSON, L.D. CHANDLER

USDA-ARS, Northern Grain Insects Research Laboratory, 2923 Medary, Brookings, South Dakota, U.S.A. 57006

S.A. CLAY, D.E. CLAY, D. MALO, T. SCHUMACHER, J. SCHUMACHER, C.G. CARLSON

Plant Science Department, South Dakota State University, Brookings, South Dakota, U.S.A. 57007

ABSTRACT

Global positioning system technology, geostatistical techniques, and geographic information systems offer new approaches to the characterization, sampling, and management of insect pest populations in relation to spatially variable mortality factors. We have characterized the spatial distribution of corn rootworm adult emergence and egg deposition patterns as semivariograms produced from seasonal adult emergence monitoring and egg sampling. Data were obtained by grid sampling from three fields of about 65 ha each, one in continuous, irrigated corn and the others in rotated, no-till corn. Emergence patterns of northern corn rootworm adults were spatially correlated in both the continuous corn and in the rotated corn. Spatial variation in western corn rootworm emergence was evident in the continuous corn field. In the case of soil-dwelling pests that do not move large distances, such as corn rootworms, geostatistical techniques should allow mapping of expected insect population densities or action thresholds.

INTRODUCTION

Western corn rootworms (*Diabrotica virgifera virgifera*) and northern corn rootworms (*Diabrotica barberi*) are major pests of corn (*Zea mays*) in the United States, accounting annually for at least $350 million in insecticide application costs (Reed *et al.*, 1991). Temperature-driven predictive models are available for survival of corn rootworm eggs (Woodson and Gustin, 1993; Woodson and Ellsbury, 1994), development of the soil-dwelling stages (Bergman and Turpin, 1986, Jackson and Elliot, 1988; Schaafsma *et al.*, 1991), and the emergence phenology of adult corn rootworms (Elliott *et al.*, 1990). Spatially variable edaphic factors, such as soil type, soil moisture, snow cover, or crop residue, influence soil temperatures and thus also affect the phenology of corn rootworm development. Since soil conditions vary spatially, distributions of corn rootworms also will be likely to vary spatially as a function of edaphic factors that mediate behavioral responses or act as mortality factors.

SPATIAL VARIABILITY OF INSECT POPULATIONS

Spatial distributions of insect populations are usually considered nonrandom or heterogeneous (Liebhold *et al.*, 1993) as the result of environmental influences and behavioral responses (Taylor, 1984) that affect distributions, particularly of the relatively nonmobile stages, of soil-dwelling insects. The result is aggregated or contagious distributions that make reliable, economical sampling or scouting strategies difficult to develop. Since the influence of abiotic environmental mortality factors may vary seasonally as well as spatially, the spatial distributions of agriculturally important insects also vary from one growing season to another. Although the spatial nature of insect population distributions can be inferred from statistical indices developed from frequency distributions, such as Moran's *I* statistic or Geary's *c* (Williams *et al.*, 1992), these techniques do not measure spatial relatedness directly and do not allow correlation of population distribution with spatially variable environmental factors.

Geostatistics and insect populations

Statistical procedures, originally developed in the mining industry for locating spatially variable ore deposits, have become known as geostatistics (Isaaks and Srivastava, 1991). Geostatistical analysis provides an alternative approach to characterization of spatially variable ecological data (Rossi *et al.*, 1992), particularly for insect pest populations (Rossi *et al.*, 1993; Liebhold *et al.*, 1993; Roberts *et al.*, 1993). Geostatistical techniques have been used to describe the spatial nature of populations of nymphs and adults of the tarnished plant bug, *Lygus hesperus* (Schotzko and O'Keefe, 1989), in lentils. Geostatistical analyses of the soil-dwelling stages of the wireworm, *Limonius californicus*, have shown aggregated spatial distributions (Williams *et al.*, 1992). Rossi *et al.*, (1993) estimated population densities of northern corn rootworms and evaluated economic risk for treatment decisions on a regional scale using geostatistical techniques applied to stochastic simulations of spatial variability in northern corn rootworms. Geostatistical techniques also have been applied to site-specific management of the Colorado potato beetle, *Leptinotarsa decemlineata* (Weisz *et al.*, 1996).

The semivariogram. A basic tool of geostatistics is the semivariogram which plots lag distance between sample pairs against a semivariance statistic calculated for all possible sample pairs at each lag distance. The semivariance is defined as

$$g(h) = [1/2N(h)]\sum [z_i - z_{i+h}]^2 \tag{1}$$

where h is lag distance and N is the number of samples for variable z. For ecological data, such as adult corn rootworm emergence patterns, the semivariance is expected to increase as the lag interval increases to a distance where spatial dependence ceases to be detectable.

Certain features of semivariograms, that are important for ecological interpretation of spatial data from insect populations, are shown in Figure 1. The nugget (C_o), when present, is the distance along the y-axis from the origin to the y-intercept and is interpreted as variability due to experimental error and/or other random effects. Presence of a nugget suggests variability on a spatial scale smaller than that detectable by the original sampling protocol. The range is the lag distance beyond which samples are considered independent. The corresponding value of the semivariance at this point is termed the sill and represents the

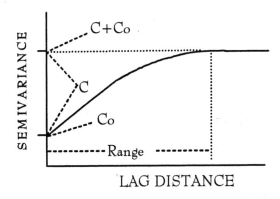

FIGURE 1. Generalized semivariogram showing the range of spatial dependence, nugget effect (C_o), variability associated with spatial dependence (C), and a sill ($C+C_o$).

combination of a nugget effect and variability, C, attributable to spatial dependence. The proportion of variability attributable to spatial dependence may be estimated as

$$\% \text{ variability} = C/(C+C_o) \tag{2}$$

where C_o is the nugget and C is variability due to spatial dependence as indicated by the distance from the nugget to the sill along the y-axis of Figure 1.

<u>Kriging</u>. Mathematical models for the semivariogram may be applied to a geostatistical technique termed kriging that provides interpolated values for response variables at points not actually sampled. Use of indicator kriging allows prediction of spatial probability that a response, expressed as an indicator variable, will exceed an established threshold. In the case of insect pest populations, geostatistical techniques should permit mapping of expected insect population density or action thresholds expressed as an indicator variable for unsampled areas of a field. This approach also should be applicable to correlation of spatial insect population patterns with spatially variable factors such as plant development, soil physical properties, or other environmental factors that influence pest population densities and may be more easily and economically sampled than are the insect populations.

Site-specific pest management depends on the availability of accurate knowledge of the spatial distributions of insect pest populations, particularly for the nonmobile stages of such pests. In the case of corn rootworms, an understanding of spatial variability in the distribution of the soil-dwelling stages of this pest is needed on a field scale for definition of the appropriate sampling procedures and to define the relationship of insect distributions to spatially variable mortality factors. Our objective was to characterize spatial variability in oviposition and emergence patterns of adult corn rootworms.

PROCEDURES

Spatial variability in corn rootworm adult emergence and egg distribution was characterized in three quarter-section (~65 ha) corn fields. Two fields, designated Elkton-South and

Elkton-North, were located near Elkton, South Dakota. These were farmed using no-till methods in a 2-year corn and soybean (*Glycine max*) rotation such that one field was planted to corn and the other to soybean each year. The third field, located near Brookings, South Dakota, had a long history of conventional cultivation in continuous corn.

Adult corn rootworm emergence was monitored during 1995 and 1996 by grid sampling each field using 0.5 m^2 cages, installed during late August in a hexagonal (offset) grid pattern. Cages were centered over corn stalks in ~1 m of row cut to within 10 cm of the ground. Numbers of emerging northern and western corn rootworm adults were counted weekly in each cage from late August through late September. Egg populations were estimated from 1 liter soil cores taken adjacent to each emergence enclosure with a golf course cup cutter. Eggs were washed from soil samples by flotation in salt solution (Shaw *et al.*, 1976) and were identified to species on the basis of chorion sculpturing (Atyeo *et al.*, 1964).

The hexagonal grid sampling pattern was chosen since this design has provided better estimates than other sampling designs for assessing the spatial structure of insect populations (Williams *et al.*, 1992). Eleven sampling transects about 75 m apart were established in each field. Sampling points within each transect also were 75 m apart and were offset from points in adjacent transects by about 32.5 m to produce the hexagonal pattern. Sampling points that occurred in waterways, wet spots, or other unplanted areas of the field were relocated whenever possible to the nearest cultivated area of the field but were placed at least 10 m in from the edge of the uncultivated area. To provide some samples at distances closer than 75 m for geostatistical analysis, cross-shaped patterns comprising 12 additional sampling points for adult emergence monitoring were established in some fields such that the minimum distance between points in the cross was about 25 m. Cross patterns were established at single sites in the Brookings and Elkton-South fields during 1995 and at three randomly selected sites per field during 1996 in the Brookings and Elkton-North fields.

Latitude and longitude of each sample point was recorded using a differentially-corrected global positioning system. Decimal degree values were converted to Cartesian coordinates in meters using the latitude and longitude of the southwest corner of each field as the origin. Where data were skewed toward small values they were transformed to log ($Z+1$). Spatial variability of adult emergence and oviposition patterns were characterized as semivariograms calculated using GS+ Geostatistics for the Environmental Sciences (version 2.3, Gamma Design Software, Plainwell, Michigan).

RESULTS AND DISCUSSION

Both northern and western corn rootworm adults were collected from continuous corn at the Brookings site. No significant western corn rootworm emergence occurred in rotated corn because this insect does not survive a two-year rotation in South Dakota. Parameters associated with semivariograms for adult emergence at these sites are presented in Table 1. Spherical models provided the best fit to the semivariograms of transformed data describing the spatial nature of western corn rootworm adult emergence during both years in the continuous corn system. Semivariograms for northern corn rootworm adults were best characterized by exponential models in both cropping systems during both study years.

The semivariogram from 1995 data for western corn rootworm adult emergence is shown

TABLE 1. Semivariogram models for spatial structure of western (WCR) and northern (NCR) corn rootworm adult emergence.

Site	Year	Species	Model, transform	r^2	Co	C+Co	Range
Brookings	1995	WCR	spherical, ln(z)	0.894	0.52	1.44	556 m
Brookings	1995	NCR	exponential, ln(z)	0.945	—	2.41	172 m
Elkton-South	1995	NCR	exponential, none	0.855	—	586	215 m
Brookings	1996	WCR	spherical, ln(z)	0.808	0.41	0.93	182 m
Brookings	1996	NCR	exponential, ln(z)	0.755	0.54	1.14	231 m
Elkton-North	1996	NCR	exponential, ln(z)	0.703	0.91	1.88	115 m

in Figure 2. Numbers of western corn rootworms per 0.5 m² sample unit in the continuous corn ranged from 0 to a maximum of about 50. Western corn rootworm emergence showed nugget effects at about 0.52 and 0.41 (y-intercepts) in semivariogram models from 1995 and 1996, respectively. This suggested that about 56% and 44% of the variability among sample pairs in those years was explained by the spatial structure of this population. One possible interpretation of the nugget effects observed in the western corn rootworm data is that spatial structure existed that was not detectable at the sampling scale used. Total numbers of adult western corn rootworms trapped were spatially correlated out to a distance of about 550 m. The spatial structure of northern corn rootworm adult emergence, typified by the semivariogram in Figure 3, was best described by exponential models (Table 1). Total numbers of northern corn rootworm adults that emerged per 0.5 m² cage unit varied from 0 to over 100 beetles in certain areas of each field. Spatial dependence was evident over a range of influence from 115 to 230 m for northern corn rootworms in both rotated and continuous corn. Adult populations of both species of corn rootworms exhibited distinctive spatial structure in their emergence patterns. Reasons for this spatial variability are unknown

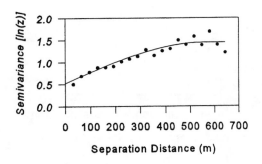

FIGURE 2. Semivariogram showing spatial variability in 1995 western corn rootworm adult emergence data taken at the Brookings continuous corn site (spherical model, r^2 = 0.894):sill at 1.44, nugget effect at 0.52 (y-intercept), and range of influence extending to 556 m.

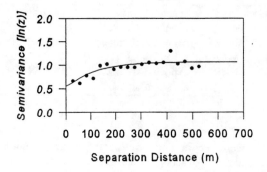

FIGURE 3. Semivariogram showing spatial variability in northern corn rootworm adult emergence data taken during 1996 at the Elkton-North rotated corn site (exponential model on ln(z), $r^2 = 0.70$): sill at 1.88, nugget effect at 0.91 (y-intercept), and range of influence extending to 115 m.

at this time, but probably relate to differential influence of mortality factors acting at several possible points in the life cycle of corn rootworms.

Semivariograms for egg samples showed no spatial dependence (Figure 4). This suggested that oviposition behavioral responses may not be related to spatial variability in crop or soil physical conditions within areas of the field that influence distribution of emerging adults.

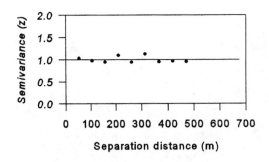

FIGURE 4. Semivariogram showing lack of spatial dependence in northern corn rootworm egg sample data taken from the Elkton-South site during 1996.

Differential overwinter mortality to eggs associated with variable snow cover, crop residue, soil moisture conditions, or topography may be more likely determinants of the spatial distribution of emerging adults. Mortality also may occur when newly-hatched larvae are affected by soil physical conditions such as texture (Turpin and Peters, 1971) or compaction (Ellsbury *et al.*, 1994) which may vary spatially depending on soil type, moisture conditions, tillage, and topography.

CONCLUSION

Geostatistical techniques, global positioning system technology, and geographic information systems offer new approaches to the characterization, sampling, and management of insect pest populations (Liebhold *et al.*, 1993). Kriging based on semivariogram properties (Isaaks and Srivastava, 1989) can provide interpolated values for response variables, such as insect population densities, at points not actually sampled. In the case of soil-dwelling pests, such as corn rootworms, geostatistical techniques should allow development of contour maps of expected insect population density in relation to spatially variable mortality factors. These thresholds may be expressed in terms of potential indicator variables, such as host plant condition or soil properties, that are correlated spatially with rootworm survival and population dynamics, and sampled more easily and economically than insect populations.

REFERENCES

Atyeo, W.T., Weekman, G. T., Lawson, D. E. (1964) The identification of *Diabrotica* by chorion sculpturing. *Journal of the Kansas Entomological Society*, **37**, 9-11.

Bergman, M. K.,Turpin, F. T. (1986) Phenology of field populations of corn rootworms (Coleoptera: Chrysomelidae) relative to calendar date and heat units. *Environmental Entomology*, **15**, 109-112.

Elliott, N.C., Jackson, J.J., Gustin, R.D. (1990) Predicting western corn rootworm beetle (Coleoptera: Chrysomelidae) emergence from the soil using soil or air temperature. *The Canadian Entomologist*, **122**, 1079-1091.

Ellsbury, M.M., Schumacher, T.E., Gustin, R.D., Woodson, W.D. (1994) Soil compaction effect on corn rootworm populations in maize artificially infested with eggs of western corn rootworm (Coleoptera: Chrysomelidae). *Environmental Entomology*, **23**, 943-948.

Isaacs, E. H., Srivastava, R. M. (1989) *Applied geostatistics*. London, Oxford University Press, 561 pp.

Jackson, J.J., Elliott, N.C. (1988) Temperature-dependent development of immature stages of the western corn rootworm, *Diabrotica virgifera virgifera* (Coleoptera: Chrysomelidae). *Environmental Entomology*, **17**, 166-171.

Liebhold, A.M., Rossi, R.E., Kemp, W.P. (1993) Geostatistics and geographic information systems in applied insect ecology. *Annual Review of Entomology*, **38**, 303-327.

Reed, J. P., Hall, F. R., Taylor, R. A. J., Wilson, H. R. (1991) Development of a corn rootworm (*Diabrotica*, Coleoptera: Chrysomelidae) larval threshold for Ohio. *Journal of the Kansas Entomological Society*, **64**, 60-68

Roberts, E.A., Ravlin, F.W., Fleischer, S.J. (1993) Spatial data representation for integrated pest management programs. *The American Entomologist*, **39**, 92-107.

Rossi, R.R., Borth, P.W., Tollefson, J.J. (1993) Stochastic simulation for characterizing ecological spatial patterns and appraising risk. *Ecological Applications*, **3**, 719-735.

Rossi, R.E., Mulla, D.J., Journel, A.G., Franz, E.H. (1992) Geostatistical tools for modeling and interpreting ecological spatial dependence. *Ecological Monographs*, **62**, 277-314.

Schaafsma, A.W., Whitfield, G.G., Ellis, C.R. (1991) A temperature-dependent model of egg development of the western corn rootworm, *Diabrotica virgifera virgifera* LeConte (Coleoptera: Chrysomelidae. *The Canadian Entomologist*, **123**, 1183-1197.

Shaw, J.T., Ellis, R.O., Luckmann, W.H. (1976) Apparatus and procedure for extracting corn rootworm eggs from soil. *Illinois Natural History Survey Biological Notes*, **96**, 4pp.

Schotzko, D.G., O'Keeffe, L.E. (1989) Geostatistical description of the spatial distribution of *Lygus hesperus* (Heteroptera: Miridae) in lentils. *Journal of Economic Entomology*, **82**, 1277-1288.

Taylor, L.R. (1984) Assessing and interpreting the spatial distributions of insect populations. *Annual Review of Entomology*, **29**, 321-357.

Turpin, F.T., Peters, D.C. (1971) Survival of southern and western corn rootworm larvae in relation to soil texture. *Journal of Economic Entomology*, **64**, 1448-1451.

Weisz, R., Fleischer, S.J., Smilowitz, Z. (1996) Site-specific integrated pest management for high-value crops: impact on potato pest management. *Journal of Economic Entomology*, **89**, 501-509.

Williams, L. III, Schotzko, D.J., McCaffrey, J.P. (1992) Geostatistical description of the spatial distribution of *Limonius californicus* (Coleoptera: Elateridae) wireworms in the northwestern United States, with comments on sampling. *Environmental Entomology*, **21**, 983-995.

Woodson, W.D., Ellsbury, M.M. (1994) Low temperature effects on hatch of northern corn rootworm eggs (Coleoptera: Chrysomelidae). *Journal of the Kansas Entomological Society*, **67**, 102-106.

Woodson, W.D., Gustin, R.D. (1993) Low temperature effects on hatch of western corn rootworm eggs (Coleoptera: Chrysomelidae). *Journal of the Kansas Entomological Society*, **66**, 104-107.

SPATIAL AND SEASONAL FACTORS AFFECTING CROP YIELD AND N REMOVAL FROM GLACIAL TILL SOILS

D.L. KARLEN, C.A. CAMBARDELLA, D.B. JAYNES, T.S. COLVIN, T.B. MOORMAN

USDA-ARS, National Soil Tilth Laboratory, Ames, IA, USA, 50011-4420

ABSTRACT

Spatial and seasonal variation in crop yield must be known and evaluated before attempting to develop site-specific management practices that are possible with the positioning and real-time control technologies associated with precision agriculture. The objective for this presentation is to examine crop yield and N removal data collected for 1992 through 1995 from fields with Typic Haplaquolls and Typic Hapludolls located in a central Iowa, USA watershed. Corn (*Zea mays* L.) and soybean [*Glycine max* (L.) Merr.] yield, grain N concentrations, relative grain yield, and annual N removal are evaluated according to soil map unit. Relationships with seasonal rainfall patterns, N fertilizer rates, and selected biological, chemical, and physical indicators of soil quality are discussed.

INTRODUCTION

Variable rate agriculture requires three primary components: (I) positioning (currently the global positioning system, GPS) to know where equipment is located, (II) real-time mechanisms for controlling nutrient, pesticide, seed, water, or other crop production inputs, and (III) databases or sensors that provide information needed to develop appropriate input response to site-specific conditions. The technologies associated with requirements I and II are well advanced compared to development and interpretation of information needed to obtain the economic and environmental benefits of precision agriculture. Building multiple-year crop yield databases to accurately define spatial variability (Karlen *et al.*, 1990; Sadler *et al.*, 1993) is one of the first field operations needed to implement precision farming practices such as site-specific fertilizer, herbicide, seeding, or irrigation rates. These multi-year databases should examine both anthropogenic (controllable) and nonanthropogenic factors which are contributing to the spatial variability encountered within a sampling area (Cambardella *et al.*, 1994; Karlen and Sharpley, 1994). Landscape position, especially through its impact on the soil water balance, is a major nonanthropogenic factor that affects not only soil fertility parameters but also crop growth and yield (Pierce *et al.*, 1995). Tillage, crop rotation, and application of manure are controllable factors that can have a major impact on several physical, chemical, and biological parameters within a soil, and should therefore be included in the overall information database.

Crop yield and N accumulation information presented herein was obtained from two fields within a Management Systems Evaluation Area (MSEA) watershed that was established in central Iowa, USA as part of the President's Water Quality Initiative

(Ward *et al.*, 1994). This 5130 ha watershed is located in Major Land Resource Area (MLRA) 103. Dominated by row crops, over 85% of the land area is managed in a corn-soybean rotation.

Soils in this Des Moines Lobe watershed were formed in calcareous glacial till deposited approximately 14000 years BP during the most recent advance of the Late Wisconsinan ice sheet into central Iowa (Steinwand, 1992). They are generally poorly drained due to their low relief and relatively young geologic development. Soil map units within depressional or 'pothole' areas are characterized by very poorly drained clay material, but other soils within the watershed are predominantly silt loam. The soils are classified as part of the Clarion-Nicollet-Webster association. Crop yield potential is high because of deep profiles, high water holding capacity, and high fertility levels. Spatial variability for several soil properties within these two fields was reported by Cambardella *et al.* (1994). Our objective is to discuss crop yields and estimates of N removal for the period 1992 through 1995.

EXPERIMENTAL APPROACH

These field-scale studies were conducted in the 36 ha 'Pothole' and 96 ha 'No-Till' fields within the Walnut Creek watershed, which is located at latitude 41° 55' N to 42° 00' N and longitude 93° 32' W to 93° 45' W. The pothole field is managed by our cooperator in a 2-yr corn-soybean rotation with conventional tillage practices. After soybean harvest and when soil temperature in the upper 20 cm was below 10° C, 137 kg N ha^{-1} was applied using anhydrous ammonia. An additional 44-49-138 kg ha^{-1} N-P-K, respectively, was broadcast prior to each corn crop. Herbicides were applied in spring while preparing the seedbed with a field-cultivator. In 1992, no-till management was initiated in a nearby field that was split to accommodate both corn and soybean each year. The east side of the field was planted to soybean in the even years and corn the odd years. The west side of the field was planted to soybean in the odd years and corn in the even years. Anhydrous ammonia was injected following soybean at rates of 122 to 137 kg N ha^{-1} but there was no other surface tillage. Granular N-P-K fertilizer was also broadcast prior to each corn crop, providing 14 to 35 kg ha^{-1} N.

Our cooperator planted commercial corn hybrids and soybean varieties and made all other cultivation and seasonal management decisions. Before grain harvest, several transects were marked within both fields. Individual grain samples were collected from areas that were approximately 15 m long and from 2.3 to 4.0 m wide depending on the crop and machine used for harvest. This experiment was initiated before reliable yield monitors were commercially available, so grain samples were collected by starting and stopping harvest operations at selected positions along each transect.

The equipment used for this operation is available on both a Gleaner K[1] and John Deere Model 4420 combine. Initially, transect location was verified using standard survey

---

[1]Mention of trademark, proprietary product, or vendor does not constitute a guarantee or warranty of the product by the USDA and does not imply its approval to the exclusion of the other products or vendors that may also be suitable.

236

methods, but more recently, a GPS was used to locate each sampling position. After harvest, maps were generated and overlaid onto a digitized Order 2 (1:15840) soil survey map. Each sampling point was assigned a specific soil map unit. Visual interpolation was used for cells with boundaries that crossed map units.

While the combine was stopped at each sampling site, GPS position, grain weight, and grain moisture were recorded. Grain samples collected for total N analysis were dried at 65° C, ground to pass a 0.5 mm screen, and stored in sealed plastic bags, until they could be analyzed using a Carlo-Erba NA1500 NCS analyzer (Haake Buchler Instruments, Patterson, NJ). This instrument measures total N using dry combustion or the Dumas method (Bremner and Mulvaney, 1982). N removal was calculated by multiplying grain yield at 0% water content by N concentration. Total N was determined for all grain samples in 1992 and 1994, but for only every fifth sample in 1993 and 1995. To provide some degree of randomization when only 20% of the samples were analyzed, grain was collected from positions 1, 6, 11, ... on transect I, 2, 7, 12, ... on transect II, and 3, 8, 12, ... on transect III, etc. Relative yield was computed by dividing the measured grain yield for each sampling cell by the anticipated grain yield for that soil map unit as published in the Story County Soil Survey Manual (USDA-SCS, 1984). Yield data is normalized for weather and crop differences when using relative yields, allowing more accurate comparisons of yields from different years and different crops. Statistical analyses were computed using the SAS statistical software package (SAS, 1985).

RESULTS AND DISCUSSION

Yield relationships

The seven predominant soil map units identified in the pothole and no-till fields are classified in Table 1. All are Mollisols, differing primarily in drainage, slope, and landscape position. They are generally considered to be very productive soils.

The average relative yield in 1993 was significantly lower (0.54) than in 1992, 1994, or 1995 (1.24, 1.24 and 1.14 respectively). The excessive rainfall which occurred throughout most of the midwestern USA in 1993 resulted in prolonged flooding of low relief areas (i.e., Okoboji soils), creating O_2 stress throughout most of July and August and resulting in reduced yields. Excluding 1993, relative yields averaged across all years in the pothole field and in the east- and west-halves of the no-till field were not significantly different (1.10, 1.24, and 1.23, respectively).

Variability in absolute yield estimates, averaged for each soil map unit at the pothole field, was generally reduced about 70% of the time when compared with whole field average yield variability for both corn and soybean (Table 2). The pattern at the no-till field was different for the two crops. For corn, the variability in yield averaged for each soil map unit was less than whole field yield variability only 50% of the time (Table 3). Soybean yield variability was reduced more than 60% of the time compared to whole field yield variability when averaged within individual map units (Table 4). This suggests that no-till management may be enhancing corn yield variability within individual map units compared with chisel plow management at the pothole field, where yield variability was consistently lower when averaged within map units. Differences in

TABLE 1. Soil map units identified in the two central Iowa, USA study sites.

Symbol	Soil name	Taxonomic classification
55	Nicollet loam 1-3% slope	Fine-loamy, mixed, mesic Aquic Hapludolls
90	Okoboji mucky silt loam 0-1% slope	Fine, montmorillonitic, mesic, Cumulic Haplaquolls
95	Harps loam, 1-3% slope	Fine-loamy, mesic Typic Calciaquolls
107	Webster clay loam 0-2% slope	Fine-loamy, mixed, mesic Typic Haplaquolls
138B	Clarion loam 2-5% slope	Fine-loamy, mixed, mesic Typic Hapludolls
138C	Clarion loam 5-9% slope	Fine-loamy, mixed, mesic Typic Hapludolls
507	Canisteo clay loam 0-2% slope	Fine-loamy, mixed (calcareous), mesic Typic Haplaquolls

variability between crops also emphasizes the importance of including weather data in addition to soil test, yield, input, management, and economic information when developing databases for precision farming operations.

N removal

Two factors encouraging development of precision farming strategies are the desire for improved economic return on production inputs and the need for environmental protection by improving the efficiencies with which nutrients and herbicides are used. Within the experimental watershed where the pothole and no-till fields are located, 85% of the land is managed in a corn-soybean rotation. Surface water quality is a primary concern (Ward et al., 1994) because tile drainage has been shown to contribute a significant amount of NO_3-N to the stream.

Grain N concentrations from year to year were very similar ranging from 11.0 to 11.8 g kg^{-1} for corn and from 61.5 to 62.4 g kg^{-1} for soybean. Standard deviations for soybean samples grouped by map unit showed small differences (±1.3 to 2.2 g kg^{-1}) provided at least 20 sites were sampled. For corn grain, standard deviation among samples ranged from 0.2 to 1.6 g kg^{-1} with no detectable loss in accuracy associated with analyzing every fifth sample. Reducing the number of samples that need to be processed is important because of its impact on fiscal and personnel resources.

Within the pothole and no-till fields, N removal by corn accounted for approximately 110 kg N ha^{-1} or 69% of the N applied, except in 1993 when yields were very low and

TABLE 2. Average grain yield and N removal by corn and soybean grown on various soil map units identified within the 'Pothole' field.

Soil symbol	Yield Avg(SD)	Yield CV	N removal Avg (SD)	Yield Avg (SD)	Yield CV	N removal Avg (SD)
	Mg ha^{-1}		kg ha^{-1}	Mg ha^{-1}		kg ha^{-1}
	1992 -- Corn			1993 -- Soybean		
55	11.8 (0.9)	7.6	107 (16)	2.16 (0.2)	9.3	137 (5)
90	10.8 (1.3)	12.0	108 (10)	0.07 (0.2)	285.7	----*
95	12.4 (1.0)	8.1	114 (8)	1.93 (0.6)	31.1	133 (6)
107	11.2 (0.7)	6.3	100 (14)	2.25 (0.1)	4.4	142 (-)
138B	11.3 (1.1)	9.7	100 (14)	2.05 (0.4)	19.5	131 (13)
138C	11.3 (1.1)	9.7	110 (22)	1.92 (0.4)	20.8	130 (9)
507	11.3 (1.8)	15.9	103 (17)	1.41 (0.9)	63.8	95 (47)
field	11.4 (1.4)	12.3	103 (16)	1.76 (0.8)	45.5	117 (36)
	1994 -- Corn			1995 -- Soybean		
55	10.4 (0.9)	8.7	131 (11)	3.20 (0.1)	3.1	203 (13)
90	11.6 (0.5)	4.3	141 (12)	3.55 (0.1)	2.8	235 (12)
95	10.4 (0.4)	3.8	120 (13)	3.36 (0.3)	8.9	218 (9)
107	10.4 (0.1)	0.96	124 (7)	3.31 (0.3)	9.1	201 (-)
138B	10.5 (0.8)	7.6	131 (12)	3.17 (0.2)	6.3	206 (19)
138C	10.4 (0.5)	4.8	132 (6)	3.16 (0.2)	6.3	----
507	10.6 (1.3)	12.3	133 (18)	3.37 (0.3)	8.9	220 (18)
field	10.6 (1.3)	12.3	131 (15)	3.27 (0.3)	9.2	214 (9)

* Missing data because map unit did not occur or because no grain samples were saved for analysis.

removal only accounted for 31% of the applied N. Soybean removed 132 kg N ha^{-1} in 1993 and averaged 208 kg N ha^{-1} other years, presumably most from fixation.

CONCLUSIONS

Spatial and temporal differences in crop yield and N removal data from two central Iowa, USA fields are examined. Relative yield comparisons across four years for the two fields emphasize the importance of collecting climate data when developing databases for precision farming. The variability pattern for absolute yield estimates

TABLE 3. Average grain yield and N removal by corn grown on various soil map units identified within the 'No-Till' field.

Soil symbol	Yield Avg(SD)	Yield CV	N removal Avg(SD)	Yield Avg(SD)	Yield CV	N removal Avg(SD)
	Mg ha^{-1}		kg ha^{-1}	Mg ha^{-1}		kg ha^{-1}
	1992 (west)			1993 (east)		
55	11.9 (0.7)	5.9	108 (8)	4.98 (0.6)	12.0	---*
90	11.7 (1.0)	8.5	111 (10)	------	-----	---
107	-------	-----	----	5.31 (0.8)	15.1	56 (16)
138B	11.8 (0.8)	6.8	110 (11)	4.86 (1.4)	28.8	44 (16)
138C	11.4 (0.4)	3.5	109 (8)	------	-----	---
507	11.5 (0.8)	7.0	106 (9)	4.73 (1.4)	29.6	49 (16)
field	11.6 (0.8)	6.9	107 (10)	4.90 (1.3)	26.5	48 (10)
	1994 (west)			1995 (east)		
55	11.0 (0.6)	5.5	112 (7)	9.62 (0.3)	3.1	113 (10)
90	11.8 (0.3)	2.5	114 (2)	-----	-----	----
107	-------	-----	----	9.79 (0.7)	7.2	96 (9)
138B	10.6 (1.0)	9.4	106 (12)	9.47 (0.6)	6.3	99 (10)
138C	11.2 (0.5)	4.5	111 (8)	------	-----	----
507	11.5 (0.6)	5.2	113 (10)	9.61 (0.6)	6.2	100 (10)
field	11.4 (0.8)	7.0	112 (10)	9.61 (0.6)	6.2	99 (10)

* Missing data because map unit did not occur or because no grain samples were saved for analysis.

averaged within map units and compared to whole field yield variability was different for the two fields. No-till management appears to affect corn yield variability differently than soybean yield variability within map units. We anticipate that correlation analysis between soil biochemical properties and crop yield will help explain the reasons for these differences.

The data suggest that when using yield transects to assess spatial variability of crop parameters, grain N does not have to be measured for every yield sample. Nitrogen removal by corn accounted for approximately 70% of the fertilizer N applied for all years except 1993 for both fields. This is relatively high nutrient use-efficiency when compared to average values for the upper Midwest corn belt (25-30%) and suggests our farmer cooperators were managing the soil resource in a highly efficient fashion.

TABLE 4. Average grain yield and N removal by soybean grown on various soil map units identified in the 'No-Till' field.

Soil symbol	Yield Avg(SD)	Yield CV	N removal Avg(SD)	Yield Avg(SD)	Yield CV	N removal Avg(SD)
	Mg ha^{-1}		kg ha^{-1}	Mg ha^{-1}		kg ha^{-1}
	1992 (east)			1993 (west)		
55	3.17 (0.1)	3.2	-----	2.45 (0.3)	12.2	140 (4)
90	------	-----	-----	0.19 (0.4)	210.52	----
107	3.38 (0.2)	5.9	-----	------	-----	----
138B	3.42 (0.3)	8.8	-----	2.57 (0.2)	7.8	152 (-)
138C	------	-----	-----	2.51 (0.2)	8.0	132 (5)
507	3.39 (0.2)	5.9	-----	1.82 (0.8)	44.0	116 (29)
field	3.20 (0.2)	6.3	-----	1.88 (0.8)	42.6	121 (28)
	1994 (east)			1995 (west)		
55	------	-----	----	3.58 (0.3)	8.4	194 (-)
90	------	-----	----	3.27 (0.2)	6.1	182 (-)
107	3.83 (0.3)	7.8	225 (16)	------	-----	----
138B	3.71 (0.2)	5.4	210 (14)	3.33 (0.1)	3.0	191 (14)
138C	------	-----	----	3.52 (0.1)	2.8	212 (4)
507	3.83 (0.3)	7.8	220 (14)	3.52 (0.2)	5.7	194 (11)
field	3.77 (0.3)	7.9	217 (16)	3.39 (0.2)	5.9	194 (81)

* Missing data because map unit did not occur or because no grain samples were saved for analysis.

REFERENCES

Bremner, J.M., Mulvaney, C.S. (1982) Nitrogen--Total. *Methods of Soil Analysis. Chemical and Microbiological Properties.* A.L. Page (Ed.), Part 2. 2nd ed. Agronomy Monograph No. 9, Madison, WI, American Society of Agronomy, pp 595-624.

Cambardella, C.A., Colvin, T.S., Jaynes, D.B., Karlen, D.L. (1996) Spatial variability analysis: a first step in site-specific management. *Proc. 8th Annual Integrated Crop Management Conference*, Ames, IA, Iowa State University, pp165-173.

Cambardella, C.A., Moorman, T.B., Novak, J.M., Parkin, T.B., Karlen, D.L., Turco, R.F., Konopka, A.E. (1994) Field-scale variability of soil properties in central Iowa soils. *Soil Sci. Soc. Am. J.*, **58**, 1501-1511.

Karlen, D.L., Sadler, E.J., Busscher, W.J. (1990) Crop yield variation associated with coastal plain soil map units. *Soil Sci. Soc. Am. J.*, **54**, 859-865.

Karlen, D.L., Sharpley, A.N. (1994) Management strategies for sustainable soil fertility. *Sustainable Agriculture Systems,* J.L. Hatfield and D.L. Karlen (Eds.), Boca Raton, FL., CRC Press, Inc., pp 47-108.

Pierce, F.J., Warncke, D.D., Everett, M.W. (1995) Yield and nutrient variability in glacial soils of Michigan. *Site-Specific Management for Agricultural Systems,* P.C. Robert, R.H. Rust, and W.E. Larson (Eds.), Madison, WI, ASA-CSSA-SSSA, Inc.

Sadler, E.J., Evans, D.E., Busscher, W.J., Karlen, D.L. (1993) Yield variation across coastal plain soil mapping units. *Site-Specific Management for Agricultural Systems,* P.C. Robert, R.H. Rust, and W.E. Larson (Eds), Madison, WI, ASA-CSSA-SSSA, Inc., pp 373-374.

SAS Institute (1985) SAS user's guide: Statistics. Version 5. Cary, NC, SAS Inst.

Steinwand, A.L. (1982) Soil geomorphic, hydrologic, and sedimentologic relationships and evaluation of soil survey data for a Mollisol catena on the Des Moines Lobe, central Iowa. PhD Dissertation, Iowa State University.

USDA-SCS (1984) Soil survey of Story county, Iowa. Washington, DC, U.S. Gov. Printing Off.

Ward, A.D., Hatfield, J.L., Lamb, J.A., Alberts, E.E., Logan, T.J., Anderson, J.L. (1994) The management systems evaluation areas program: Tillage and water quality research. *Soil & Tillage Research,* **30**, 49-74.

YIELD VARIATION AND CROP WATER USE: CAUSE OR EFFECT?

E.I. LORD

ADAS, Woodthorne, Wergs Road, Wolverhampton, WV6 8TQ, UK

M.A. SHEPHERD

ADAS, Gleadthorpe Farm, Meden Vale, Mansfield, Nottinghamshire, UK

P.M.R. DAMPNEY

ADAS Boxworth, Boxworth, Cambridgeshire, UK

ABSTRACT

On two clay fields with 21 measurement points in each, yield was positively correlated with soil moisture deficit in July. However both measurements were more strongly correlated with early crop growth, and with take-all scores, than with any measured soil properties. This suggests that the differences in crop water use were related to crop growth affecting crop demand for water and possibly rooting effectiveness. On a sandy site, in a drought year, yield was strongly positively correlated with soil available water capacity, indicating that in this case water restriction was the cause of the reduced yield. The implications of the results for identification of the causes of low yield, and appropriate management responses, are discussed.

INTRODUCTION

Precision farming offers the opportunity to manage inputs within a field in response to measured spatial variation in the crop or its environment. Such measurements can influence management only if they are made sufficiently early in the cropping year, or reflect permanent features of the site with predictable effects.

Within-field variation in soil type may affect yield through variation in moisture supply to the crop (Gales, 1983). However the effect of soil type on the timing and severity of restrictions on water uptake; and the implications of this for the crop's response to other inputs; are not sufficiently understood. Such understanding is essential before appropriate management responses can be formulated.

METHODS

Studies from two sites with contrasting soils are reported here. At Boxworth, two fields within the same farm had combine yield maps, and detailed measurements from 21 sampling points, for harvest years 1995 and 1996 (Clarke *et al.*, 1996). The points were evenly distributed across the field (Figure 1). At each of the 21 points neutron probe measurements of soil volumetric moisture content were made in April and July, and calibrated against topsoil and subsoil samples. Crop growth measurements (including

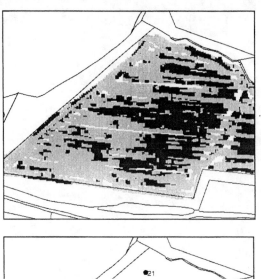

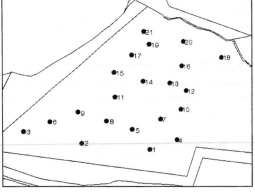

FIGURE 1. a) Combine yield map, Knapwell field, 1995. Darkest areas > 5 t/ha, pale areas < 5 t/ha wheat at 85% dry matter. b) Location of 21 sampling points.

plant or tiller numbers in spring; above-ground dry matter at Growth Stage 39; take-all scores and hand harvested yields) and soil properties (including nutrient content, pH, calcium carbonate content, and % clay) were also measured at each point. The soil at Boxworth is mapped as mainly Hanslope series (clay loam developed over chalky boulder clay), with a small area of Wickham series (with a greater depth of clay loam over the chalky boulder clay) in Knapwell field. Due to the similarity of the soil physical parameters, the close proximity of the sites (within less than 500 m), and their uniform management, yield variation is discussed between as well as within these two fields.

At Gleadthorpe, a field containing a mixture of loamy sand (Newport series) and sandy loam (Wick series) showed severe lodging in 1992 after heavy rain. A detailed soil map was used to analyse differences in yield and lodging measured at 88 points, on a regular grid. Yields were measured with a plot combine (cuts approximately 2 m by 20 m), and the percentage of lodging was estimated by eye for each cut area.

Crop water use at each site was modelled using the Irriguide model (Bailey and Spackman, 1996; Bailey *et al.*, 1996). This model has been used for many years for irrigation scheduling, and has been validated against a range of crops. The model, based on the principles of the Penman-Monteith equation, assumes normal growth rates without checks other than those due to lack of water, and allows details of crop management, dates of specified crop cover percentage, date of senescence and soil properties to be specified. Modelled evapotranspiration declines progressively relative to potential as the soil moisture deficit exceeds a critical value, the critical deficit being defined in terms of soil moisture characteristic and crop rooting depth. The soil properties assumed are given in Table 1.

TABLE 1. Site details. Soil available water contents are expressed as volumetric percentages for the range 0.05 to 15 kPa; values for easily available water (0.05 to 2 kPa) are in brackets.

Site	Years	Crop	Soil Texture	Available water %	
				Topsoil	Subsoil
Boxworth: Top Pavements and Knapwell fields	1995, 1996	W. Wheat	Clay loam	18(11)	16(10)
Gleadthorpe	1992	W. Barley	Sandy loam	17(11)	15(11)
			Loamy sand	15(10)	9 (6)

RESULTS

Boxworth sites. These two fields were mapped as having similar clay loam soils, and this was reflected in very similar soil moisture contents in early spring, as determined by neutron probe measurements at the 21 points. However water removed by the crop was greater throughout the subsoil, and continued to a greater depth, in Top Pavements than in Knapwell field, in both years. The deficit measured in July was significantly greater at Top Pavements than at Knapwell in both years (124 v 94 mm in 1995, $p < 0.01$; 122 v 111 mm in 1996, $p < 0.05$; subsoil deficit 102 v 77 mm in 1996, $p < 0.001$) (Figure 2).

The soil profile water available to a winter wheat crop is stated to be 115 mm for the Hanslope series and 125 mm for the Wickham series (Hodge *et al.*, 1984). The soils are classified as moderately and slightly droughty respectively for winter wheat at Boxworth. Thus it would be expected that moisture stress would be not uncommon at these sites, with perhaps slightly less risk of drought on the Knapwell field.

Irriguide model simulations for a clay loam soil for the two years indicated a maximum deficit for a crop with effective rooting depth of 1.20 m (reflecting depth of water extraction at Knapwell) of 113-114 mm in both years. For an effective rooting depth of 180 cm, reflecting observations at Top Pavements, the maximum calculated deficit was 156-158 mm. Deficits were well developed by May, causing modelled actual evapotranspiration to fall below potential from May onwards (Table 2). Unfortunately neutron probe measurements were not available for this intermediate period to allow comparison of measured with modelled deficit development.

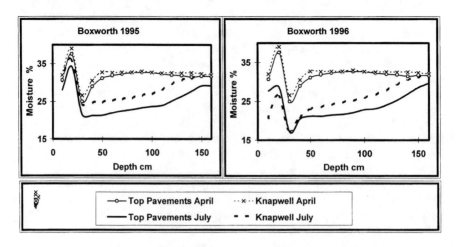

FIGURE 2. Soil moisture content measured by neutron probe at Boxworth, early spring and July, 1995 and 1996, showing greater depth and quantity of water extraction at Top Pavements than Knapwell field.

TABLE 2. Modelled crop water use: Boxworth and Gleadthorpe sites.

Site	Year	Soil	Modelled water use mm	
			May	June
Boxworth	1995	Clay loam	45	26
	1996	Clay loam	75	31
Gleadthorpe	1992	Loamy sand	43	58
		Sandy loam	80	67

The yields at Knapwell were less than those at Top Pavements, especially in 1995 (Figure 3). Yield (hand harvested) for 1995 for the 21 measurement points per field was found to be highly significantly correlated ($p < 0.01$) with soil moisture deficit measured in July ($r^2 = 0.4$), tiller numbers in spring ($r^2 = 0.64$), and dry matter at growth stage 39 ($r^2 = 0.74$). When data for the two sites for 1995 were plotted on the same axes, the results appeared to be part of the same distribution (Figure 3).

Yield was negatively correlated with take-all scores ($r^2 = 0.33$). Take-all would have had the effect of reducing rooting, and hence potentially causing drought stress in the summer. However the correlation was weaker than those with early growth. Calcium carbonate content of the topsoil, which varied from 0 to over 10%, could be an indictor of soil structural stability, which could affect seedbed conditions and root development. However the correlation between calcium carbonate content and yield was not significant. There was also no consistent link between yields in 1995 or 1996 and the yield zones defined for previous harvests.

The relative difference between the two fields persisted in 1996, but was less marked. The correlations with explanatory variables, although still present, were also weaker in 1996 (Figure 3).

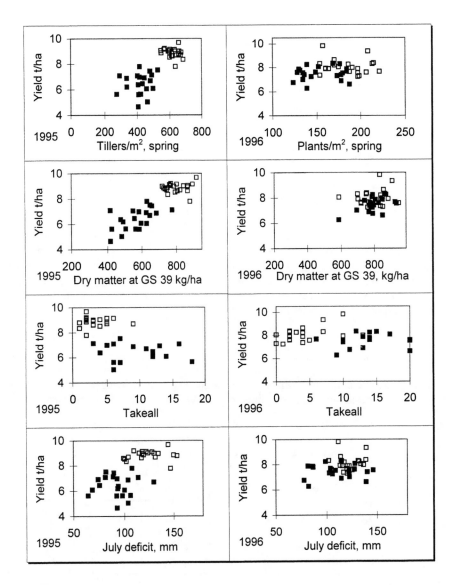

FIGURE 3. Yield of winter wheat at Boxworth (□ Top Pavements, ■ Knapwell field) as a function of early growth, take-all and water use.

Gleadthorpe site. This sandy field is on a farm where large responses to irrigation are common. In the summer of 1992, after heavy rains, severe lodging occurred in the areas identified as being of sandy loam texture (Figure 4), but not where the soil was a loamy sand. Despite the lodging, yields were greater in the area of sandy loam (5.1 t/ha (± 0.1) compared with 3.6 t/ha (± 0.2); p < 0.001). The greater water holding capacity of the sandy loam soil had supported greater crop growth and yield, and the lodging did not outweigh this yield advantage.

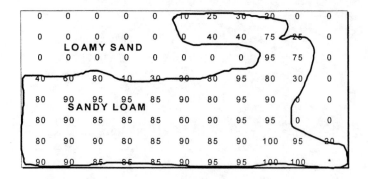

FIGURE 4. Map showing link between soil type and lodging (%): Gleadthorpe 1992.

Modelled actual evapotranspiration during the critical months of May and June was 49 mm greater on the sandy loam than the loamy sand soil (Figure 5 and Table 2), indicating considerable stress in the droughtier areas. The lodging suggests that the nitrogen inputs were in excess of the optimum for the more retentive soil, despite its greater yield potential relative to lighter parts of the field.

DISCUSSION

Both sites studied confirm a link between water use and yield which has been well documented by other authors (e.g., Green *et al.*, 1983). However the mechanism of that

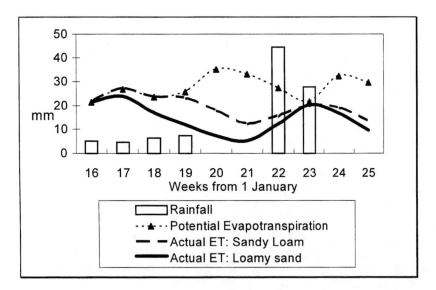

FIGURE 5. Modelled potential and actual evapotranspiration of a Winter Barley crop at Gleadthorpe in 1992, for a loamy sand and sandy loam soil. Severe moisture stress developed in early May, causing reduced yields on the droughtier soil.

link, and hence the appropriate management response, appears to differ. At Boxworth, measured soil physical properties and soil descriptions imply similar soil water supply for both fields and within the fields. Measured water use and yield were however consistently less within Knapwell field. Strong correlations were found between early growth and yield in 1995. These growth measurements were made before or in the early stages of any modelled crop water stress, and therefore imply some other limitation to growth. Neutron probe measurements showed equal soil water content across the sites in spring, which is consistent with a similar ability to supply moisture to the crop, yet moisture abstraction was smaller, and reached a shallower depth, in the Knapwell field in both years. It is possible that poor soil structure, take-all disease, or both contributed to reduced growth, restricting crop demand for water. Reduction in water use due to other limitations to growth, such as inadequate nutrient supply, was reported by Scott *et al.* (1994). It is equally possible that poor soil structure and take-all may have restricted the density and depth of rooting into the subsoil, thereby effectively reducing soil water available to the crop.

At Gleadthorpe, in contrast, variation in soil water supply appeared to be the direct cause of variation in yield. It is estimated that during May and June both crops suffered drought, but the crop on the sandy loam soil had access to almost 50 mm more water than on the loamy sand, resulting in an increase in yield of 1.6 t/ha. The severe lodging on the sandy loam soil suggested that nitrogen fertiliser inputs exceeded the optimum on this soil. This is consistent with MAFF fertiliser recommendations, which assume 20 kg/ha greater soil nitrogen supply on a loamy than a sandy soil (MAFF, 1994).

The site reported by Lark and Stafford (1996; 1997) provides interesting data on the effect of soil texture, since the same field contains both sands and clays. In the first four seasons (1991-4) yields on the clay soils were no better or slightly worse than on the sandy soils. In the fifth year (1995), the sandy soils yielded less. Furthermore, within the sandy areas, yields in 1995 were stated to be positively correlated with soil moisture content in early spring. We modelled crop water use for May and June of 1991 to 1995, using Irriguide, as 104, 133, 148, 92 and 40 mm respectively, compared with 101 mm for the Gleadthorpe field discussed above. This series of results suggests that in years of moderate to good water availability, clay soils have no advantage over sandy soils, and may indeed yield less. Reduced yields on clay soils may be caused by waterlogging restricting rooting, or more commonly perhaps by poor structural and seedbed conditions on these difficult soils affecting establishment and early growth, and these problems will often be greater in wet years.

One interesting feature of the modelled data is that, using published values for soil water availability (Hodge *et al.,* 1984), drought stress is predicted to be almost as great on clay as on sandy soils, with only 10-20 mm more water taken up during May and June on a clay soil than on a loamy medium sand. This is less than the year-to-year variation. Yet irrigation responses are generally deemed to be unlikely on clays (Tinker and Widdowson, 1982). Modelled water use at Gleadthorpe in May and June 1992 totalled 101 mm on the loamy sand soil, compared with 147 mm on the sandy loam. The equivalent value for the clay at Boxworth in 1995 was 66 mm, yet yields averaged about 8 t/ha grain compared with only 3.6 - 5 t/ha at Gleadthorpe. Soil moisture measurements show that total water abstraction from the clay was no greater than that modelled.

Weir and Barraclough (1986) demonstrated the critical importance of rate and depth of root development under drought conditions, which may be an important factor both on sandy soils and heavy clays, especially where wetness impedes early root growth. However it is possible that other factors, including the rapidity and severity of stress onset, may be important in explaining the differing responses to drought of crops on sandy and clay soils.

CONCLUSIONS

Increased yields were correlated with increased water use; but the latter was not always the cause of the former. Appropriate management response to variation in water uptake or yield requires an identification of the true limitation to yield and the degree to which it can be overcome. Management decisions will need to take into account also confounded factors, such as the effect of soil type on both yield potential and nutrient availability.

Modelled water use partly explained differences between sites and years in sensitivity to soil type. However the small differences between sands and clays in soil available water capacity did not fully explain the greater sensitivity to drought of the sandy soils. Comparison of the model predictions of deficit development with direct measurement, and with measurements of crop growth, should allow some distinction to be made between causes, and timing of onset, of reduced water use. Such detailed studies may also help us to understand more clearly how the timing and severity of drought stress interact with the building of yield potential, on different soils.

ACKNOWLEDGEMENTS

Financial support from the Ministry of Agriculture, Fisheries and Food; and from Massey Ferguson UK; is gratefully acknowledged. The authors are grateful to many colleagues, particularly James Clarke, for their part in managing and carrying out the scientific work which is discussed here.

REFERENCES

Bailey, R., Spackman, E. (1996) A model for estimating soil moisture changes as an aid to irrigation scheduling and crop water-use studies. 1. Operational details and description. *Soil Use and Management,* **12(3),** 122-129.

Bailey, R., Groves, S.J., Spackman, E. (1996) A model for estimating soil moisture changes as an aid to irrigation scheduling and crop water-use studies. II Field test of the model. *Soil Use and Management,* **12(3),** 129-134.

Clarke, J.H., Froment, M.A., Dampney, P.M.R., Goodlass, G., Stafford, J.V., Lark, R.M. (1996) An investigation into the relationship between yield maps, soil variation and crop development in the UK. *Third International Conference on Precision Agriculture,* Minnesota, USA.

Gales, K. (1983) Yield variation of wheat and barley in Britain in relation to crop growth and soil conditions - a review. *Journal of the Science of Food and Agriculture,* **34**, 1085-1104.

Green, C.F., Vaidyanathan, L.V., Hough, M.N. (1983) An analysis of the relationship between potential evapotranspiration and dry-matter accumulation for winter wheat. *Journal of Agricultural Science, Cambridge,* **100**, 351-358.

Hodge, C.A.H., Burton, R.G.O., Corbett, W.M., Evans, R., Seale, R.S. (1984) *Soils and their use in Eastern England.* Harpenden, Soil Survey of England and Wales, Bulletin 13.

Lark, R.M., Stafford, J.V. (1996) Consistency and changes in spatial variability of crop yield over successive seasons: methods of data analysis. *Third International Conference on Precision Agriculture*, Minnesota, USA.

Lark, R.M., Stafford, J.V. (1997) Classification as a first step in the interpretation of temporal and spatial variation of crop yield. *Ann. App. Biol.,* **130,** (in press).

MAFF (1994) *Fertiliser recommendations for arable crops.* Reference Book 209, London, HMSO.

Scott, R.K., Jaggard, K.W., Sylvester-Bradley, R. (1994) Resource capture by arable crops. *Resource Capture by Crops,* J.L. Monteith, R.K. Scott, and M.H. Unsworth, (Eds), Nottingham, Nottingham University Press, 469 pp.

Tinker, P.B., Widdowson, F.V. (1982) Maximising wheat yields and some causes of yield variation. *Proceedings no. 211*, London, The Fertiliser Society.

Weir, A.H., Barraclough, P.B. (1986) The effect of drought on the root growth of winter wheat and on its water uptake from a deep loam. *Soil Use and Management,* **2(3),** 91-97.

MAPPING OF POTATO YIELD AND QUALITY

S.M. SCHNEIDER, R.A. BOYDSTON

U.S. Department of Agriculture, Agricultural Research Service, Systems Research Group, Prosser, WA, 99350, USA

S. HAN, R.G. EVANS

Irrigated Agriculture Research and Extension Center, Biological Systems Engineering, Washington State University, Prosser, WA, 99350, USA

R.H. CAMPBELL

HarvestMaster, Inc., Logan, UT, 84321, USA

ABSTRACT

Precision crop management requires knowledge of the spatial variation within and among fields. Crop yield and quality are the end result of the integration of many processes over the entire growing season. The most valuable function of yield mapping is to indicate the variability of yields across fields for further analyses. Potato yields were mapped in commercial irrigated fields in 1995, 1996, and are planned for 1997. In 1996 several other field characteristics were also mapped. Soils were sampled for pre-plant potassium and phosphorus. Weed populations were mapped just prior to harvest. Field topography was mapped for all fields. Tuber quality samples were taken at harvest and evaluated according to processor standards. All data are being analyzed to determine which spatially variable factors significantly contributed toward potato yield and quality and therefore are the most critical to precision management.

INTRODUCTION

Crop yield and quality are the end results of the integration of all factors that affect crop growth and development over the entire growing season. Maps of yield and quality are indicators of the spatial variability present in the field and can be used as starting points for site-specific management decisions. The overall goal of precision management is to reduce the effect of spatial variability, resulting in optimum yield and uniform quality. Areas of reduced yield or quality are evaluated to determine what factors limit crop performance and how to manage these areas in the future to optimize crop performance. In cases such as rock outcrops, the limitations cannot be practically or economically removed, and unnecessary inputs should be eliminated.

The Site-Specific Management Research Team in Prosser, WA, USA is comprised of U.S. Department of Agriculture and Washington State University scientists, commercial growers, processors, variable rate fertilizer applicators, and a yield monitor manufacturer. The scientific team is multidisciplinary, including expertise in weed science, irrigation, computerized controls, GIS and GPS, micrometeorology, remote sensing, crop modeling, soil fertility, plant pathology, and systems engineering. By working as a coordinated team with industry partners, the contribution of each component to the overall picture can be explored, within the context of the whole picture, rather than in isolation. Research was conducted on one commercial farm in 1995, two commercial farms in 1996, and at a USDA research site in 1994 and 1995. Work is planned for a commercial farm and the USDA research site in 1997. This paper will present data from one of the 1996 commercial farms.

METHODS AND MATERIALS

Site description and data collection

A 53.9 ha field equipped with a center pivot irrigation system in eastern Washington was selected for study in 1996. Soil texture in the field ranged from a fine, sandy loam to a sandy loam. Average annual rainfall is less than 250 mm. Pre-plant soil samples were taken in January 1996 on a 61 m x 61 m grid to a depth of 30 cm for a total of 145 samples. Samples were analyzed by a commercial soil test lab for phosphorus and potassium and the results used to create residual potassium and phosphorus maps using an inverse distance weighting algorithm. A commercial processor specified the desired level for P and K. Using the maps of soil test P and K values, and the specifications of the processor, the variable rate applicator developed application maps for P and K fertilizer with the goal of uniform availability of P and K.

A short season potato variety, 'Shepody', was planted in March 1996. Water and nitrogen were applied through the irrigation system in accordance with standard grower practice. A late spring frost injured part of the field, especially those plants in the low areas. Little disease or insect pressure was observed during the season, although minor occurrences of *Sclerotinia* were present. The four primary weeds present in the field were common lambsquarter (*Chenopodium album*), prickly lettuce (*Lactuca serriola*), russian thistle (*Salsola iberica*), and nightshade (*Solanum sarrachoides*).

Mapping - weeds, yield, quality, and topography

Weed mapping. Weeds were mapped in two strips of the field. The day before harvest began, an all-terrain vehicle was driven through the field, and each plant of the four most prevalent weeds was recorded using a Trimble Pro/XL GPS with sub-meter accuracy and a hand-held computer. Ten passes, each two rows wide (1.7 m), were made through the field, resulting in a total mapped area of 1.36 ha.

Yield mapping. HarvestMaster HM-500 yield monitors were installed on two 4-row Lockwood potato diggers. Yield data were recorded at a 3 s time interval. Each record included a time stamp, GPS position (latitude, longitude, and elevation), potato weight, load cell values, belt speed, and digger identification number. The time required for a potato mass to move from where it was dug to where it passed over the load cells, based on the total travel distance of the potato mass and belt speed, was determined. If the belt speed had been constant, this method was equivalent to using a fixed time delay; however, when the belt speed was variable, using the distance and the belt speed provided a better estimate of location. Using the time delay, the actual location where the potatoes were dug was determined and linked with the corresponding yield measurement. This process created a point data file with yields and corresponding positions. Additional details of the operation of the yield monitor are given in Schneider *et al.* (1996).

Each digger loaded into a truck with an approximate capacity of 14 t. The start and stop time for each filling of each truck, along with the truck identification were recorded. Each truck was then weighed with a portable field scale, emptied, then weighed again, to get the net weight of the potatoes. The net weights were used to calibrate the yield monitor. The weights recorded by the yield monitors were known to include some soil along with the potatoes. More of the soil was removed before the potatoes were loaded into the trucks. Comparison of the truck weights to the yield monitor values was used to estimate the amount of soil passing over the load cells. The yield data were then adjusted to remove the weight of the soil.

The final step in yield mapping was to convert the point yield measurement data into raster (cell by cell) format. Raster format is compatible with most GIS software and is easy to analyze. If there was sufficient area coverage by all of the points within a cell, the cell yield value was calculated as the average of all the point values within the cell. If there was insufficient area coverage, a non-parametric distance-weighting algorithm (Han *et al.*, 1993) was used to estimate the cell yield value. At normal digger speeds, the 3 s sampling interval resulted in points being approximately 3 m apart. If a small cell size is desired, the original point data can be discretized into smaller distance intervals and used in the yield calculations. A 6.1 m x 6.1 m (20 ft x 20 ft) cell size and a 0.3 m (1 ft) discretization interval were used for the maps presented in this paper.

Quality mapping. One quality sample, ca. 18 kg, was collected from each 14 t truck load after the truck was weighed. Although the exact growing location of the sample was not known, the swath containing the growing location was known. The quality samples were submitted to a commercial potato processing plant for standard quality analyses. Recorded data included total weight of the sample, number of tubers in the sample, washed weight, number of potatoes in each of four weight categories, specific gravity, average tuber size, number of potatoes that were misshapen, and additional measures of specific internal and external defects. To create the quality maps, the value for a given quality characteristic was assumed to uniformly represent the entire swath loaded into a given truck. A 0.3 m discretization interval was used to generate a point data file, which was then used to generate raster quality maps as described in the yield mapping section.

Topographic mapping. Two topographical features were considered: elevation and slope. Elevation data were measured from a DGPS system (4000RS[TM] and Pathfinder ProXL by Trimble Navigation Limited, Sunnyvale, CA) with sub-meter accuracy. The slope data were derived with a TIN (Triangulated Irregular Network) model from ARC/INFO GIS software (Environmental Systems Research Institute, Inc., Redlands, CA). Yield, elevation, and slope were converted into grids in ARC/INFO database, with a 6.1 m x 6.1 m cell size. The larger 61 m x 61 m cell size used for yield masked too much of the elevation and slope variability, necessitating the smaller grid size.

Statistical analysis. The SAS statistical software package (SAS, 1988) was used to calculate the mean, standard deviation, minimum and maximum values and to conduct correlation analyses between yield, soil P and K, and the quality characteristics. Yield measurement data were averaged over 145 cells, each 61 m x 61 m. The center of each yield cell was chosen based on the soil sampling locations. Specific gravity and tuber weight data were averaged over 125 values, one sample per 14 t truck. Phosphorus and potassium values were averaged over 145 points, one sample per 61 m x 61 m grid. Elevation and slope data were averaged over 12430 cells, each 6.1 m x 6.1 m. Weed data were averaged over 106 cells, each 24 m x 24 m.

Correlation analysis of the very large elevation and slope datasets resulted in little useful information. A modified iterative optimization clustering procedure in ARC/INFO was used to separate the elevation and slope cells into a user specified number of distinct unimodal groups. Elevation was divided into 5, 10, 15, and 20 classes and slope was divided into 5 and 10 classes. The correlation analyses were then conducted on the classes of yield, elevation, and slope.

RESULTS

Basic statistics for yield, specific gravity, tuber weight, residual P and K, elevation, and slope are given in Table 1. The yield map, presented as a percentage of the field average, shows a great deal of variability across the field (Figure 1). The white areas in the map represent areas of missing data. Statistical analysis showed no significant correlation between yield and soil test values for P and K. However, if only the low yielding cells (the first quantile with yield < 93.1%) were considered, a weak correlation between yield and P ($r = 0.266$, $P = 0.10$, $n = 38$) and a strong correlation between yield and K ($r = 0.325$, $P = 0.05$, $n = 38$) were discovered.

Spatial variability was present in all quality characteristics measured. Two maps, specific gravity (Figure 2) and average tuber weight (Figure 3), are shown. The striping

TABLE 1. Basic statistics for yield and field characteristics.

Variable	# Data Points	Mean	Std. Dev.	Min.	Max.
Yield (% field ave.)	145	100.0	10.1	75.4	129.6
Specific Gravity (g/cm^3)	125	1.082	0.004	1.069	1.097
Tuber Weight (g/tuber)	125	15.8	2.2	10.7	22.6
Phosphorus (mg/kg)	145	11.4	7.4	4.0	88.0
Potassium (mg/kg)	145	211.7	58.3	85.0	540.0
Elevation (m)	12430	243.0	9.6	219.2	260.3
Slope (%)	12430	4.4	3.8	0.0	38.3

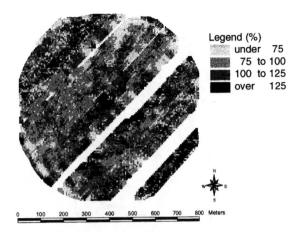

FIGURE 1. 1996 Potato yield map.

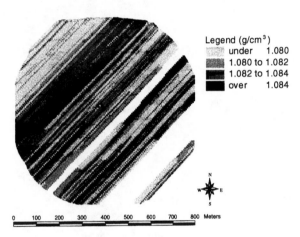

FIGURE 2. 1996 Potato quality map - specific gravity.

effect is due to having only one sample per truck load. Specific gravity was low in the far northwest and extreme southeast sections of the field and was higher midway between the far northwest edge and the center of the circle. Average tuber weight was higher in the far southeast and northwest corners and low midway between the far northwest edge and the center of the circle. Specific gravity was negatively correlated with average tuber size (r = -0.202, P = 0.026, n = 122) for this field in this year. Because the exact locations of the quality samples were unknown, correlation analysis between yield and quality characteristics was not performed.

The field topography was variable having more than a 35 m elevation change (Figure 4). Yield was found to be positively correlated with elevation (P = 0.0001) at all four elevation class sizes (5, 10, 15, 20). Slope, when grouped into either 5 or 10 classes, was found to be negatively correlated with yield (P = 0.02 and 0.008

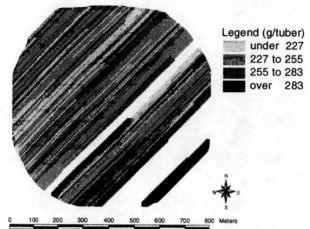

Legend (g/tuber)
under 227
227 to 255
255 to 283
over 283

0 100 200 300 400 500 600 700 800 Meters

FIGURE 3. 1996 Potato quality map - average tuber weight.

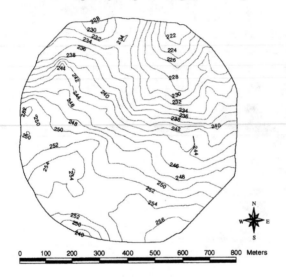

0 100 200 300 400 500 600 700 800 Meters

FIGURE 4. Field topography.

respectively). The combined effect of elevation and slope on yield was tested based on six class sizes for elevation/slope combinations (5, 10, 15, 20, 25, 30). Elevation was strongly positively correlated with yield (P = 0.0001 in all cases). Slope was slightly negatively correlated with yield. Results of stepwise multivariate regression analysis did not warrant inclusion of the slope variable into the model at P = 0.10 level.

Distribution patterns differed among the four weed species. Lambsquarter was fairly uniformly distributed over the area surveyed, with the exception of a few less dense patches. Prickly lettuce was densely distributed over several long streaks, but there were also areas with few or no occurrences of this weed. Prickly lettuce density tended to be inversely correlated with lambsquarter density. Russian thistle was more sparsely distributed than lambsquarter or prickly lettuce. Nightshade was present in one dense patch with several other sparse, isolated patches. There was a weak positive correlation between yield and total weed density ($r = 0.170$, $P = 0.08$, $n = 106$).

DISCUSSION

The results presented in this paper were taken from one field. The variability present in this field was greater than in some of the other fields in the study, due largely to greater variability in topography. Although specific correlation coefficients would be expected to differ between fields, the overall insights and conclusions were similar across the range of other fields studied.

Adequate supplies of P and K are necessary for proper growth and development of the crop. In this study, the lack of correlation between yield and soil test P and K values suggest that in most of the field, variable rate fertilizer application was successful and P and K were not yield limiting factors. However, when the low yielding areas were analyzed separately, the data suggest that low soil test P and K were contributing factors. Two of the lowest yielding cells were in areas where an error in the fertilizer application map resulted in fertilizer K not being applied, but the soil test K values indicated that the inherent soil K was insufficient to meet the crop need. These results illustrate that analyzing low and high yielding areas of a field separately, in addition to whole field analyses, can lead to knowledge that might be missed by whole field analyses alone.

Although the data suggest a positive correlation between weed populations and yield, this is an artifact. Visual observations during harvest documented more soil being carried onto the harvester and over the load cells in the areas of the field with higher weed densities, especially in areas with high soil moisture. The greater weight recorded by the yield monitor was attributed to soil, not potatoes. Modifications to the yield monitor will be made to address this problem.

The impact of elevation and slope on yield is related to drainage and microclimate factors including soil water and nutrient deficit or excess, early season soil warming, danger of frost, and favorability of conditions to disease, insect, and weed development. In this study the positive correlation between yield and elevation, indicated that conditions in low lying areas had a negative impact on yield in 1996. A late spring freeze caused greater injury in the draws. During harvest, more rotten tubers were encountered in the low areas of the field, possibly due to excess soil moisture. Site-specific application of water at critical times might reduce the negative impact of elevation and slope related factors. In a different situation, for example an irrigation system which applied too little water, the correlation between elevation and yield might be negative, i.e., low-lying areas would have greater yield due to less water deficit. Although the correlation between yield and elevation might change from positive to negative, the underlying principle did not; apply the optimal level of water for each part of the field.

High specific gravity and large tubers are both desirable characteristics. However, in this field in 1996, specific gravity and tuber weight are inversely related. This is not unreasonable in a short season cultivar, but would be less likely in a long season cultivar. Trade-offs between specific characteristics and between performance and cost will have to be made. The overall goals for a specific field, year, and market will have to be decided before management strategies can be selected. These goals could include yield and quality, economic goals, reduced pesticide use, and environmental concerns.

Once yield and quality goals have been set, management strategies to achieve those goals must be developed. Delivery systems for the site-specific application of inputs are commercially available or under development. Application of inputs such as P and K through ground-based systems is available from several commercial vendors. Computerized control of banks of sprinklers on self-propelled irrigation systems for the site-specific application of water and nitrogen has been developed and tested (Evans *et al.*, 1996). Computerized control of individual nozzles on secondary booms for site-specific application of pesticides is under development. Inputs can be delivered based on the spatial and temporal variation present in a field. The missing link is the prescription. What is the optimal, site-specific level of a given input, or combination of inputs, under a given set of site-specifically variable field conditions with specified production goals? The use of artificial intelligence methods such as case-based reasoning, coupled with statistical techniques should be considered (Hess and Hoskinson, 1996). Development of site-specific prescriptions is an area in which much research is still needed.

CONCLUSIONS

Technologies such as yield mapping have clearly illustrated the spatial variability present in agricultural fields. Mapping of field characteristics such as topography, soil nutrient status, pest status, and soil texture can help identify which combinations of factors are most likely to be contributing to especially high or low yield and quality characteristics. Some delivery systems for the application of site-specific inputs are already commercially available, while others are under development. Prescriptions which use the information in the field and yield maps, and the production objectives of the grower to specify the site-specific application of inputs are still incomplete. Management strategies must include consideration of the rotation crops grown between potato crops. Whole field research sites will have to rotate crops, adding another layer of variability to be characterized. Additional efforts are needed before the full potential of site-specific management is reached in potato production systems.

ACKNOWLEDGMENTS

The authors thank CENEX/LAND 'O LAKES, and Lamb-Weston, Inc. for their cooperation on this project and Doug Burt and Mike Mahan for technical assistance.

REFERENCES

Evans, R.G., Han, S., Schneider, S.M., Kroeger, M.W. (1996) Precision Center Pivot Irrigation for Efficient Use of Water and Nitrogen. *Proceedings of the 3rd International Conference on Precision Agriculture,* P.C. Robert (Ed.), Madison, WI, ASA, CSSA, SSSA.

Han, S., Goering, C.E., Cahn, M.D., Hummel, J.W. (1993) A robust method for estimating soil properties in unsampled cells. *Transactions of the ASAE,* **36(5),** 1363-1368.

Hess, J.R., Hoskinson, R.L. (1996) Methods for Characterization and Analysis of Spatial and Temporal Variability for Researching and Managing Integrated Farming Systems. *Proceedings of the 3rd International Conference on Precision Agriculture,* P.C. Robert (Ed.), Madison, WI, ASA, CSSA, SSSA.

Schneider, S.M., Han, S., Campbell, R.H., Evans, R.G., Rawlins, S.L. (1996) *Proceedings of the 3rd International Conference on Precision Agriculture,* P.C. Robert (Ed.), Madison, WI, ASA, CSSA, SSSA.

SAS (1988) *SAS Procedures Guide, Release 6.03.* Cary, NC, SAS Institute, Inc., 441 pp.

DYNAMIC WEIGHING FOR ACCURATE FERTILIZER APPLICATION

J. VAN BERGEIJK, D. GOENSE, L.G. VAN WILLIGENBURG, L. SPEELMAN

Wageningen Agricultural University, Department of Agricultural Engineering and Physics, Bomenweg 4, 6703 HD Wageningen, The Netherlands

ABSTRACT

The mass flow of fertilizer spreaders must be calibrated for different types of fertilizer. To obtain accurate fertilizer application, manual calibration of actual mass flow to the theoretical mass flow has to be repeated frequently. An automatic calibration is possible by estimation of the actual mass flow based on dynamic weighing of the spreader. In this research, a dynamic weighing system is designed and tested. This resulted in weight readings under field conditions that had a standard deviation of 20 N, over a measurement range of 6 kN to 20 kN, with a time delay through filtering of a maximum of 1 second. Time stamped data logging of spreader weight, theoretical application rate and position information, gives the opportunity to relate the realized spatial application of fertilizer to the prescribed application rate.

INTRODUCTION

Application of mineral fertilizer in a precision agriculture system depends on site specific crop requirements and is based on the crop-soil-weather interaction. To address spatial variability of available nutrients in the soil, three functional requirements for fertilizer spreading can be formulated:

a. the prescribed application rate varies for different locations in a field,
b. the prescribed application rate has to be realized through a correct setpoint mass flow,
c. the mass flow obtained has to be precisely distributed.

For granular fertilizers, the spreading technique should be able to vary the application rate at a scale corresponding to the variations in crop and soil. Application rate errors due to mass flow deviations in the spreader have to be minimized and the realized spatial application rate must be reported to the management system to analyze site specific crop response and to adjust future applications within a growing season.

Mass flow determination for fertilizer spreaders is needed to calibrate the flow control device for different types of fertilizer with different physical properties (Hofstee and Huisman, 1990). Due to the hygroscopic properties of fertilizer, the calibration for a single type of fertilizer might even change during spreading. The flow control device on a spreader is equipped with a sensor measuring either the opening state of a valve or the rotational speed of studded roller feeders, depending on the type of spreader used. The information from this sensor is called the theoretical mass flow, based on the calibration currently used for the flow control device. Conventional calibration is carried out by collection of a small sample of about 25 kg from the fertilizer spreader during a test run. This sample is weighed and compared with the theoretically applied amount and yields a new calibration valid for the current fertilizer conditions. To reduce application rate errors, several manufacturers have started to mount load cells in the spreaders to monitor the applied amount of fertilizer. The semi-automatic way of calibrating these type of spreaders is to weigh the loaded and unloaded spreader during stand stills on a level surface. The measured load difference must be equal to the amount of fertilizer theoretically spread. If a difference occurs the calibration is adjusted. This calibration is an improvement over the manual method because of the larger measurement range and because it is less labor and time consuming.

A next step is the continuous measurement of the spreader weight. This method will be referred to as 'dynamic weighing'. Dynamic weighing of the fertilizer spreader enables automatic calibration of the flow control device. In this paper, the design and results of a dynamic weighing system to automate flow control device calibration and to create a report of the amount of fertilizer applied will be discussed.

OBJECTIVES

The implementation of an automatic calibration algorithm for a flow control device requires an accurate and reliable weight measurement system. Disturbances due to uneven terrain, mechanical vibrations of the tractor and spreader, and the operation of the spreader on slopes have to be compensated for or suppressed. The first objective, therefore, is to design a dynamic weighing system suitable for operation under field conditions in agriculture.

The second objective is the automatic calibration of the mass flow control mechanism of a granular fertilizer spreader. Accurate measures of driving speed and effective working width are needed. In this research, the working width is specified by the type of spreader used and the driving speed is supplied by an integrated 'position-speed-heading' system on the tractor (van Bergeijk et al., 1996).

The third objective is to store both the theoretical application rate and the weight measurements during spreading to be able to reconstruct the amount of fertilizer applied.

LITERATURE REVIEW

Several methods for weight measurement of implements in the three point hitch already exist. Separate derricks or weighing frames used to weigh silage handling implements or livestock carriers are available for static weighing (Knechtges, 1988). These frames are mounted between the tractor three point hitch and the standard implement. Although the design is a versatile solution, the major drawbacks to its use for dynamic weighing are the displacement of the implement away from the tractor and the static weight reading restriction. More rigid design resulted in a derrick suitable for dynamic use which has been released on the market recently (Lely, 1993).

Integration of dynamic weighing in the tractor or implement itself offers a more compact solution. An overview of measuring forces in the three point hitch system is given by Auernhammer and Stanzel (1990) and by van den Heuvel and van Meeteren (1991). They conclude that implement weighing by measurement of forces in the three point hitch is technically feasible but in practice disturbed by different hitch geometries. On the implement, experiments with three extended octagonal sensors at the individual three points of the hitch resulted in weighing accuracies of ± 2% over a range from 6 kN to 20 kN (van Noort, 1995).

Under field conditions, compensation is needed for deviations due to vertical accelerations, and for vibrations due to engine, pto and implement shaft rotations and operation on slopes. Inclination sensors can be used to determine slope angle and vertical accelerations can be measured by an acceleration sensor. When separate information on inclination and vertical acceleration is not needed, a single sensor which measures a reference weight mounted parallel to the unknown weight (Böttinger, 1989) can be used.

MATERIALS AND METHODS

Spreader

A standard pneumatic spreader with a working width of 12 m was used in the experiments (Amazone Jet 1504). Three characteristics which make a pneumatic spreader suitable for precise application of fertilizer are the relatively small effective working width, the rapidly adjustable rate and the small application depth. The load of the spreader is approximately 6 kN and the maximum additional fertilizer load is 12 kN.

Sensor mountage

The original hitch studs of the spreader were replaced by three spring blades on a triangular subframe. Figure 1 illustrates the use of the spring blades, one at the top middle position of the subframe and two at the lower sides, to construct a reversed parallellogram. The subframe has mounting positions for two strain gauge force sensors. The vertical forces of the entire spreader act on the weight sensor while the horizontal forces run parallel through the spring blades. The reference sensor measures a known weight. The mounting orientation of the reference sensor is similar to the weight sensor to ensure that both sensors are affected by the same disturbances. Two strain gauge sensors, each with different dynamic behaviour, were tested as reference sensors. The first reference sensor was loaded with 1.776 N while the second sensor had a reference load of 38.40 N. Table 1 lists the specifications of the sensors used in the experiments.

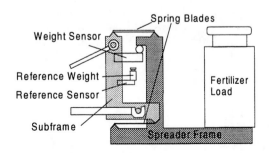

FIGURE 1. Sensor mounting positions.

TABLE 1. Strain gauge sensor specifications.

	reference sensor I	reference sensor II	weight sensor
type	Celtron LPS2	Celtron LOC50	Tedea Huntleigh 3510
	single point load cell	single point load cell	shear beam load cell
range	0 .. 20 N	0 .. 500 N	0 .. 50 kN
total error*	0.025 %	0.02 %	0.02 % of applied load

*according to manufacturer

Data acquisition

Two different data acquisition systems for the strain gauge sensors were examined. The first method consisted of a strain gauge amplifier connected through a low pass filter to an analog to digital conversion card in a personal computer system. Although the quality of the individual data acquisition components was high, an alternative was needed to minimize noise sources in the weighing system. The second data acquisition method therefore uses a single integrated circuit directly connected to the strain gauge sensor. This AD7715 integrated circuit combines an amplifier, a low pass filter and a sigma delta analog to digital converter in a single package. To connect the spreader weight sensor and the reference sensor to the data acquisition system two AD7715 integrated circuits were used. The digitized sensor readings were time stamped and stored on harddisk together with the studded roller speed readings for later analysis.

Experimental setup

Several characteristics of the strain gauge sensors were investigated. Before mounting the reference sensor on the spreader, resonance frequency, temperature influence, linearity and hysteresis were determined. For the weight sensor the temperature influence was measured before mounting. Linearity and hysteresis were tested by applying calibrated loads on the spreader while it was fixed to the three point hitch of a tractor. Deviations due to spreader operation on slopes were simulated by adjustment of the top link of the three point hitch.

After the static calibration series, dynamic measurements with various loads in the spreader were carried out, consisting of measurements while driving with a known load on a road with traffic obstacles and acquired during fertilizer spreading on a 6 ha field. The results were used to implement a filter algorithm to estimate the spreader weight. Data from two field trials where the spreader was used to spatially vary the fertilizer application rate were available to check the conventional calibration of the studded roller feeder and to design an automatic calibration routine. In the field trials, a calcium ammonium nitrate 27% fertilizer was applied.

RESULTS

Sensor analysis

The static weight sensor response was determined by applying calibrated loads to the sensor over a range similar to the maximum fertilizer load of the spreader. Table 2 lists the ranges and characteristics of the acquired responses of the different sensors and data acquisition methods. Both reference sensors were linear over the specified range. The weight sensor, built into the spreader, was slightly nonlinear. The standard deviation of the residues of this sensor decreased significantly with a second order polynomial fit. Static weighing accuracy also improved slightly with the replacement of the strain gauge amplifier with separate A/D conversion by an integrated circuit data acquisition method.

Sensor hysteresis was not detected by the data acquisition equipment. Comparison of the response of the weight sensor and reference sensor on different slopes showed small errors in the sensor mounting positions. The weight sensor was mounted at -0.29° away from the horizontal plane. Results from a similar cosine fit through the signal of the reference sensor resulted in a -0.75° offset. The temperature influence on the weight sensor over a range from 5 to 25 °C was negligible. According to the specifications of this sensor, a temperature range from -10 to 45 °C was compensated for correctly. Reference sensor I was slightly influenced by temperature fluctuations according to the specified 0.06% over 20 °C.

TABLE 2. Residue characteristics of sensor responses from static load/unload experiments.

	reference sensor I	reference sensor II	weight sensor	weight sensor	weight sensor
data acquisition	strain gauge amplifier, low pass filter and A/D card				AD7715 i.c.
regression method	------------------ linear ---------------------			- 2nd order polynomial -	
r^2	0.9999983	0.99999993	0.9999773	0.9999955	0.9999978
load range	0 - 5 N	0 - 100 N	6 - 16.50 kN	6 - 16.50 kN	6 - 16.50 kN
standard deviation	2.45×10^{-3} N	7.95×10^{-3} N	11.7 N	5.69 N	4.42 N
minimum	-4.12×10^{-3} N	-13.7×10^{-3} N	-30.3 N	-17.7 N	-10.9 N
maximum	3.92×10^{-3} N	11.8×10^{-3} N	26.1 N	22.0 N	9.32 N

An important aspect of the reference sensors were the resonance characteristics. According to previous research (van den Heuvel and van Meeteren, 1991), particularly low frequency (< 2Hz) vibrations during driving affect the weight readings. The typical resonant frequency for the chosen combination of sensor material and reference load must therefore be at least a decade above these low frequencies.

Dynamic weighing

Three different combinations of sensors and data acquisition systems were investigated with respect to the dynamic weighing performance. Data were acquired on roads with traffic obstacles at 3 m/s driving velocity. For series A, reference sensor I was used in combination with the strain gauge amplifier and PC A/D card. Series B used the same data acquisition method with reference sensor II. In series C the data acquisition method was replaced by the AD7715 integrated circuits while the sensors were identical to the setup in series B. Table 3 summarises the experimental setup and lists the measured weight characteristics of the three series. To evaluate the compensation algorithm, data series with comparable spreader load and disturbance magnitude were analysed. The weight sensor signal row lists the characteristics of the dynamic disturbances. The last row of Table 3 lists the weight characteristics after compensation and filtering. The filter was a second order low-pass butterworth filter with 0.5 Hz cutoff frequency.

The power spectral density graph of reference sensor I had a resonance peak near 80 Hz. Although compensation of the weight signal is only needed for lower frequencies and the signal is low pass filtered at 0.5 Hz (-3 dB) the excitations of reference sensor I due to this resonance led to limited compensation capacity. The power spectral density graph of reference sensor II showed a similar peak near 40 Hz but required more energy to excite. As a result, the compensation based on reference sensor II in series B and C reduced the standard deviation of the weight readings to 19.6 N (Table 3). The specifications of the low pass filter in the compensation algorithm had to match a required update rate of weight readings and maximum allowed standard deviation. Increase of weight accuracy decreases the response time of the weight determination.

TABLE 3. Dynamic weighing system performance (N) of three experiments.

		series A	series B	series C
data acquisition method		separate strain gauge amplifier, low pass filters and A/D card		AD7715 sigma delta A/D convertors
reference sensor (load)		I (1.776 N)	------------ II (38.40 N) ------------	
total load on weight sensor		9823.7	11312.9	9827.7
weight	standard deviation	948.6	1090.9	1063.4
sensor signal	minimum	5971.4	6826.8	6593.3
	maximum	12545.0	15378.2	12921.7
compensated	standard deviation	168.7	18.6	19.6
and filtered	minimum	9314.6	11241.3	9811.0
signals	maximum	10063.1	11396.3	9903.2

Automatic calibration

The fertilizer flow control algorithm of a single section fertilizer spreader is graphically represented by a data flow diagram (Figure 2). A conventional fertilizer spreader with a velocity dependent fertilizer flow incorporates the processes 'Flow Control Device Setpoint Calculation' and 'Closed Loop Flow Control Device Controller'. To enable automatic update of the flow control device calibration, the processes 'Dynamic Weighing' and 'Automatic Calibration' are added. The process 'Dynamic Weighing' produces a weight figure with accuracy according to the results of the previous section. A method for implementing the process 'Automatic Calibration' will be discussed in this section.

The theoretical mass flow is calculated from the data flow 'flow control device actual value' and 'spreader calibration'. Integration of theoretical mass flow over time yields the theoretical applied amount of fertilizer which can be compared with the measured weight decrease over the same time period. Although this procedure is straightforward, any modification of the current calibration has to be justified by certain criteria. A correct initial calibration should not be altered due to less accurate weighing results while an incorrect calibration should be detected and modified at an early stage.

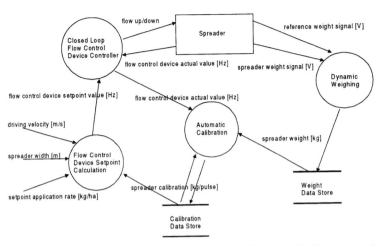

FIGURE 2. Data flow diagram for automatic calibration of a flow control section.

A simulation study, based on the data obtained in the previous section, has been carried out to evaluate spreader calibration update algorithms. The assumptions were the use of a 12 m wide spreader, a velocity of approximately 2.8 m/s and an application rate of 300 kg/ha which resulted in a mass flow of 1 kg/s. The measured weight update frequency was reduced to 1 Hz to avoid autocorrelation in the weight signal. A linear regression model of theoretical weight decrease and measured weight decrease was analysed. The time window for linear regression was varied between 20 and 120 s. A correct calibration should produce a regression line according to equation 1 with coefficient $A_i = 0$ and coefficient $B_i = 1$. Calibration deviations were recognized when theoretical weight differed significantly from measured weight and/or when the B coefficient was significantly different from 1.

$$W_{m,i} = A_i + B_i * W_{t,i} \tag{1}$$

Where: $W_{m,i}$ = measured weight at time i
$W_{t,i}$ = theoretical weight at time i
A_i = linear regression offset at time i
B_i = linear regression gain at time i

Confidence intervals for coefficient B and for predicted weight at each time step are possible criteria to justify calibration adjustment. Figure 3 shows the confidence limit of coefficient B over linear regression windows ranging from 20 to 120 s in length at a mass flow rate of 1 kg/s. According to this figure, a calibration accuracy of 4.3% could be obtained with 99% reliability after the application of 60 kg of fertilizer (i.e., after 60 s application). Figure 4 shows the confidence interval of the predicted weight $W_{m,i}$. Given a certain calibration accuracy at the start of a fertilizer application, correction of the calibration started when the weight measurements reached a better accuracy than the initial calibration and continued to a chosen level of desired calibration accuracy.

Figure 5 shows the results of calibration updates from an algorithm which modified the calibration based on the accuracy of the B coefficient. Initial calibration accuracy was

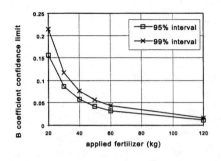

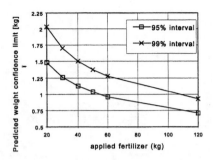

FIGURE 3. Reliability of B at 1 kg/s.   FIGURE 4. Reliability of pred. weight at 1 kg/s.

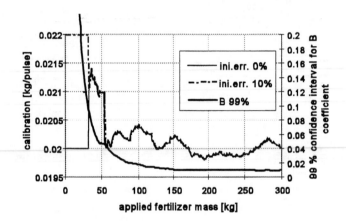

FIGURE 5. Calibration updates based on B reliability.

set to 10% and the calibration accuracy goal was 1%. The data set covered a fertilizer application range from 1000 kg down to 0 kg at a constant mass flow of 1 kg/s. The 99% confidence interval for regression coefficient B reached an accuracy of 10% (0.1 at the right hand axis in Figure 5) after application of 30 kg of fertilizer. The calibration, initially set at either the correct 0.02 kg/pulse or 0.022 (10% error) kg/pulse, was then modified. With increasing applied fertilizer mass, the calibration accuracy increased until 1% accuracy with 99% reliability for the B coefficient was reached.

During the application of the first 160 kg of fertilizer, the linear regression window was enlarged and comprised all weight samples available. The regression window was kept at the size needed to maintain the same accuracy after the accuracy goal was reached. Although accuracy might be improved by further enlargement of the regression window, a quick response to calibration changes is also needed. This implies a regression window which is as small as allowed by the accuracy criterion and which depends on the actual dynamic weighing accuracy.

DISCUSSION

The non-linearity encountered in the response of the main weight sensor might be caused by bending of the weighing derrick or by slightly non-parallel mounted spring plates. In both situations, a small portion of the vertical force is not measured by the weight sensor. The measurements showed an increasing error with load and a reduced sensor response.

Integration of the required data acquisition and signal processing into a single integrated circuit and digital signal processor was a move in the direction of practical implementation of the dynamic weighing system. In addition to a more compact implementation, accuracy improved slightly. One drawback of the use of an integrated data acquisition method was the required hardware and software for the interface, which had to be designed specifically for our experimental setup. Another drawback in the use of an AD7715 i.c. is reduced flexibility in sample rate settings and low pass filter methods compared to the capabilities of an analog data acquisition card mounted in a personal computer.

Based on dynamic weighing, several flow control device calibration update methods can be defined. The fertilizer mass flow is, in practice, not constant over time due to changing application rates and fluctuations in driving speed. Both measured weight and theoretical weight should therefore not be compared as time series but related directly to each other.

In addition to the confidence interval of the B coefficient, other criteria for calibration have to be checked. For predicted weight $W_{m,i}$ (equation 1), the confidence interval might indicate significant deviations between measured weight and theoretical weight. A reason for not using this criterion is that correction of the calibration relies entirely on the B coefficient. Even when predicted weight differs significantly from theoretical weight an inaccurate or less reliable B coefficient may not alter the calibration. Although both the B coefficient confidence interval and the predicted weight confidence interval are closely related, a criterion based on the B coefficient confidence interval could be used directly to modify the calibration.

CONCLUSIONS

Fertilizer application errors due to mass flow calibration deviations can be reduced by spreader weight measurement during fertilizer application. The use of integrated data acquisition method, specifically designed for load cell measurements, improved performance over a separate strain gauge amplifier and data acquisition card.

A second load cell with known weight to compensate for operation on slopes and for the dynamic weight fluctuations, equipped with a low pass filter to remove resonance frequency influences further reduced standard deviation of dynamic weighing.

The flow control device calibration can be monitored and eventually modified when accurate spreader weight measurements are available. Based on the dynamic weighing results, a calibration accuracy goal of 1% can be maintained under field conditions.

REFERENCES

Auernhammer, H., Stanzel, H. (1990) Wiegemöglichkeiten im schlepperdrheckkraftheber und in transportfahrzeugen, (Weighing opportunities in rear tractor hitch and in transport vehicles), *Landtechnik*, **10(43)**, 414-419.

Böttinger, S. (1989) Durchsatzbestimmung für eine Regelung der Mähdrescher-Reinigungsanlage, *Proceedings of the 11th international congress on agricultural engineering*, V.A. Dodd and P.M. Grace (Ed.), Balkema, Rotterdam, pp 1857-1862.

Hofstee, J.W., Huisman, W. (1990) Handling and spreading of fertilizers Part 1: Physical properties of fertilizer in relation to particle motion. *Journal of agricultural engineering research*, **47**, 213-234.

Knechtges, H. (1988) Behälterwägung im dreipunktanbau, (Implement weighing on three point hitch mounting). *Landtechnik*, **5(44)**, 218-219.

Lely (1993) European Patent Application 0537857A2, An implement for spreading material.

van Bergeijk, J., Goense D., Keesman, K.J., Speelman, L. (1996) Enhancement of global positioning system with dead reckoning, *Proceedings of EurAgEng*, Madrid 1996.

van den Heuvel, C.L.H., van Meeteren, C.M.A. (1991) Dynamische gewichtsmeting voor de bewerking kunstmeststrooien, (Dynamic weighing for fertilizer application), Msc thesis agricultural university Wageningen, Department of agricultural engineering and -physics.

van Noort, J. (1995) On the improvement of actual applied rates of fertilizer spreaders, Msc thesis agricultural university Wageningen, Department of agricultural engineering and -physics.

SITE SPECIFIC MANAGEMENT OF WATER AND CHEMICALS USING SELF-PROPELLED SPRINKLER IRRIGATION SYSTEMS[1]

H.R. DUKE, G.W. BUCHLEITER, D.F. HEERMANN

USDA-Agricultural Research Service, AERC-CSU, Fort Collins, Colorado 80523, USA

J.A. CHAPMAN

Valmont Industries, Inc., Valley, Nebraska 68064, USA

ABSTRACT

A linear-move sprinkler irrigation system has been equipped to apply predetermined, variable amounts of both water and chemicals along the pipeline and has been successfully used for irrigation for three seasons. Sprinkler heads are pulsed in a one-minute cycle to reduce the application of water where needed. A second low volume, pulsed application system, independent of the irrigation water supply, has been developed to apply chemicals over the canopy at rates ranging from 35 to 1800 l/ha. Base station software for a commercial radio telemetry control system has been modified to allow the irrigator to program the desired pattern of both water and chemical application to automatically implement the desired spatial application patterns. Future modifications will allow automatic interpretation of a GIS map into control commands to implement the desired variable application.

INTRODUCTION

Irrigated lands provide a disproportionate fraction of the world food supply. Aside from its importance in food production, irrigation of arid and semi-arid lands provides the opportunity for more precise management of water application than does rainfed agriculture. This potential for careful water management gives the irrigator an advantage in controlling the transport of chemicals and subsequent contamination of water supplies. Well designed self-propelled sprinkler irrigation systems provide additional advantages of substantially reducing the effect of soil characteristic variability on irrigation system performance, and of being inherently adaptable to automation. Center pivot and linear-move irrigation systems are widely adopted machines, usually dedicated to a single field, which can operate day or night under a wide range of weather conditions. The operator need not necessarily be present, if permitted by pesticide label, thereby minimizing the risk from exposure to chemicals.

Variability of soil properties, topography, plant density and growth stage, and precipitation patterns result in a lack of uniformity of water infiltration, storage, and subsequent plant water use across a field. It may be desirable to vary water application under self-propelled sprinkler systems to account for these factors (particularly in

---

[1]A contribution of USDA-Agricultural Research Service, Northern Plains Area, Fort Collins, CO and Valmont Industries, Inc., Valley, NE, USA.

regions where rainfall probabilities are significant), or to avoid application to unproductive zones such as rock outcrops or waterways. The ability to vary irrigation water application across a field may substantially reduce the potential for leaching chemicals into the water supply.

Perhaps even more significant than variable application of irrigation water is the ability to variably apply agricultural chemicals. Much research has been conducted on the areal variability of soil fertility. In irrigated areas, some effort has been expended on temporal nutrient application to minimize the amount of plant nutrients available for leaching or erosion, both during and between growing seasons. The ready availability of an irrigation-system-transported chemical application system capable of selective areal application will negate the need for broad scale prophylactic chemical applications in anticipation of a problem, and allow both spatially and temporally optimum application as determined by plant needs. As an example of the potential chemical reduction, studies conducted in the central USA (Wiles et al., 1992) showed that 71% of the field was free of broadleaf weeds and 94% free of grass weeds following pre-emergence herbicide application. Thus, the potential for reducing herbicide application by applying only when and where needed is quite large. Similar reduction in chemical application is anticipated for the range of pesticides used for agricultural production. Variability of plant nutrients is likewise well documented, and management of soluble nitrogen fertilizer applications is particularly promising for managing water quality.

VARIABLE IRRIGATION WATER APPLICATION

Fraisse et al. (1992; 1995) developed the concept of pulsing conventional irrigation spray heads to modulate the application depth as the machine moves across the field. Laboratory tests showed little distortion of the sprinkler pattern due to pulsing, with the pattern radius reduced by only about 5% and only for very short cycle periods (4 s). They showed in field tests, pulsing the 40 mm solenoid valves at 1 cycle per minute, that coefficients of uniformity exceeding 90 percent could be achieved, with the particular heads used, at virtually any solenoid duty cycle so long as the machine percent timer was set at 50 percent or less. Thus, the irrigation application for various segments can be controlled by first setting the percent timer to give the desired application depth on the most heavily irrigated segment, then varying the duty cycle of those segments to receive lesser applications such that the fraction of each minute that each segment is ON corresponds to the desired fraction of maximum application determined by the percent timer setting. The solenoid energization period is somewhat shorter than the percentages of each minute cited above to compensate for the fact that the solenoid closes more slowly than it opens (about 2.5 s to close).

A linear-move sprinkler irrigation system is located at Colorado State University's Agricultural Research, Development, and Education Center (ARDEC), about 10 km northeast of Fort Collins, CO, USA. The 177 m long, four-span system is divided into eight manifold segments, each 21 m long, with a 40 mm solenoid valve connecting between the mainline and manifold as shown in Figure 1 (Buchleiter et al.,1995; Duke et al., 1992; Duke et al., 1994). Each manifold is fitted with drop hoses on 1.5 m centers, pressure-relief check valves (to prevent drain down between pulses), and

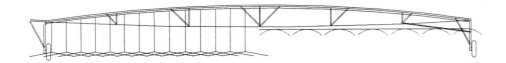

FIGURE 1. Manifold systems attached to sprinkler span for variable application of water and chemicals.

pressure-regulated (40 kPa) in-canopy spray heads (Senninger Quad-Spray[2]) mounted 60 cm above the soil surface. At that height, these heads have a wetted radius of about 3 m.

The machine has been equipped with a secondary computerized controller, operated by the irrigation system control panel, to pulse the 40 mm solenoid valves at a preselected duty cycle, thus pulsing the irrigation spray heads. This system has been used to apply predetermined variable water treatments to research plots at ARDEC during four seasons, with each solenoid valve having been operated for approximately 50,000 cycles without failure.

VARIABLE CHEMICAL APPLICATION

Self-propelled sprinklers provide a convenient method for application of chemicals, ranging from nitrogen fertilizer to a wide range of pesticides. Chemigation has been used for more than two decades as an effective technique to apply some chemicals. The necessity of applying irrigation water as a carrier for the chemical raises issues of whether the crop needs water, the possibility of backflow contamination of the water supply, and the need for registration of hazardous chemicals specifically for chemigation. The typical sprinkler irrigation system applies a minimum of about 2 mm water (about 20,000 l/ha), which creates special problems with foliar applied chemicals. This large water volume results in high dilution rates of the chemical, and in much of the applied water and chemical running from the crop canopy onto the soil. To avert the problem, immiscible oil carriers are often used as a means of attempting to preferentially hold the chemical on the intended target.

The chemical application system chosen here is adapted from a unique low volume system developed for orchard irrigation (Intertec Corp, Lynchburg, VA). The original system is constructed from extruded thin-wall polyethylene tubing (9 mm) which connects molded PVC spray heads. The original chemical system operated by expansion of the tubing which interconnected discharge heads in series. When water under pressure was introduced, a kevlar disc covered the discharge port, allowing water to pressurize the downstream tubing.

---

[2]Brand names are mentioned for the convenience of the reader and do not imply endorsement by USDA-ARS or Valmont Industries.

Under pressure, the thin-wall tubing expanded and no water was discharged. When water pressure was relieved at the inlet by a specially designed automatically pulsing valve, the discharge valve sealed the inlet side of the first spray head, allowing depressurization of the downstream segment of tubing and discharge of the water recovered from contraction of that tubing. This discharge, in turn, released the pressure against the discharge valve of the next spray head in sequence, and discharge progressed downstream. Since discharge volume is controlled by the volume of expansion of the tubing rather than an orifice, the diameter of the sprinkler nozzle is irrelevant, and can be made sufficiently large that filtration of fine particles is not necessary. The discharge orifice of the Intertec heads is about 3.5 mm in diameter.

Due to the fact that the black tubing was exposed to air temperature fluctuations as well as solar radiation, its temperature changed widely during the day. The large change in modulus of elasticity of the low density polyethylene tubing resulting from this temperature change caused significant temperature sensitivity of the discharge volume. The diurnal fluctuation of discharge per pulse was on the order of two-fold, making it difficult to achieve uniform application of chemicals.

Therefore, we redesigned the original Intertec system to minimize expansion, and replaced the thin wall tubing with a more rigid translucent tubing (14.5 mm o.d., 2.4 mm wall). The discharge head with its kevlar disk, large discharge port, and normally-closed-when-pressurized characteristic was retained. We designed and fabricated a small hydraulic accumulator (patent pending by Valmont Industries), into which the discharge head is incorporated, to provide the elastic storage originally contained within the expanding tubing. This accumulator has an adjustable displacement piston, sealed with a rolling diaphragm, which is forced by liquid pressure against a compression spring. Displacement of the piston is limited by a threaded piston stop and can be adjusted from zero to about 14 ml. Although there is still a small amount of expansion in the tubing when operated at 275 kPa, temperature sensitivity is greatly reduced. The adjustable displacement allows for increased discharge volume from heads within a manifolded segment to compensate for the larger area covered by heads at greater radius from the pivot. We used an electrically operated three-way solenoid valve to control pressurization of the spray heads and accumulators and to initiate discharge.

The new, like the original, system is designed to give zero discharge when pressurized, which gives it capabilities for a wide range of application rates. Each segment of the chemical application system is controlled independently, which allows varying the pulsing frequency to apply greater volumes from segments at larger radius from the pivot. This controlled pulse chemical spray system is capable of delivering volumes ranging from 35 l/ha to 1750 l/ha, depending on the frequency of pulsing and speed of the parent machine. These spray volumes compare favorably with the range of applications typical of low-volume aerial systems as well as high volume ground sprayers, and should overcome the problems of applying more chemical solution than the crop canopy can retain. As it is composed of injection molded plastic components, the chemical spray system is expected to be sufficiently low cost that it can be permanently attached to the irrigation machine.

Although most tests have been conducted with water, tests conducted with UAN (urea ammonium nitrate), a commonly used liquid nitrogen fertilizer, showed that application of this viscous, dense (specific gravity, 1.35) fluid was only about 20% less than that of water for a given set of operating conditions. The remainder of the results presented are for water.

The maximum application rate is limited by the minimum speed of the irrigation machine and the pulse frequency. Laboratory tests for a single segment string of eight heads spaced at 2.5 m showed that the system was fully pressurized (2.4 mm solenoid valve orifice) within about 3.5 s of valve closure. To assure pressurization, we assumed that the maximum pulse frequency should be 5 s. The minimum application depends on the maximum time between pulses that will give an acceptable coefficient of uniformity. Computer simulations showed that, assuming a percent timer setting of 30%, a controlled pulse head spacing of 2.5 m along the pipeline would give a uniformity along the pipeline of 97.6%. These simulated uniformities were confirmed by fluorometer evaluation of lengths of cotton string mounted both parallel and perpendicular to the path of the machine. Fluorescent dye (Rhodamine) was mixed with water, and the system pulsed as it passed over the strings. Assuming that pulsing the head once every 1.25 m of machine travel (heads have approximately 4 m pattern diameter) would give a satisfactory areal uniformity, then the range of application rates practical with this system is from about 35 l/ha to 1750 l/ha. These low spray volumes will allow chemical application with much less dilution than required for existing sprinkler-transported spray systems or by chemigation. In addition, pressurization can be provided with a very inexpensive, low volume pump fitted with a simple pressure relief valve, rather than the precision pumps required for chemical injection by conventional means.

CONTROL SYSTEM

The linear-move irrigation system is fitted with the C.A.M.S. (Valmont Industries' Computer Aided Management System) computerized control panel and the radio telemetry option. The system is monitored and controlled by a personal computer at the ARDEC headquarters. The C.A.M.S. panel has the capability to control sixteen output relays, of which seven are used for control of this machine (reversing relays, speed controls, etc.). Thus nine relays are available for additional control. These output relays are used to provide a binary input to a secondary controller, which in turn controls the eight 40 mm normally-closed solenoid valves on the water manifolds and eight segments of the chemical application system.

Five of the C.A.M.S. output relays are used as binary inputs to select one of 32 (2^5) irrigation sets, each of which consists of predetermined application percentages to each of the eight segments along the pipeline (Figure 1). Several methods are used to determine when a boundary between sets is reached, including passive antennae designed for buried utility locators, above-ground switch actuating markers, and inexpensive GPS units. On reaching the boundary between sets, the C.A.M.S. panel is instructed to set the output relays corresponding to the desired water pattern to apply. These relay closures are interpreted by the secondary controller (GE Fanuc Series 90 Programmable Logic Controller). The timing sequence of each of the valves is stored in secondary controller memory for each of the possible irrigation sets to be selected. This secondary controller operates self-contained relays to pulse the water control solenoid valves, opening the valve for the portion of each minute required to result in the predetermined fraction of the maximum application depth (as determined by the linear machine's percent timer setting). Four control ports from the C.A.M.S. panel give sixteen chemical application patterns which can be selected at any field position. Chemical controls are implemented by the secondary controller in a manner similar to water control, except that the length of the pulse is held constant (less than 500 ms pulse is required to activate the sequential discharge) and the pulse frequency is controlled to adjust the chemical application.

Figure 2 illustrates how a particular water application pattern is programmed into the base station computer. Each selected segment for a particular water application pattern is programmed for the desired percentage of each minute (pulse ratio) water is to flow. Selection of the *Chemical* button in the upper right side of the window opens a similar window to allow programming of the desired chemical application patterns.

Once the desired application patterns have been entered, the location and sequence of desired application patterns is selected from the *Program* window as shown in Figure 3. First, the field location is chosen at which a change of application pattern is desired, then the appropriate application pattern to start at that location is selected. For the fixed plot treatments used at the ARDEC study site, this method is quite satisfactory for entering desired changes in application patterns, but the procedure is obviously cumbersome when application patterns change frequently as would be the case if responding to real-time sensing of water needs or pest infestations. Future development of this base station software will concentrate on allowing the irrigator to develop, using GIS techniques to process available data, maps of desired application and to automatically interpret these maps into control commands which will be sent to the C.A.M.S. remote terminal unit when the machine reaches the appropriate position in the field.

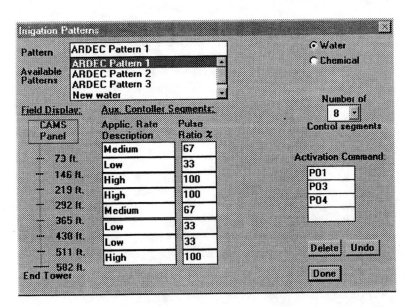

FIGURE 2. Window used for programming water application pattern along sprinkler.

278

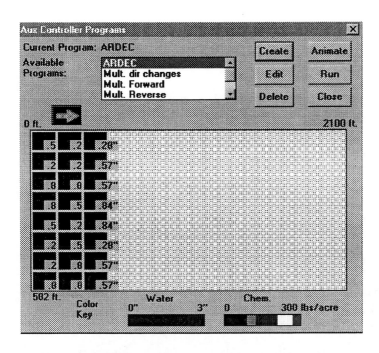

FIGURE 3. Window used for programming sequence of water and chemical application patterns to be applied as sprinkler moves across the field.

SUMMARY AND CONCLUSIONS

The pulsed irrigation system has been implemented at CSU's ARDEC using commercially available components. Field tests indicate that it will provide a very flexible means of applying variable water treatments while delivering acceptable uniformity of water application (CU > 90%). The variable water application system has operated successfully for four irrigation seasons.

The variable chemical application system has been thoroughly tested in the laboratory, and the chemical application system has been installed on the linear irrigation system. The controlled pulse chemical spray system can deliver water at rates comparable to both conventional and low-volume sprayers. Chemical solutions can be delivered with very high uniformity in a system that is easily controlled and potentially very inexpensive.

We will continue development of the chemical application system with the intent of providing a low cost system that can be attached to a self-propelled sprinkler to provide variable chemical application on demand. Software development will also continue to make it easier for the irrigator to convert real-time remotely sensed data into sprinkler control commands for automated implementation of spatially variable water and chemical application.

REFERENCES

Buchleiter, G.W., Heermann, D.F., Duke, H.R. (1995) Automation of variable irrigation water and chemical applications. *Proc. Clean Water-Clean Environment-21st Century. Vol. III: Practices, Systems, and Adoption,* St. Joseph, MI, ASAE, pp 49-52.

Duke, H.R., Heermann, D.F., Fraisse, C.W. (1992) Linear move irrigation system for fertilizer management research. *Proc. from the Int'l. Expo. and Tech. Conf.,* The Irrigation Assoc., pp 72-81.

Duke, H.R., Carnahan, B., Buchleiter, G.W., Heermann, D.F. (1994) Spatially variable sprinkler irrigation and chemical application. *ASAE Paper 94-2096,* St. Joseph, MI, ASAE, 11 pp.

Fraisse, C.W., Heermann, D.F., Duke, H.R. (1992) Modified linear move system for experimental water application. *Proc., Advances in Planning, Design, and Management of Irrigation Systems as Related to Sustainable Land Use,* Center for Irrig. Engr., K. U. Leuven, Leuven, Belgium, 14-17 September, pp 367-376.

Fraisse, C.W., Heermann, D.F., Duke, H.R. (1995) Simulation of variable water application with linear-move irrigation systems. *Trans. of the ASAE,* **38(5),** 1371-1376.

Wiles, L.J., Wilkerson, G.G., Gold, H.J., Coble, H.D. (1992) Modeling weed distribution for improved postemergence control decisions. *Weed Sci.,* **40,** 546-553.

DETERMINING APPROPRIATE SEED SOWING RATES

R.H. ELLIS

Department of Agriculture, The University of Reading, Earley Gate, P.O. Box 236, Reading RG6 6AT, UK

ABSTRACT

Seed sowing rate and seedling emergence and establishment affect plant population density, and so the yield of combinable crops. Yield-density equations enable target plant population densities to be estimated. Appropriate sowing rates can then be derived. Spatial variation in emergence within cereal crops can be large, but analyses combining this variation with the yield-density equations suggest that the direct effect of such variation on yield-sowing rate relations is negligible. Thus, altering sowing rate spatially in order to obtain a uniform plant population density may not be necessary, unless there are indirect reasons for such action (e.g. weed suppression, or because yield potential varies spatially).

INTRODUCTION

For the most widely-grown combinable crops in the UK, seed represents 20% (winter wheat) to 43% (spring oats) of the variable costs of production (Nix, 1994). Given such low values, for autumn-sown crops at least, and the risk averse tendency of many farmers to oversow in order to insure against crop emergence failure, it is not surprising that seed sowing rates have received less attention than fertilizer and spray application rates. I consider here a quantitative approach to determining appropriate sowing rates, and whether or not sowing rate should be deliberately varied within a field in order to reduce spatial variation in seedling emergence.

DISSERTATION

Sowing rates for combinable crops are usually specified as seed weight per unit area. Assuming row width is fixed, combinable crop growers require information on three factors in order to decide upon the seed sowing rate (s, g m^{-2}): mean seed weight (m, g), target plant population density (t, plants m^{-2}), and expected field emergence (e, fraction of seed population sown from which plants are established), where

$$s = mt / e \qquad (1)$$

Mean seed weight

Most growers of combinable crops sow (and purchase) seeds by weight. Since mean seed weight within a cultivar can vary by a factor of two among seed lots, the direct effect of m on s is considerable; m is determined during seed lot certification. Indirect effects of m on s as a result of variation in e with m, see for example Pieta Filho and Ellis (1991), are not considered here since, where detected, such indirect effects are generally negligible in comparison with the direct effect of *m* on *s*.

Target plant population density

The target is essentially that value of plant population density at which the grower believes economic returns will be maximised. How can this value be estimated? Historically, empirical approaches have tended to proliferate. An approach advocated here is to quantify first relations between yield and plant population density. Relations between yield and plant population density in the temperate cereals and in oilseed rape are asymptotic. This arises because increasing competition reduces yield per plant (w, g plant^{-1}), with a linear relation (Figure 1a, b) between the reciprocal of yield per plant ($1/w$)

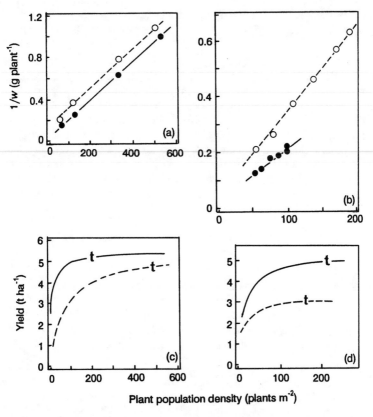

FIGURE 1. Linear relations between the reciprocal of seed yield (g d.m.) per plant ($1/w$) and plant population density (ρ): (a) spring barley (cv. Doublet) sown on 31 March (solid circles) and 21 April (open symbols) 1988 (from Ghassemi-Golezani, 1992); (b) winter oilseed rape (cv. Jet Neuf, data from Ogilvy, 1984) sown on 31 August 1979 (open symbols) and 25 August 1981 (solid symbols), and the resultant asymptotic relations (with target densities, t) between yield and density in these crops of barley (c) and oilseed rape (d).

and plant population density (ρ, plants m^{-2}), i.e.

$$1/w = a + b\rho \tag{2}$$

where a and b are the intercept and slope of this relation (Willey and Heath, 1969). Seed yield per unit area (Y, g m^{-2}) is the product of w and ρ, thus

$$Y = \rho/(a + b\rho) \tag{3}$$

Examples of the asymptotic relations described by Equation 3 are shown in Figure 1c, d. Note that in the earlier-sown barley crop high yields are obtained at comparatively low densities (Figure 1c), because at lower densities a higher proportion of tillers survive, are fertile, and so contribute to grain yield. Similarly, branching in oilseed rape results in high yields being obtained at comparatively low densities (Figure 1d).

If yield-density relations are quantified in this way, then it is possible to determine optimum plant population densities for different scenarios. One estimate of the economic optimum is that point on the asymptotic relation where the increase in the value of the crop equals the cost of extra seed. Figure 1c shows the effect of one agronomic variable, sowing date, on target plant population densities (t) in barley. In this example, e is assumed to be 0.8, m 0.050 g, and the cost of 1 kg of seed is 2.5 times the value of 1 kg grain. In this case the target for the later sowing (480 plants m^{-2}) is 2.4 times that (200) for the earlier (because the later the sowing, the greater the density required to ensure that crop canopy and root volume develop sufficiently rapidly in order to maximise resource capture). The relations for oilseed rape (Figure 1d) show the effect of another factor, inter-seasonal variability. Calculations similar to those described above for barley provide targets of 156 plants m^{-2} in 1979/80, but 215 in 1981/82. Thus differences between seasons affect potential yield, and so estimates of t.

Expected field emergence

Dead seeds cannot produce seedlings; neither can abnormal seedlings survive in the field. Accordingly, the viability of the seed lot (v, defined here as the fraction of the seed lot able to germinate normally in a standard laboratory germination test) will affect field emergence directly. Not every seed able to germinate normally in a laboratory germination test under optimum conditions will germinate, emerge and establish in the seedbed,

TABLE 1. Effect of variation in seed lot quality (estimated by seed lot viability, v) and seedbed conditions on the likely emergence (% of seeds sown) of spring wheat and winter oilseed rape at one site (From Khah et al., 1986; Ellis and Dolman, 1988).

v	Spring wheat		Winter oilseed rape	
	Poor seedbed	Good seedbed	Poor seedbed	Good seedbed
	Likely field emergence (%)			
0.85	29	67	51	70
0.90	37	75	61	78
0.95	52	85	74	87
0.98	67	93	85	94

however. Bleasdale (1984) used the term field factor (f) to describe the proportion of viable seeds from which seedlings will emerge and establish in the field. Thus

$$e = vf \qquad (4)$$

The field factor is affected by the seedbed environment, not just at sowing but also subsequently (i.e. until the crop is established). Moreover, it is also affected by the quality of the seed. Nevertheless, the combined effect of seedbed and seed lot factors on emergence can be quantified. For example, in one experiment with spring wheat (in one

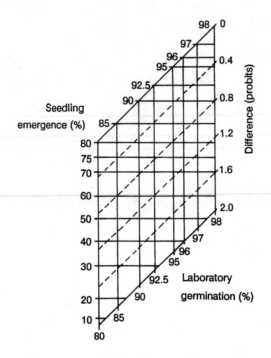

FIGURE 2. Nomogram to estimate likely field emergence of different seed lots in different conditions. Plotting historical results on this grid will show the difference in probits (diagonal lines) between laboratory germination (vertical axis) and seedling emergence (horizontal axis) for both the best and poorest seedbed conditions experienced locally. From the resultant pattern of responses, estimates of likely field emergence can then be obtained for each seed lot to be sown. For example, if the pattern of previous experience extends between the broken lines at 0.4 and 1.6 probits, then these values represent the best and poorest seedbeds (least and greatest stress for seeds) encountered. If the grower has a seed lot with 98% laboratory germination and expects seedbed conditions to be some of the poorest experienced (i.e. 1.6 probits), then emergence would be likely to be about 70%. Whereas, for a seedlot with the legal minimum value for certification of 85% laboratory germination less than 30% emergence would be likely from that same seedbed (From Wheeler and Ellis, 1992).

particular soil type)

$$E = K_i + 0.042 - 0.136M + 0.066T \tag{5}$$

where E is probit percentage field emergence, K_i is probit percentage viability (normal germination in a standard laboratory test), and M and T are mean soil moisture content (%) and temperature (°C) between sowing and emergence (Khah *et al.*, 1986).

Information on the normal germination of a seed lot is available from seed lot certification documents, while soil moisture contents and temperatures can be determined simply. However, this approach cannot be used directly to predict field emergence, because the seedbed environment can only be quantified retrospectively. Nevertheless, this approach has enabled a simple solution to the problem of determining the effects of seed lot and environment on emergence to be developed. Table 1 summarises seedbed environments as

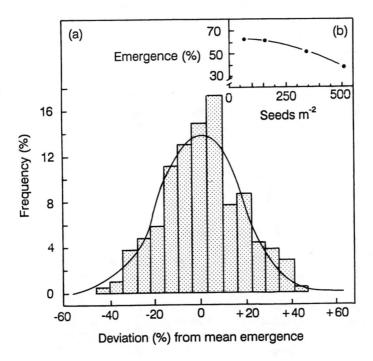

FIGURE 3. Histograms show the variation in emergence recorded in 1995/96 among 208 plots in an 11 ha winter wheat crop (cv. Estica) sown into a light sandy loam soil overlying gravel at Sonning Farm, The University of Reading (a). The solid curve in (a) is a normal frequency distribution with a standard deviation of 17%. The inset (b) shows the relation between establishment and the number of seeds sown (mean of the cvs. Axona, Baldus, Chablis, Tonic of spring wheat sown into a light sandy loam soil overlying gravel at Sonning Farm in 1996).

merely good or bad (i.e. the best and poorest seedbed conditions likely to be encountered), and seed quality by the results of the standard germination test. This guide is only useful for the site at which it was developed, but a nomogram (Figure 2) has been developed which enables growers to develop similar guidelines from their own results.

Variation in emergence within a field and its consequences

Figure 3a shows that the variation in establishment recorded (histograms) within one cereal crop (away from headlands) conforms to a normal frequency distribution (the solid bell-shaped curve). Is such variation of any practical significance? Given asymptotic yield-density relations (Figure 1), the consequences may depend on sowing rate. Moreover, sowing rate itself can also influence establishment; above a certain value increase in sowing rate may reduce establishment (Figure 3b).

The solid line in Figure 4 shows the relation between yield and sowing rate with no spatial variation in emergence (or yield potential) in the model, whereas the broken lines show these relations for several different scenarios in which spatial variation in emergence has been incorporated into the model. In all these scenarios, Equation 3 quantifies yield-density relations and the normal frequency distribution shown in Figure 3a quantifies variation in emergence within the field. The impact of the latter variation in emergence on yield is negligible (Figure 4). Similarly, Figure 4 shows that there is a negligible effect on yield-sowing rate relations of adjusting sowing rate spatially (assuming this could be done precisely using drills now being developed combined with an adequate prior knowledge of spatial variation in emergence) so that in those areas where emergence is lower than average (i.e. below the mean in Figure 3a) more seeds are sown in order to obtain the average plant population density, or on the other hand fewer seeds are sown in those areas where emergence is higher than average (i.e. above the mean in Figure 3a). In other words, although varying sowing rate spatially would affect yield the effect on the relation between these two variables would be small. The only scenario which alters yield-sowing rate relations substantially is when a decline in emergence at high sowing rates (quantified by the relation shown in Figure 3b) is included in the model.

There are good reasons for paying very close attention to the factors which affect mean emergence and target plant population density when determining sowing rate but, at least for the example reported here, the direct consequences for yield of spatial variation in emergence *per se* are unimportant. Nevertheless, there might be indirect reasons for deliberate spatial variation in sowing rate. For example, in order to suppress weeds, or because there may also be spatial variation in yield potential. But such possibilities need rigorous evaluation before investing in the technology to make deliberate variation in sowing rate. Thus, in conclusion, deliberate spatial variation in seed sowing rate is more likely to be justified by spatial variation in potential yield than by spatial variation in potential seedling emergence.

ACKNOWLEDGEMENTS

I thank M. Cooney, G.F.S. Dolman, K. Ghassemi-Golezani, S.A. Jones, E.M. Khah, C. Lye, M. Salahi, and T.R. Wheeler for their help.

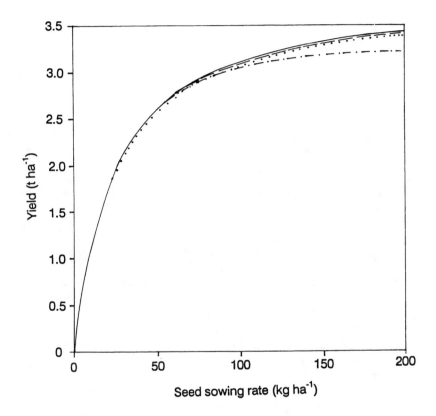

FIGURE 4. Relation between yield (d.m.) and sowing rate for spring wheat (cv. Timmo) sown on 26 April 1982 provided by Equation 3 where a = 0.1025 and b = 0.0026 (from Khah *et al.*, 1989), e = 0.65 and m = 0.04 g (solid line). Broken lines show the predicted effect of different scenarios for variation in emergence: (i) normal variation in emergence (about a mean of 65%) shown in Figure 3a (— — — —); (ii) this variation combined with the decline in emergence at high seed rates described in Figure 3b (– • – • – •); (iii) the normal variation in emergence (Figure 3a) combined with deliberate spatial variation in sowing rate so that it is increased in areas with less than the mean emergence in order to achieve the same plant population density as the mean (•••••••••); or (iv) sowing rate is reduced in areas with above mean emergence in order to achieve the same plant population density as the mean (— — — —, N.B. curve iv is coincident with i).

REFERENCES

Bleasdale, J.K.A. (1984) *Plant Physiology in Relation to Horticulture*, London: Macmillan, 143 pp.

Ellis, R.H., Dolman, G.F.S. (1988) The germination and emergence of seeds of winter oilseed rape stored and sown in a mixture with pelleted methiocarb. *Annals of Applied Biology*, **112**, 555-561.

Ghassemi-Golezani, K. (1992) Effects of seed quality on cereal yields. PhD Thesis, The University of Reading.

Khah, E.M., Ellis, R.H., Roberts, E.H. (1986) Effects of laboratory germination, soil temperature and moisture content on the emergence of spring wheat. *Journal of Agricultural Science*, Cambridge, **107**, 431-438.

Khah, E.M., Roberts, E.H., Ellis, R.H. (1989) The effects of seed ageing on the growth and yield of spring wheat at different plant-population densities. *Field Crops Research*, **20**, 175-190.

Nix, J. (1994) *Farm Management Pocketbook, 25th Edition*, Ashford: Wye College, 220 pp.

Ogilvy, S.E. (1984) The influence of seed rate on population, structure and yield of winter oilseed rape. *Aspects of Applied Biology*, **6**, 59-66.

Pieta Filho, C., Ellis, R.H. (1991) The development of seed quality in spring barley in four environments. II. Field emergence and seedling size. *Seed Science Research*, **1**, 179-185.

Wheeler, T.R., Ellis, R.H. (1992) Taking the guesswork out of seed sowing. *Grower*, August 27 1992, pp 16-17.

Willey, S.B., Heath, S.B. (1969) The quantitative relationships between plant population and crop yield. *Advances in Agronomy*, **21**, 281-321.

THE POTENTIAL FOR PRECISION FARMING IN PLANTATION AGRICULTURE

A. EMMOTT, J. HALL, R. MATTHEWS

School of Agriculture, Food and Environment, Cranfield University, Silsoe, Bedfordshire MK45 4DT, UK

ABSTRACT

Applying precision farming to plantation agriculture requires an understanding of differences between farming systems so that potential applications can be identified. Examining precision farming advances in arable crops such as wheat provides some guidance to applying precision farming in plantation agriculture. Care is needed, however, to identify appropriate technologies which complement the systems for which precision farming is being developed.

Precision farming aims to improve the spatial and temporal management of variable inputs in order to increase profit margins. By using gross margins, it is possible to identify which variable inputs to concentrate on. The analysis of variable inputs used by a farming system provides key indicators for precision farming developments. The level of technical improvement in input efficiency that can be achieved is influenced by spatial and temporal variables such as soils and climate. Crop models such as GUMTEA (Matthews and Stephens, 1995) are one way to establish whether economic input efficiency is achievable.

INTRODUCTION

Precision farming harnesses Global Positioning Systems (GPS) technology in combination with activities such as harvesting, fertiliser and agrochemical application. "Precision farming is a developing technology that modifies existing techniques and incorporates new ones" (Blackmore, 1994), in order to increase efficiency in the management of agriculture. Developing precision farming is as much about matching technologies to farming systems as it is about developing the right tools. For example, GPS has been used to provide detailed maps showing the spatial variability of yield in annual crops. Yield maps for plantation crops must also allow for the temporal nature of the harvest in order to show seasonal fluctuations and year on year changes. When the differences between farming systems have been taken into account, precision farming concepts, methodologies and technologies such as yield mapping, are likely to be applicable to plantation crops.

Nevertheless, high levels of control in precision farming require a sophisticated systems approach to managing information (Blackmore, 1994). A significant amount of computing and electronics equipment is required, some of which has been developed by organisations with little or no connection to agriculture. These technologies might be regarded as 'alien' to farmers who have limited exposure to them. For example, when discussing precision farming with plantation crop managers, the use of yield maps produced from a combine harvester fitted with a GPS has been used to illustrate the potential benefits of the technology. This approach often leads to negative comments such as, "precision farming is interesting but not relevant to my situation, you can't provide every tea plucker with a GPS". The extent, therefore, to which technologists take into account farmer's perceived needs and priorities will undoubtedly influence the successful transfer of new technologies such as precision farming (Kowalski *et al.*, 1996).

Successful transfer of precision farming technology between farming systems ultimately depends on whether the additional benefits exceed additional costs over time as calculated in a cost-benefit analysis. Before any detailed cost-benefit analysis is carried out, it is useful to identify specific areas where precision farming is likely to have the greatest impact on technical input efficiency. Gross margin analysis is an effective tool for providing an insight into areas which will improve profit margins. Areas identified then need to be assessed on their potential to improve economic efficiency. In order to improve economic efficiency, precision farming must ensure that the cost of application of variable inputs is exceeded by the additional revenue from the production and sale of the extra output.

PRECISION FARMING

The management of variability

Precision farming technology provides farmers with the opportunity to improve the management of spatial and temporal variability. Farmers are constantly refining their management of variability as demonstrated by changes in the distribution and timing of fertilisers and agrochemical applications. Precision farming has been developed to improve the management of information about variability, in order that more attention to detail can be used to influence decisions.

Farmers' decisions on agronomic practice take into account the costs, benefits and risks associated with those decisions. Agronomic practices are linked, therefore, to a wide range of variables such as:

1. Spatial variability: climate, infrastructure, soils, topography and yield distribution;
2. Temporal variability: climate, nutrient requirement, pests, diseases, weeds and yield distribution;
3. Projected variability: plans such as harvest interval based on crop and weather forecasts and physical variables such as soils.

Precision farming solutions designed to manage variables will differ between farming systems. The technology used will be influenced by the size of farms, their profitability and the scale of the problem to be addressed. Monitoring of variables will need to be modified to suit the management of those variables.

Annual and perennial crops are all influenced by soils, nutrition, water and climate, but the management of crop production is a good example of where different crops need different monitoring procedures. Plantation crops are harvested over extended periods during the year and over many years, while annual crops are harvested over short periods in the year and tend to be part of a rotation. Both systems require management information relating to spatial and temporal variability. An example of similar temporal information needs would be the way variations of nutrition requirement through a season can affect the final yield. The way the crop harvest is distributed through the season illustrates the difference between the two systems. Arable crops, with a single harvest, tend to be harvested mechanically allowing GPS to be used to identify spatial variability within fields. Digital hand-held weigh scales are available for use in plantation crops such as tea, which improves the collection of temporal yield data but falls short on the measurement of spatial variability. Identifying systems for the collection of information on variables, such as yield data, will be important for the development of precision farming in plantation crops.

Gross Margin Analysis

Gross margin analysis is a tool used to assess variable input efficiency between and within farming systems. The components of gross margins can, therefore, be used as pointers to where precision farming technology transfer is likely to be cost effective. The gross margin of an enterprise is its enterprise output less its variable costs. The enterprise or gross output is the total production of the enterprise multiplied by the price received for the produce, while variable inputs include cost items such as fertilisers, chemicals, and casual labour.

Gross margin maps place a financial value on variations within fields which can be enormous as shown in Figure 1. In order to improve gross margins, farmers seek the optimal use of inputs, such as fertilisers, in order to realise marginal improvements in gross outputs. Precision farming has focused attention on the relevance of this approach by demonstrating the extent and magnitude of variation within fields. Plantation crops managed over large areas might not initially use precision farming technologies to identify in-field variability. Gross margin comparisons between fields could be used to

Mercia winter wheat harvested - 13 August 1994
Mean yield - 8.18 t/ha
Mean gross margin - £548.29 /ha

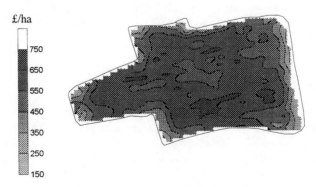

FIGURE 1. Gross margin map showing in-field gross margin variations (£/ha) for a winter wheat crop.

291

identify the areas where differences are greatest and point to where additional data are required. Precision farming technologies can then be developed to collect and manage that data.

Input Efficiency

If the transfer of precision farming technology is to be successful, farmers will expect to see improvements in farm incomes. This will come from a combination of extra revenue and cost savings resulting from improved input efficiency, as illustrated in Fig. 2.

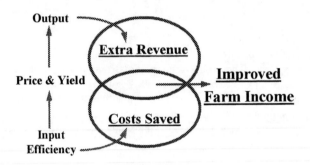

FIGURE 2. Flow diagram showing relationships between input efficiency and improved farm income.

Precision farming does offer the possibility of reducing input costs. Parkin and Blackmore (1995) suggest that considerable savings can result from selectively applying herbicides. These savings do, however, need to be balanced against changes in output or revenue because gross margins are usually more sensitive to change in output than to changes in individual inputs. An input such as fertiliser is only one of many factors affecting yield and as such has less influence on the gross margin than either price or yield. A simple comparison of gross margin components gives an indication of their relative importance. However the underlying relationships between these components need to be clearly understood. In the case of yield and fertiliser, yield response curves are used to describe the relationship between changes in fertiliser application and yield and to assess the efficiency of use within a farming system.

Gross Output

Taken that gross margins and farm incomes are particularly sensitive to changes in gross output, then focusing on factors which affect gross output should point to appropriate areas for precision farming development. Gross output is a combination of crop production (yield) and the price received for production for a given area.

Variables affecting yield and price include:
1. Yield (climate, soils, fertilisers, chemicals, variety and timing of cultivations, applications and harvest);
2. Price (product supply, demand, quality and time of sale).

The variables identified above fall into several categories. There are:
- Input variables such as chemicals and fertilisers;
- Environmental variables such as climate and soils;
- Variations in farming practice such as when to cultivate or harvest; and,
- Market variables affecting supply and demand.

The relative importance of each variable will change between crops. Precision farming can influence some but not all of the variables affecting yield and price. Gross margins are a useful guide to the relative importance of yield and price for a particular crop. Using a gross margin sensitivity analysis, distinctions can be made between the relative importance of price and yield for selected crops.

The matrix format of the sensitivity analysis (Agro Business Consultants, 1996) gives typical ranges for yields and prices against which gross margins are calculated. Gross margin calculations for dessert apples, as shown Table 1, are based on yields ranging from 6.0 to 16.5 t/ha and prices ranging from 285 to 365 £/ha. Gross margins shown in the matrix are the gross output (price x yield) less the variable costs. The resulting gross margins vary from 23 to 3065 £/ha for the complete matrix. Comparing the gross margins for a low price against a low yield illustrates the relative importance of price and yield which helps to identify variable inputs which are likely to have the greatest impact on gross output.

Fig. 3 shows typical ranges for price and yield in dessert apples as given in Table 1. These are then compared by showing the distribution around their mean values. For apples grown in the UK, yield appears to fluctuate more than price. Similarly, when replanting tea Emmott (1992) showed that return on capital (RoC) invested was more sensitive to changes in yield than to changes in price of the processed tea. The underlying reasons for these differences need to be understood in order to identify where precision farming may make the most significant impact on gross margins.

If replanting is a key issue in plantation crops, then identifying where yield improvements can be achieved in order to realise a satisfactory return on investment might be an appropriate area for precision farming development. Yield maps could be used to identify where variability exists. Additional information on soils and climatic variables could then be used in crop models to show the extent to which yields fall short of the potential and whether replanting is likely to achieve satisfactory levels of RoC.

TABLE 1. Gross margin (£/ha) sensitivity matrix for dessert apples (price x yield).

		Yield (t/ha)							
		6.0	7.5	9.0	10.5	12.0	13.5	15.0	16.5
Price	285	23	269	515	761	1007	1253	1499	1745
(£/t)	295	83	344	605	866	1127	1388	1649	1910
	305	143	419	695	971	1247	1523	1799	2075
	315	203	494	785	1076	1367	1658	1949	2240
	325	263	569	875	1181	1487	1793	2099	2405
	335	323	644	965	1286	1607	1928	2249	2570
	345	383	719	1055	1391	1727	2063	2399	2735
	355	443	794	1145	1496	1847	2198	2549	2900
	365	503	869	1235	1601	1967	2333	2699	3065

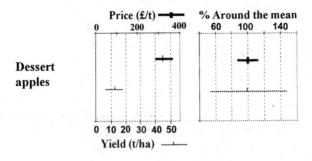

FIGURE 3. Comparison of price and yield variability in dessert apples.

Variable Costs

Since cost effective precision farming requires an understanding of technical and economic issues, then variable inputs used by different farming systems highlight different development needs. By comparing variable inputs between crops, such as wheat, apples, oil palm and tea (Figure 4), it is possible to see where precision farming has been developed and where there is potential for future applications.

Precision farming in wheat has targeted all of the three main variable costs. The variable cost breakdown suggests there is little scope to reduce fertiliser costs in apples, but pest and disease management offers more room for improvement.

Two of the main variable costs in oil palm are fertiliser application and harvesting. Fertiliser costs can account for 12-32% of operating costs and efficiency of use is paramount (Wood, 1986). Matching fertiliser applications to requirements is a complex issue particularly when extreme weather conditions can influence yields two years later. In Malaysia, where the distribution of rainfall is fairly regular, monthly crops can vary

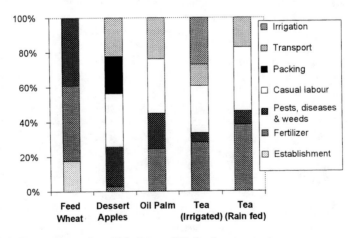

FIGURE 4. Comparison of variable inputs (%) for four crops.

from 5-12% of the annual mean. Where seasonal variations are more pronounced, yield variability is considerably greater. Precision farming applications for oil palm production should focus on variations in crop production which will lead to efficiency gains in harvesting and fertiliser applications.

In tea the main variable cost components are also fertiliser application and harvesting. However where tea is irrigated, as in Tanzania, then irrigation is a significant operating cost. Improving the economic efficiency of irrigation use should result in improved profit margins.

Fig. 5 shows the yield response curve to irrigating tea in Tanzania (Burgess, 1995). Burgess suggests that beyond 700 mm, additional irrigation gives no further yield unless irrigation application is not uniform. Spatial variability of irrigation might arise if the sprinkler system is arranged or operated incorrectly resulting in water being applied at different rates across the field. Burgess concludes that irrigation can substantially increase tea production in the absence of other constraints, but this increase can only be realised if the irrigation is properly planned and managed.

The economics of improving the technical efficiency of irrigating tea in Tanzania will depend on assessing the additional equipment required to achieve greater uniformity, and the extra running costs, against the likely improvements in output. Estimating yield improvements will need to take into account other variables such as the yield potential of estate tea compared to the research plots. The price range for made tea also needs to be built into any detailed cost benefit analysis. Crop models, such as GUMTEA (Matthews and Stephens, 1995), which predict yield potential and distribution in response to environmental and management factors, are likely to play a crucial role in understanding key variables, their variability and consequent precision farming developments.

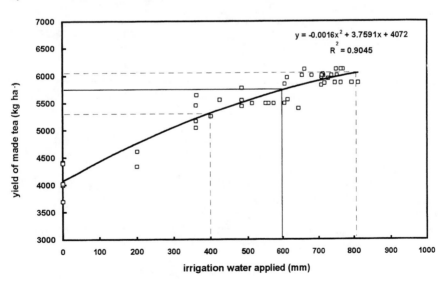

FIGURE 5. Yield response to irrigation in mature tea.

CONCLUSIONS

The use of current precision farming practice in arable crops can be used to provide an insight into possible applications for precision farming development in plantation crops. Precision farming technologies applied to plantation crops need to compliment the systems that they are being developed for. Using gross margin analysis will provide a focused approach to assessing potential areas for technology transfer. Improving the technical efficiency of inputs such as irrigation is likely to be feasible. The successful transfer of precision farming will depend ultimately on whether improvements can be made on the economic efficiency of production systems. In both tea and oil palm, harvesting and fertiliser application are likely to be key areas for developing precision farming.

ACKNOWLEDGMENTS

The authors are grateful to Massey Ferguson and Shuttleworth Farms for the use of their data and to Cranfield University staff for their support and guidance.

REFERENCES

Agro Business Consultants (1996) *The Agricultural Budgeting & Costing Book*, No. 42 May 1996.

Blackmore, B.S. (1994) Precision Farming -An Introduction. *Outlook on Agriculture*, **23(4)**, 275-280, CAB International.

Burgess, P.J. (1995) Evaluation of irrigation needs and benefits. *Proceedings of the First Regional Tea Research Seminar*, A.M. Whittle and F.R.B. Khumalo (Eds), Tea Research Foundation of Central Africa, pp 215-224.

Emmott, A.J. (1992) A Logistical and Financial Appraisal of Replanting Tea (Camellia Sinensis): A Comparison of Techniques. MSc in Plantation Management, Cranfield University, Silsoe College.

Kowalski, R., Velez, R., Lowe, J.C. (1996) Knowledge is Not Enough - the Need for Reflection in Technology Transfer, A Case Study. *Agricultural Progress*, **71**, 65-76.

Mathews, R., Stephens, W. (1995) GUMTEA: A Model Describing The Growth and Harvest of Tea. *Ngwazi Tea Research Unit Quarterly Report*, **19**, 24-26.

Parkin, C.S., Blackmore, B.S. (1995) A Precision Farming Approach to the Application of Agrochemicals, Cranfield University, Centre for Precision Farming.

Wood, B.J. (1986) A Brief Guide to Oil Palm Science, The Incorporated Society of Planters.

THE IMPACT OF VARIABLE RATE N APPLICATION ON N USE EFFICIENCY OF FURROW IRRIGATED MAIZE

G.W. HERGERT

University of Nebraska WCREC, RT. 4 Box 46A, North Platte, NE, 69101, USA

R.B. FERGUSON

University of Nebraska SCREC, Box 66, Clay Center, NE, 68933, USA

ABSTRACT

Variable rate application (VRA) of N may improve nitrogen use efficiency (NUE) and reduce nitrate leaching by minimizing high nitrate areas in the field. The objective of this research was to compare grain yield, NUE and residual soil nitrate between uniform and variably applied anhydrous ammonia for furrow irrigated maize (*Zea mays* L.). Treatments included anhydrous ammonia applied uniformly, VRA ammonia or VRA minus 15% replicated five times. VRA produced significantly more grain than uniform application at two locations in the second year. VRA minus 15% produced the lowest yields. Initial soil sampling showed highly variable soil nitrate at two of the three sites. VRA did not change soil nitrate distribution at the more uniform site which showed a normal distribution with a low coefficient of variation (CV). At a second site VRA minus 15% decreased residual nitrate significantly, but the CV was similar for the different application methods. At the third site, VRA significantly decreased variability in soil nitrate compared to uniform application and significantly decreased the CV and skewness of the soil nitrate frequency distribution. NUE was similar for all application methods. The research showed that VRA may reduce nitrate leaching potential by creating uniformly low nitrate concentrations throughout the field.

INTRODUCTION

Nitrogen fertilizer will continue to be used to sustain USA grain production (Mengel, 1990; Swanson, 1982), however, current environmental and economic considerations demand improved N use efficiency (NUE). Furrow irrigated soils in Nebraska, USA, river valleys have shown increasing nitrate-N levels since the mid 1950s (Engberg and Spalding, 1978) where land is cropped to continuous maize. These soils are generally shallow with coarse textured subsoils and groundwater 2 to 3 m below the soil surface. Nitrate levels have been reduced recently in parts of the central Platte River valley of Nebraska (Ferguson *et al.*, 1994) due in part to the adoption of Best Management Practices (BMPs) based on University of Nebraska research and education, monitoring and regulation from Natural Resource Districts (local tax-supported, elected governing board, government agency). To continue reducing ground water nitrate, improved practices will be needed for the next generation of BMPs.

Depending upon furrow length, the upper end of a furrow irrigated field is often over-irrigated and the lower end is under-irrigated. Root zone and intermediate vadose zone samples collected from the mid-Nebraska water quality demonstration project show a similar influence of furrow irrigation on accumulated nitrate (Ferguson *et al.*, 1992).

In Nebraska about half of the 1.7 million hectares of irrigated maize are grown under furrow irrigation. Almost 75% of the N fertilizer applied to maize is from anhydrous ammonia (1994-1995 Nebraska Agricultural Statistics, 1995). Since Nebraska is in the western part of the USA corn belt, rainfall is low and soil testing for residual nitrate is a commonly used and effective practice for improving N fertilizer management (Hergert, 1987). Recent research has shown that the spatial variability of nitrate-N in 43 Nebraska maize fields sampled on a 30 m grid spacing to 1.2 m had an average CV of 52% (Hergert et al., 1995b). As soil nitrate is one of the key components in the University of Nebraska N recommendation algorithm for maize, this degree of variability suggested that further increases in NUE could be attained if N fertilizer was variably applied. In this way N fertilizer application would more closely match N needs of the crop.

The overall objective of this project was to establish relationships among soil chemical and physical properties, irrigation parameters, grain yield, soil nitrate and NUE. The specific objective was to compare grain yield, NUE and soil nitrate between uniform versus variably applied anhydrous ammonia based on current N recommendations using expected yield, soil nitrate and organic matter on a spatially variable basis versus a whole field management (WFM) approach.

EXPERIMENTAL METHODS

Three VRA N experiments were initiated in 1994 on furrow irrigated sites: one in west central Nebraska (Lincoln county) on a Cozad silt loam (coarse, silty, mixed mesic Fluventic Haplustoll), one in south central Nebraska (Clay county) on a Hastings silt loam (fine, montmorillonitic, mesic Udic Argiustoll) and one in central Nebraska (Buffalo county) on an Alda loam (coarse, loamy, mixed, mesic Fluvaquentic Haplustoll).

Three N management regimes replicated five times were used: (1) a fixed uniform N rate based on an average expected yield goal for the field, average soil organic matter content, and the average soil nitrate of the field using the University of Nebraska N recommendations (Hergert et al., 1995a); (2) variable rate N based on an average expected goal, varying soil nitrate, and varying soil organic matter determined from grid sampling and (3) variable rate N calculated as in (2) minus 15%.

Treatment strips were one planter width (5 to 6 m wide depending on location) for the full field length. Field length was 280 m at the Lincoln county site, 503 m at the Clay county site and 312 m at the Buffalo county site. A transect of soil samples was taken along the middle of each 5 to 6 m wide treatment strip using a fixed spacing ranging from 20 to 30 m depending on the field. A regular rectangular grid was used at the Lincoln county site whereas an offset triangular grid was used at the other two sites. A single 5 cm core was taken at each sampling point to a 0.9 m depth. Samples were taken in the autumn of 1993 then yearly each autumn.

Grain was combine harvested from three rows 10 m long at the Lincoln county site and a subsample of grain was taken for N analysis to coincide with soil sample locations. The Clay and Buffalo county sites were harvested with an 8-row John Deere® combine equipped with an Ag Leader® yield monitor. Grain subsamples for N were not taken at these two sites.

Initial average soil test parameters based on samples for the three sites are given in Table 1.

TABLE 1. Soil analyses for the three sites.

Location	Texture	Samples	pH	Organic Matter g/kg	Bray 1 P mg/kg	DTPA-Zn mg/kg
Lincoln Co.	silt loam	225	8.0	15	20	2.0
Clay Co.	silt loam	255	6.1	27	15	1.0
Buffalo Co.	loam	158	8.1	19	24	0.9

Anhydrous ammonia was applied at the 6 to 8 leaf stage at all sites using an AgChem Falcon® controlled ammonia applicator. Applied N rates are shown in Table 2. N rates were calculated from the University of Nebraska N algorithm for maize (Hergert *et al.*, 1995a) which uses expected yield, soil nitrate and soil organic matter. For the uniform N rate, the above three parameters were averaged for the treatment strips. For the VRA N rates, spatial data was used to generate N values which were then kriged to develop a data set at twice the sampling density which was sent to AgChem for map development (an inverse distance squared interpolation method) which drove the applicator. For the VRA minus 15%, the computer was set to simply apply 15% less than was read from the electronic variable N application map.

The influence of variable rate N application on NUE was measured in several ways. The first was a comparison of the change in soil nitrate from spring to fall for the three treatments. The second was a comparison of apparent N use efficiency (ANUE) values. Data were available from all three sites for the first comparison but only from the Lincoln county site for ANUE. ANUE was calculated by a simple N balance technique for each soil sample grid and yield point at the Lincoln county site using equation (1):

$$\text{ANUE} = (\text{N uptake}_{grain}/(\text{N applied} + \text{NO}_3 \text{ previous fall} - \text{NO}_3 \text{ after harvest})) \times 100\% \quad (1)$$

where N uptake$_{grain}$ was total N in harvested grain and N applied was fertilizer N. The value of NO$_3$ previous fall is nitrate from the previous fall before spring planting. The NO$_3$ after harvest term is soil nitrate-N in a 0.9 m depth after grain harvest.

RESULTS AND DISCUSSION

The variability in residual nitrate in the 0-0.9 m depth was high, especially for the Lincoln and Buffalo county sites (Figure 1). The map shows interpolated values (Surfer®-inverse distance squared) based on all sampling points. The average N rates did not differ greatly between uniform and variable rate, however (Table 2). This is not

TABLE 2. N fertilizer application rates.

Treatment	Lincoln County 1994	1995	1996	Clay County 1994	1995	1996	Buffalo County 1994	1995	1996
	kg N/ha								
Uniform	165	270	200	165	160	170	110	130	160
Variable	160	250	190	170	165	170	120	135	180
Variable-15%	135	220	160	145	135	145	110	120	140

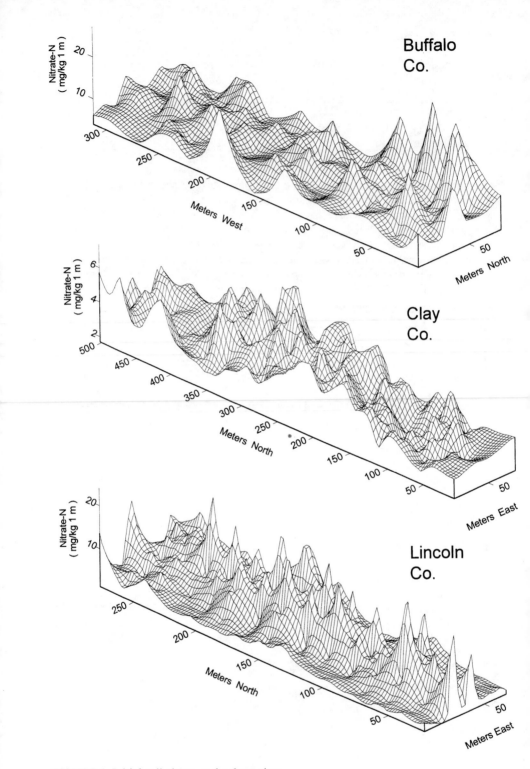

FIGURE 1. Initial soil nitrate at the three sites.

too surprising considering how the 'average' N rate for a treatment strip is calculated. The mathematics of simple averaging explain much of the similarity. The lower N applied to the variable rate minus 15% was apparent.

There was no consistent significant difference in grain yield between uniform and variable rate N application (Table 3). The variable N rate minus 15% did produce significantly lower yields than VRA, however, showing that current N recommendations were closely matched to N requirements of the yields produced.

The change in residual soil nitrate as influenced by treatments showed an encouraging pattern (Table 4). In 1994 at the Lincoln county site, both variable rate treatments produced significantly lower nitrate than the uniform N treatment. During 1995 N was over-applied due to a malfunction of the radar gun on the applicator. The over-application was 60 kg ha^{-1} for uniform application, 44 kg ha^{-1} for variable and 36 kg ha^{-1} for variable minus 15% application. The higher rates and lower yields were reflected in increased soil nitrate. Following high yields in 1996, soil nitrate levels again decreased to levels near 4.5 mg kg^{-1} (Table 4).

At the Clay county site there was little change in soil nitrate in either year, but since the initial nitrate level was low (Table 5), the treatments reflect the influence that proper N management can have on keeping soil nitrate low.

At the Buffalo county site there was a significant decrease in residual nitrate in 1994 with the uniform and variable rate minus 15% treatments showing the largest decreases (Table 4). In 1995 the variable rate minus 15% was significantly lower than the other two treatments as nitrate levels continued to decline (Figure 2). With improved irrigation and N management a goal of decreasing residual nitrate levels to 3 to 4 mg kg^{-1} uniformly throughout the field could help reduce nitrate leaching significantly.

TABLE 3. Average grain yields at the three sites.

Treatment	Lincoln County			Clay County			Buffalo County		
	1994	1995	1996	1994	1995	1996	1994	1995	1996
				Mg/ha*					
Uniform	13.2 a	11.1 b	13.9 a	10.3 a	9.0 a	11.2 a	6.9 a	4.4 b	9.0 a
Variable	13.1 a	11.3 a	13.9 a	10.4 a	9.0 a	10.7 a	6.9 a	4.9 a	9.1 a
Variable - 15%	13.1 a	11.1 b	13.9 a	10.1 b	8.9 b	10.5 b	6.9 a	4.4 b	9.0 a

*Values followed by the same letter are not significantly different at the 5% level.

TABLE 4. Yearly change in soil nitrate (kg ha^{-1} in 0-0.9 m) at the three sites.

N Treatment	Lincoln County			Clay County		Buffalo County	
	93-94	94-95	95-96	93-94	94-95	93-94	94-95
				kg/ha			
Uniform	-7	+78	-78	+11	-8	-31	-7
Variable	-25	+63	-64	+11	-4	-18	-1
Variable - 15%	-30	+64	-65	+11	-7	-27	-24

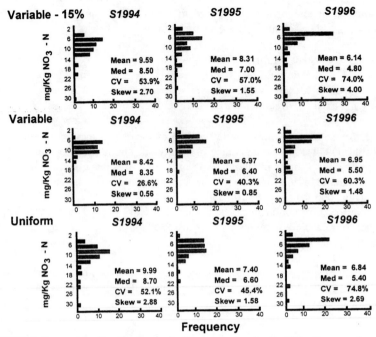

FIGURE 2. Frequency distributions of soil nitrate for the Buffalo County site.

TABLE 5. Soil nitrate (mg kg^{-1}) statistics for the 255 sample cores from Clay county.

Treatment	Year	Mean	Median	CV	Skewness
Variable - 15%	1993	3.99	3.80	29.4%	0.57
	1994	4.85	4.60	28.2%	1.03
	1995	4.32	4.30	23.5%	0.21
Variable	1993	3.88	3.80	28.6%	0.57
	1994	4.85	4.55	29.0%	0.63
	1995	4.47	4.50	39.4%	0.64
Uniform	1993	4.13	4.00	24.4%	0.70
	1994	5.21	4.90	30.6%	0.73
	1995	4.53	4.40	37.3%	0.55

Frequency distributions and statistics of soil nitrate were also analyzed to determine the impact of variably applied N on accomplishing the goal of improving NUE. Soil nitrate concentrations at the Clay county site were fairly low and uniform at the beginning of the experiment (Table 5). The CVs for nitrate and the skewness values were moderately low for soil nitrate. There was no significant influence of variable rate application compared to uniform application on the change in soil nitrate (Table 4) or in the frequency distributions (data not shown).

The Buffalo county site had high initial soil nitrate that was quite variable based on CVs and skewness (Figure 2). Initial soil nitrate concentrations were reduced significantly by all application methods the first year and by variable rate minus 15% in 1995 (Table 4). Variable rate application, however, did not significantly influence the frequency distributions compared to uniform application based on changes in CV and skewness (Figure 2).

High variability in soil nitrate at the Lincoln county site was reduced more by variable rate application than by uniform application (Figure 3). Even with the over-application of N in 1995, the variation in both variable rate application treatments was significantly lower than for the uniform application as shown by lower CVs and skewness. This data suggests that VRA could decrease N leaching if the effect were similar on other highly variable sites.

Apparent NUE values for the Lincoln county site were high (Table 6). Growing conditions during 1994 were excellent and yields and N removed in grain were high. Yields in 1995 (and N removed in grain) were good, but the over-application of N decreased ANUE. The values still reflect high NUE values, however. The recovery of residual nitrate in the fall 1995 sampling helped the overall NUE, especially for the variable minus 15% application (Table 6). Since N applied for the variable minus 15% treatment was closer to an optimum N rate than other treatments, the effect on NUE was evident. Data for 1996 were not available at publication time. The data do confirm the potential of variable rate N application for improving NUE, and decreasing long-term potentially leachable nitrate and nitrate-N in Nebraska ground water.

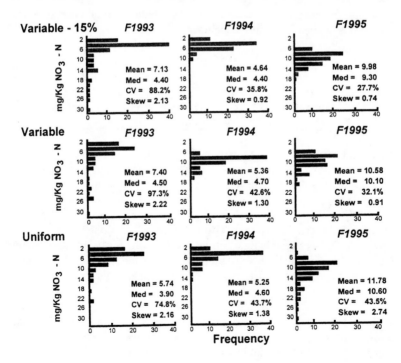

FIGURE 3. Frequency distributions of soil nitrate for the Lincoln County site.

303

TABLE 6. ANUE for treatments at the Lincoln County site.

N Treatment	1994	1995
	------ % ------	
Uniform	88.8 a	64.4 b
Variable	84.2 a	69.7 b
Variable - 15%	88.4 a	80.0 a

CONCLUSIONS

Variable rate application did not have as great an impact as we expected (initial euphoria about the potential of VRA in 1993). Soil nitrate variability and quantity was high at 2 or the 3 sites and VRA (optimum and reduced rates) on average decreased the quantity and variability more than uniform application. At the other site with low and fairly uniform initial nitrate, no significant changes occurred in nitrate quantity or variation which would be expected. In spite of significant nitrate variation, VRA did not consistently produce more grain than uniform N application. The VRA minus 15% treatment always yielded less than the other two treatments at the low nitrate site indicating that the N recommendation algorithm is meeting N demands of the yields produced and cannot be reduced much more without affecting yields and profit. Both VRA treatments tended to show improved ANUE compared to uniform N application and confirm the potential of VRA to improve NUE. VRA will have its greatest impact on fields with high nitrate variability and as shown by the data, can decrease nitrate hot spots and long-term potentially leachable nitrate. On fields with fairly uniform and low nitrate, uniform N application is still recommended.

REFERENCES

Engberg, R.A., Spalding, R.F. (1978) Groundwater quality atlas of Nebraska. *Resource Atlas No. 3*, Conservation and Survey Division, University of Nebraska.

Ferguson, R.B., Christiansen, A.P., Gosselin, D., Kuzila, M., Barnes, E., Murphy, T. (1992) Root zone and intermediate vadose zone nitrate accumulation as influenced by nitrogen and irrigation management. *Agron. Abstr.*, Madison, WI, Am. Soc. Agron, pp 277.

Ferguson, R.B., Schepers, J.S., Hergert, G.W., Peterson, T.A., Cahoon, J.E., Gotway, C.A. (1994) Variable Rate Nitrogen Application for Irrigated Agriculture: Opportunities for Groundwater Protection. *National Symposium on Protecting Rural America's Water Resources*, Washington D.C., Partnerships for Pollution Solutions, pp 207-216.

Hergert, G.W. (1987) Status of residual nitrate nitrogen soil tests in the United States of America. *Soil testing: Sampling, correlation, calibration, and interpretation,* J.R. Brown (Ed.), SSSA Special Publication 21, Madison, WI, USA, SSSA, pp 73-88.

Hergert, G.W., Ferguson, R.B., Shapiro, C.A. (1995a) Fertilizer Suggestions for Corn. *NebGuide G74-174 (Revised)*, University of Nebraska Cooperative Extension Service.

Hergert, G.W., Ferguson, R.B., Shapiro, C.A., Penas, E.J., Anderson, F.B. (1995b) Classical statistical and geostatistical analysis of soil nitrate. *Site-Specific Management for Agricultural Systems-Second International Conference Minneapolis, MN. March 1994,* P.C. Robert, R.H. Rust and W.E. Larson (Eds), Madison, WI, ASA, CSSA, SSSA, pp 175-186.

Mengel, K. (1990) Impacts of intensive plant nutrient management on crop production and environment. *Transactions of the 14th International Congress of Soil Science, Plenary Papers, Contents, Author Index,* Kyoto, Japan Intl. Soc. Soil Sci., pp 42-52.

1994-1995 Nebraska Agricultural Statistics (1995) NE Department of Agriculture, NE Agricultural Statistics Service, Lincoln, NE.

Swanson, E.R. (1982) Economic implications of controls on nitrogen fertilizer use. *Nitrogen in agricultural soils,* F.J. Stephenson (Ed.), Agronomy 22, ASA, SSSA, pp 773-790.

PATCHY WEED CONTROL AS AN APPROACH IN PRECISION FARMING

H. NORDMEYER, A. HÄUSLER, P. NIEMANN

Federal Biological Research Centre for Agriculture and Forestry,
Institute for Weed Research, Messeweg 11-12, D-38104 Braunschweig, Germany

ABSTRACT

Field observation showed that there is a marked tendency of several weed species to form aggregated spatial patterns. On one hand the uneven distribution can be related to certain aspects of weed biology, e.g., weeds forming perennial root systems. On the other hand weed distribution on field scale is influenced by specific site factors. An irregular distribution implies that a portion of the field is free of weeds or with weed densities below economic threshold levels. So confining spraying to weed patches may allow more or less considerable savings of herbicides. Weed distribution was recorded by field walking with digital registration using a Differential Global Positioning System. Geostatistical methods were used to quantify the spatial dependence of weed distribution data. Maps of weed abundance were generated and coupled with a Geographical Information System. Based on these weed maps, a local weed management system is recommended and patch spraying was carried out.

INTRODUCTION

Weed control is an important part of crop production. Since the availability of herbicides chemical weed control became widespread. But with the long term use of herbicides environmental contamination occurred. Many herbicides have been found in surface and ground water (Pionke and Glotfelty, 1989; Ritter, 1990; Spalding *et al.*, 1989). These contaminations caused a very intensive discussion about the necessity of herbicide use in agriculture. Consequently, many attempts were made to reduce herbicide inputs in order to minimize environmental contamination (e.g., mechanical weed control, reduced dosages, new low dosage herbicides, reglementations, herbicide mixtures).

Patch spraying is a new approach to minimize the amount of herbicides in order to reduce farmer's costs and environmental contamination. Patch spraying can be done by using different herbicides or no herbicides at all according to spatial weed distribution and economic thresholds. Using variable dosages according to growth stage of weeds or soil conditions is an additional tool in this concept.

Field observations showed that there is a marked tendency of several weed species to form aggregated spatial patterns (Cousens and Mortimer, 1995; Nordmeyer *et al.*, 1996; Walter, 1996). They often grow in clumps or patches of varying densities. Spatial weed distribution can be permanent over many years with varying densities (Wilson and Brain, 1990). This uneven distribution of weeds can be related to certain aspects of weed biology, e.g., weeds forming perennial root systems like *Cirsium arvense*, to spatial variability of specific site factors and to field management. An irregular distribution implies that a portion of the field is free of weeds or with weed densities below economic thresholds. Mortensen *et al.* (1995) reported that based on spatial analyses of weed populations in 12 Nebraska farm fields the postemergence herbicide application could be reduced by 71% and 94% for broadleaf and grass weeds respectively. This indicates a high potential of herbicide reduction under specific circumstances. Studies of weed distribution and patch spraying are becoming more significant in future as there is an increasing public pressure to reduce the use of herbicides.

The realisation of the patch spraying concept in agricultural practice requires the solution of two problems: weed identification and sprayer control with high spatial precision. There are two concepts of approach (Miller and Stafford, 1991; Nordbo *et al.*, 1995). One can be described as the real-time approach (online-concept). A weed sensing system mounted on the tractor controls herbicide application in real-time. Today's technique allows to distinguish between photosynthetic and non-photosynthetic material. This is an approach to control weeds in direct drilling system and in row crops. Systems which are able to distinguish between weed species or weeds and crops are not available in practice yet. Such a system requires an image recognition system combined with a data pool of weeds in different growth stages. Problems arise from background structure, variability of light, crop and weed development and overlapping of plants. Especially for weeds with low economic threshold a high accuracy in weed detection is necessary and it is most likely that the required hardware for the detection of weeds in real-time won't be available in the near future.

Another approach is the mapping concept which includes two separate steps. The first step is the generation of weed maps and the second is the weed control according to these maps. The concept requires a positioning system with high accuracy for weed location and sprayer control (e.g., Differential Global Positioning System). One of the most important questions in this concept is how to determine weed distribution with little time and costs for the farmer. In the beginning of the growing season farmers usually have not enough time to walk across their fields to estimate weed species and weed distribution in a sufficiently small grid. But at the moment there seems to be no other realistic possibility than weed mapping by field observations. So there is a strong need to look for new techniques which allow a reduction of expenditure of weed recording. The required data for generating weed maps can be derived from actual field walking (Stafford and le Bars, 1996), weed maps from previous years (Walter, 1997), aerial photography (Thurling *et al.*, 1985), remote sensing data (Everitt and Deloach, 1990), registration of weeds during routine field work (e.g., soil management, harvesting), soil properties and farmer's experience. All these data should be recorded in a weed data base and prepared for map generation by a Geographical Information System (Prather and Callihan, 1993). Finally herbicide application maps could be generated with less annual work for weed recording. This concept requires data collection over several years.

The objective of our work was to determine the distribution patterns of relevant weed infestations (weed densities above economic threshold levels) on selected agricultural fields and to demonstrate patch spraying based on these weed distribution data.

MATERIALS AND METHODS

Two experimental fields (Sickte VII, Sickte IX) were part of the research station in Sickte which is located in the south of Braunschweig (North Germany). The third field

(Edesbüttel) with a total area of 13 ha is situated north of Braunschweig. Sickte and Edesbüttel belong to an arable area with cereals and sugar beet as main crops. The soil is a brown earth/pseudogley (soil type: loamy sand/sandy loam). In 1995 winter barley was cultivated on Sickte VII, winter wheat on field Sickte IX and sugar beet on Edesbüttel. In 1996 winter wheat was cultivated on Edesbüttel.

In 1995 and 1996 weed distribution was recorded by field walking with a hand held data logger with integrated real-time Differential Global Positioning System (DGPS). In Sickte weed density and weed coverage were monitored in April 1995 (5 to 12 April). Weed densities were counted on 0.1 m^2 at every grid point and specified in plants m^{-2}. Grid spacings were 30 m. Counts were recorded for individual weed species. In Edesbüttel *Cirsium arvense* patch positions and shoot densities were determined on 26 May 1995, 22 August 1995, 24 and 29 May 1996.

The real-time DGPS consists of a base station (reference station) and a portable remote station (back pack receiver) connected with a data logger (Figure 1); data from the remote station were differentially corrected for sufficient accuracy by using data simultaneously collected by the reference station. The accuracy of the localisation was 1 to 3 m. Weed data and co-ordinates of grid points were linked and stored in one file.

Geostatistical methods were used to quantify weed distribution by using GS[+] software. Based on semivariograms interpolation (kriging) was done for creating contour maps of weed species and the total weed infestation as a basis for patch spraying. Contour maps of weed distribution and herbicide application maps were generated using a mapping software (SURFER) and the Geographical Information System (GIS) PC ARC/INFO. GIS was also used for determining the total area with weed densities exceeding threshold levels.

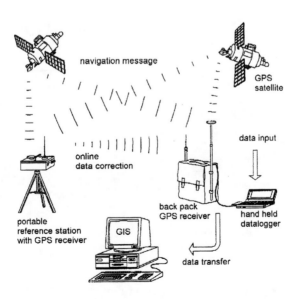

FIGURE 1. Differential Global Positioning System (DGPS) system configuration.

Patch spraying of *Cirsium arvense* in Edesbüttel (6 June 1996) was done with MCPA (as U 46 M-Fluid) by using a commercial tractor and sprayer with a boom width of 21 m. The tractor was equipped with real-time DGPS. The field was separated into subfields (5 x 5 m). Every subfield which includes or touches a *Cirsium arvense* patch (shoot density ≥ 1 m^{-2}) was defined as area to be treated with herbicide. Based on the co-ordinates of these subfields a manual on/off control of the sprayer was carried out while travelling along the tramlines.

RESULTS

All examinations of weed distribution patterns showed a marked tendency to cluster leaving large areas of the field relatively free of weeds altogether. This holds true for single weed species as well as for the total weed infestation. Contour maps were created for grass and broadleaf weeds and for weeds with extremely low economic threshold levels (e.g., *Galium aparine*). According to Kees *et al.* (1993) the economic threshold for grass weeds can be specified with 20 plants m^{-2}, for broadleaf weeds with 40 plants m^{-2} and for *Galium aparine* with 0.2 plants m^{-2}. In the subsequent contour maps only weed infestations are presented which exceed economic threshold levels. Figure 2 shows the distribution of grass weeds, broadleaf weeds and *Galium aparine* on field Sickte VII (1995, winter barley) combined in one figure. The different areas partly overlap. Areas with weed infestations below economic threshold were found in the northern part of the field.

Based on these weed distribution data an application map was generated (Figure 3). The area with herbicide spraying corresponds to the portion of the field with weed densities above threshold levels. Approximately 60% of field Sickte VII has to be treated with herbicides.

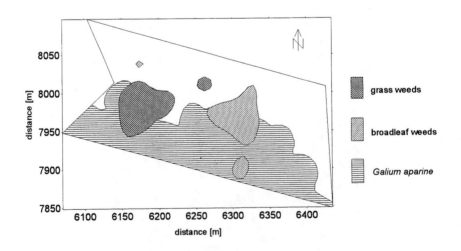

FIGURE 2. Distribution of grass weeds, broadleaf weeds and *Galium aparine* for Sickte VII (spring 1995, winter barley, only areas with weed densities above economic threshold).

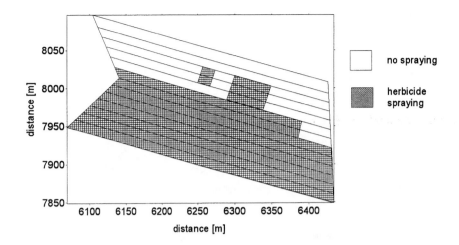

FIGURE 3. Herbicide application map for Sickte VII according to weed infestation in 1995 and tramlines.

Figure 4 gives an impression of the weed infestation of *Galium aparine* and broadleaf weeds on Sickte IX (1995, winter wheat). In general weed densities were higher than on Sickte VII. On this site local variations in weed populations` densities were also determined. But in case of the broadleaf weeds the low spatial heterogeneity and the high weed densities did not justify patch spraying. In contrast to the distribution of broadleaf weeds, grass weeds showed a more irregular distribution pattern (Figure 5). The area to be treated with herbicide against these weeds can be specified with nearly 70%. So assuming a separate control of grass weeds also on this site patch spraying is recommended.

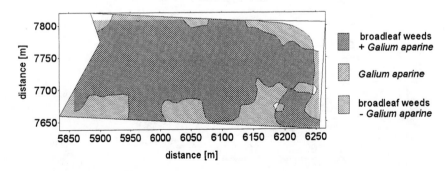

FIGURE 4. Distribution of *Galium aparine* and broadleaf weeds for Sickte IX (spring 1995, winter wheat, only areas with weed densities above economic threshold).

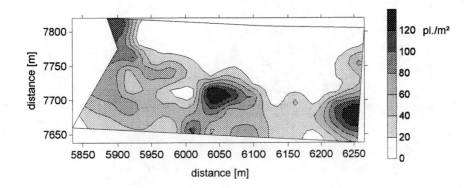

FIGURE 5. Distribution of grass weeds for Sickte IX.

Figure 6 shows the application map for control of *Cirsium arvense* for field Edesbüttel. The spray area is greater than weed patches to have security zones around the weeds in order to avoid misapplication caused e.g. by inaccuracy of positioning. A comparison of treated and total field area gives an impression on the potential of herbicide reduction. In this case the portion of treated area can be specified with 21% (total field area: 13 ha). The success of patch spraying was estimated by monitoring the weed density in the growing period until harvest.

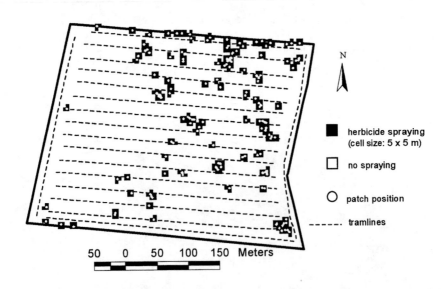

FIGURE 6. Herbicide application map for field Edesbüttel and positions of *Cirsium arvense* patches.

DISCUSSION

These investigations showed an uneven distribution of weed species and total weed infestations. This is in accordance with results of many authors (e.g., Cousens and Mortimer, 1995; Walter, 1996; Wiles *et al.*, 1992; Wilson and Brain, 1990). Often a patchy distribution pattern can be recorded but nearly on the whole field weed densities exceed the economic threshold levels. In further investigations it should be clarified under which conditions (field size, distribution pattern of relevant weed infestations, weed densities above economic threshold levels, monetary loss due to weed competition, costs and benefits of patch spraying) patchy weed control can be successfully integrated in precision farming. Nevertheless, under specific circumstances patch spraying will in the long run reduce costs and can contribute to a sustainable agriculture.

CONCLUSIONS

- There is a marked tendency of weed species to occur in patches.
- Based on geostatistical methods weed maps can be created for agricultural fields.
- For both environmental and economic aspects, patch spraying seems to be a successful way of reducing herbicide use.
- In future, the patch spraying concept will become more important in agricultural practice.

ACKNOWLEDGEMENTS

The authors wish to express their appreciation to the German Federal Environmental Foundation (Deutsche Bundesstiftung Umwelt) for their financial support.

REFERENCES

Cousens, R., Mortimer, M. (1995) *Dynamics of weed populations*, Cambridge, Cambridge University Press.

Everitt, J.H., Deloach, C.J. (1990) Remote sensing of Chinese Tamarisk (*Tamarix chinensis*) and associated vegetation. *Weed Science*, **38**, 273-278.

Kees, H., Beer, E., Bötger, H., Garburg, W., Meinert, G., Meyer, E. (1993) *Unkrautbekämpfung im Integrierten Pflanzenschutz*, Frankfurt/Main, DLG-Verlag.

Miller, P.C.H., Stafford, J.V. (1991) Herbicide application to targeted patches. *Proc. BCPC-Weeds*, Volume 3, 1249-1256.

Mortensen, D.A., Johnson, G.A., Wyse, D.Y., Martin, A.R. (1995) Managing spatially variable weed populations. *Site-Specific Management for Agricultural Systems*, ASA-CSSA-SSSA Publications, 397-415.

Nordbo, E., Christensen, S., Kristensen, K. (1995) Teilflächen Unkrautmanagement. *Zeitschrift für Pflanzenkrankheiten und Pflanzenschutz*, **102**, 75-85.

Nordmeyer, H., Häusler, A., Niemann, P. (1996) Weed mapping as a tool for patchy weed control. *Proc. Second International Weed Control Congress, Copenhagen,* Denmark, Volume 1, 119-124.

Pionke, H.B., Glotfelty, D.E. (1989) Nature and extent of ground water contamination by pesticides in an agricultural watershed. *Water Research,* **23**, 1031-1037.

Prather, T.S., Callihan, R.H. (1993) Weed eradication using Geographical Information Systems. *Weed Technology,* **7**, 265-269.

Ritter, W.F. (1990) Pesticide contamination of ground water in the United States - A review. *Journal Environmental Science Health,* **25**, 1-29.

Spalding, R.F., Burbach, M.E., Exner, M.E. (1989) Pesticides in Nebraska's ground water. *Ground Water Monitoring Review,* **9**, 126-133.

Stafford, J.V., le Bars, J.M. (1996) A hand-held data logger with integral GPS for producing weed maps by field walking. *Computers and Electronics in Agriculture,* **14**, 235-247.

Thurling, D.J., Harvey, R.N., Butler, N.J. (1985) Aerial photography of field experiments using a remotely-piloted aircraft. *Proc. BCPC-Weeds,* Volume 1, 357-363.

Walter, A.M. (1996) Temporal and spatial stability of weeds. *Proc. Second International Weed Control Congress, Copenhagen,* Denmark, Volume 1, 125-130.

Walter, A.M. (1997) Patch spraying using weed maps from previous years. *Proc. 10th EWRS Symposium, Poznan,* Poland, in press.

Wiles, L.J., Oliver, G.W., York, A.C., Gold, H.J., Wilkerson, G.G. (1992) Spatial distribution of broadleaf weeds in North Carolina soybean (*Glycine max*) fields. *Weed Science,* **40**, 554-557.

Wilson, B.J., Brain P. (1990) Weed monitoring on a whole farm - patchiness and the stability of distribution of *Alopecurus myosuroides* over a ten year period. *Proc. EWRS Symposium Integrated weed management in cereals,* Helsinki, Finland, 45-52.

AGRONOMIC CONSEQUENSES OF VARIABLE N FERTILIZATION

H.S. ØSTERGAARD

The Danish Agricultural Advisory Centre, Department of Plant Production, Udkaersvej 15, Skejby, 8200 Aarhus N, Denmark

ABSTRACT

The contents of different nutrients in soil were measured in four fields. Soil samples were taken either in sub-fields or in a fixed grid with a side length of 50 m. The individual sub-fields were identified by yield maps, the farmer's and advisers' experience and the topography of the field. Fertiliser trials were carried out in a number of the sub-fields, making it possible to evaluate the economic effect of applying nitrogen fertilizer in the individual sub-fields. Based on results from a three year experiment it was estimated that the direct potential economic benefit in the fields examined from precision agriculture was \$40 - \$50 ha^{-1} compared to an even application of N, P, K and lime.

INTRODUCTION

For the sake of both the environment and the economy, it is important to balance the allocation of consumable inputs against the requirements as precisely as possible. In principle, all inputs should be based on the requirements of the specific sites of the individual fields but some are more important than others.

Among the nutrients, the most interesting is nitrogen. In an average Danish soil, it is estimated that nitrogen fertilizing of grain crops results in increased yields of 1000 - 5000 kg per ha at costs corresponding to between 300 and 800 kg per ha. As regards phosphorus, potassium and magnesium the potential short-term increased yields due to application are much lower, only 0 - 100 kg per ha at costs corresponding to 100 - 300 kg per ha. From an environmental angle, nitrogen is also the most interesting nutrient, primarily due to nitrate leaching caused by fertilizing above the optimum level.

MATERIALS AND METHODS

The environmental and economic effects of precision agriculture were estimated in five fields. The size of the fields varied from 10 ha to 12 ha and were chosen because of a considerable variation in soil type and topography. Based on variation in soil type, yield, topography, aerial photographs and the farmers' and the advisers' experience, each field was divided into 12 - 17 sub-fields. In each of the sub-fields, soil samples were taken to determine pH, P, K, texture and N-min (nitrate- + ammonium nitrogen). Based on the collected data and the expected yields the fertilizer requirements were calculated for the individual sub-fields.

In a number of sub-fields, trials were conducted with increasing nitrogen applications. A yield curve and the economically optimum N rates were determined. Based on the N-min measurements in the individual sub-fields and in accordance with the rules of the

Danish Plant Directorate the nitrogen requirements were also calculated to determine the potential effect of balancing the nitrogen rates against the requirements of the individual fields.

RESULTS

Economic effect

The results of the trials with increasing nitrogen application rates in the fields 1, 2, 3, 4 and 5 are shown in Table 1. The table compares the economic effect of optimum fertilization and the economic effect of applying nitrogen in accordance with the N-min method (Danish Agricultural Advisory Centre, 1995) and the standards laid down by the Danish Plant Directorate (Danish Plant Directorate, 1995). The calculations imply that every trial reflects a sub-field and that all the sub-fields are of the same size.

In field 1, the 1994 crop was perennial ryegrass for seed production. The experimental field was divided into fifteen sub-fields and, in four of these fields, trials were conducted where increasing N rates were applied. According to Table 1, the economically optimum nitrogen application ranged from 134 - 178 kg N per ha, corresponding to net yields of 893 - 1656 kg seed per ha. An average of the four trials shows that the net yields were nearly identical when using the standard recommendations or the N-min method and were about 450 DKK lower per ha compared to optimum fertilization in all four trials. The nitrogen consumption was approximately 25 kg lower per ha if using these two methods compared to the economically optimum method. If applying the average rate of the measured optima, the economic loss averaged 135 DKK per ha.

In field 2, the 1994 crop was winter wheat and four trials were conducted where rising nitrogen rates were applied. Table 1 shows that the economically optimum nitrogen rate in the field ranged from 127 - 234 kg N per ha, corresponding to a variation in the net yields of 4300 - 8600 kg grain per ha. The table also shows the N application rates and the corresponding net yields both for the standards fixed by the Danish Plant Directorate and for the N requirements determined by means of the N-min method. An average of the four trials shows that the same net yield was achieved no matter whether the standards or the N-min method was used. The net yield was about 200 kg lower than that achieved from optimum fertilization in all four trials. The N consumption was the same in all three situations.

In 1995, field 2 was sown to winter wheat and the economically optimum N rate ranged from 97 - 300 kg N per ha. The economic gain of varied fertilization was just below 200 DKK per ha. The N-min method did not work properly in field 2, especially not in two of the trials where the reduced economic gain of using the method varied between 400 and 700 DKK per ha.

In field 3, the 1995 crop was spring barley and the measured economically optimum N rate ranged from 0 - 129 kg N per ha within the same field. The reduced economic gain of fertilizing in accordance with the standards of the Danish Plant Directorate amounted to 280 DKK per ha. If applying the average of the measured optima the average economic loss was 200 DKK per ha. Using the N-min method in the individual trials made the reduced economic gain drop to 63 DKK per ha on average.

TABLE 1. Results of N trials** in 1994 and 1995. The table shows the N application rates and the reduced economic effects of N fertilization following the standards of the Danish Plant Directorate, the N-min method and as average of the measured optima.

Site	Crop	Soil type 0–25 cm	Yield applying 0 N	Economically optimum N application, kg per ha	Excess yield with optimum application, t/ha	Nitrogen application, kg/ha			Reduced economic gain, US$ per ha		
						Av. of optimum	Plant Directorate	N-min method	Av. of optimum	Plant Directorate	N-min method
Field 1, 1994											
Trial 1	peren. ryegrass	humus	3.1	178	7.6	154	125	92*	26	125	304*
Trial 2	peren. ryegrass	sand	6.4	155	10.9	154	125	144	0	141	21
Trial 3	peren. ryegrass	sand	3.3	149	6.3	154	125	143	2	40	2
Trial 4	peren. ryegrass	sand	3.1	134	8.2	154	125	138	62	10	1
Av.			4.0	154	8.2	154	125	129	23	79	69
Field 2, 1994											
Trial 1	winter wheat	clay	38.0	234	57.6	193	188	190	47	48	48
Trial 2	winter wheat	sand	21.7	232	47.0	193	188	188	17	16	16
Trial 3	winter wheat	sandy loam	28.9	178	40.0	193	188	189	2	0	0
Trial 4	winter wheat	loamy sand	18.9	127	29.5	193	188	185	32	32	16
Av.			26.9	193	43.5	193	188	188	24	24	20
Field 3, 1995											
Trial 1	spring barley	clay	37.7	129	27.2	71	125	118	60	1	2
Trial 2	spring barley	humus	45.2	36	4.3	71	125	68*	24	96	21
Trial 3	spring barley	sandy loam	37.1	110	19.2	71	125	115	23	3	1
Trial 4	spring barley	sandy loam	52.8	81	18.3	71	125	84	3	40	0
Trial 5	spring barley	humus	55.6	0	0.0	71	125	35	57	94	29
Av.			45.7	71	13.8	71	125	84	33	47	11

317

Site	Crop	Soil type 0-25 cm	Yield applying 0 N	Economically optimum N application, kg per ha	Excess yield with optimum application, t/ha	Nitrogen application, kg/ha			Reduced economic gain, US$ per ha		
						Av. of optimum	Plant Directorate	N-min method	Av. of optimum	Plant Directorate	N-min method
Field 4, 1995											
Trial 1	winter barley	clay	29.7	199	27.5	175	160	160	4	11	11
Trial 2	winter barley	clay	31.1	135	24.0	175	160	167	13	6	9
Trial 3	winter barley	clay	32.0	116	18.5	175	160	161	26	16	17
Trial 4	winter barley	clay	24.9	205	29.7	175	160	155	7	15	18
Trial 5	winter barley	clay	34.8	149	12.7	175	160	163	3	1	1
Trial 6	winter barley	clay	40.8	180	18.0	175	160	166	1	3	2
Trial 7	winter barley	clay	30.1	235	20.3	175	160	167	10	16	13
Trial 8	winter barley	clay	30.5	182	16.1	175	160	162	0	1	1
Av.			31.7	175	20.9	175	160	163	8	9	9
Field 5, 1995											
Trial 1	winter wheat	loamy sand	24.8	196	46.5	210	200	184	2	0	2
Trial 2	winter wheat	clay	35.1	97	40.8	210	200	190	103	113	115
Trial 3	winter wheat	clay	29.0	223	49.9	210	200	194	2	6	10
Trial 4	winter wheat	sandy loam	24.6	251	51.3	210	200	192	16	24	33
Trial 9	winter wheat	clay	37.0	205	35.2	210	200	153	0	0	32
Trial 10	winter wheat	sandy loam	23.5	300	46.7	210	200	194	100	61	68
Trial 12	winter wheat	clay	27.8	200	50.5	210	200	208	2	1	2
Av.			28.8	210	45.8	210	200	188	32	29	37

* It is not recommended to use the N-min-method on this soil type.

** In the N trials the rates of nitrogen were increased in steps of 50 kg N per ha from 0 to 250 kg N per ha.

In field 4 the 1995 crop was winter barley and the economically optimum N rate ranged from 135 - 235 kg N per ha. The economic effect of using graduated N fertilization was about 50 DKK per ha.

<u>Nitrate leaching</u>

Potential nitrate leaching due to varied application of N to different parts of the field can be determined by measuring N-min at harvest. Other things being equal, increased N-min content at harvest will result in increased nitrate leaching (Ostergaard *et al.*, 1995).

The results collected in the experimental field at the Risø National Laboratory are shown in Figure 1. It shows that in nearly all cases the N-min content increased with increasing N rates but, in general, the largest increase was recorded when the N rate exceeded the economically optimum N rate.

DISCUSSION AND CONCLUSION

Varied fertilization will benefit the environment and reduce resource consumption because, if a correct application rate is used everywhere in the field, excessive application to those parts of the field where the requirements are low, can be avoided. On the basis of this study, it is not possible to indicate the size of the environmental impact.

It is extremely difficult to make a correct determination of the nitrogen requirements before the growing season. In this series of trials, the economically optimum N rates in the individual trials were compared with different methods of determining the requirements before the growing season. Therefore, the difference between the measured N rates and the results of the different methods of predicting the N

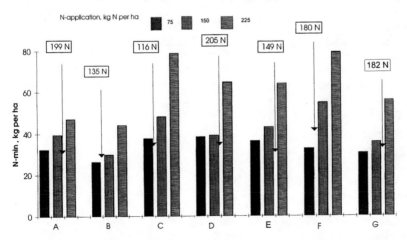

FIGURE 1. Results of N-min measurements at harvest, 1995, in the experimental field of the Risø National Laboratory. The N-min measurements were carried out in three treatments concerning economically optimum fertilization which is indicated in the figure by means of arrows.

requirements indicates the ideal situation where the correct N rate predictions can be made before the growing season. For this reason, the quoted economic effects of varied N fertilization are uncertain and to be considered as potentials which can be realized to a greater or lesser extent. It is very important to find out how the N requirements of the sub-fields are best determined on the basis of all the available information.

Besides, when evaluating the results it is important to bear in mind that the results originate from fields chosen because of the great variation within the fields. In fields with less variation, the effect of graduated N fertilization will be correspondingly lower.

There are significant differences between the five fields in the potential economic effect of graduated N fertilization within the field. In field 4, the potential economic effect was especially small due to low yield response to N application. The results show that the largest economic effect of graduated fertilization can be achieved by a correct determination of the N requirements in the humus-rich parts of the field. Therefore, it is necessary to develop a method designed to determine the N requirements of humus-rich soils. Furthermore, if the fertilizer requirements are based on the N-min content and the expected yields, then the results show that the highest and the lowest calculated N requirements will often be lower than 40 kg N per ha in mineral soil. Only where the humus content is high, are larger differences in the calculated N requirements recorded.

Based on two years' experience from five fields and given the subdivision of the fields used, it can be concluded that the direct economic gain of varied N fertilization will range from $15 - $35 per ha in most agricultural crops grown in mineral soils.

The total economic gain of varied input of nutrients and lime will probably amount to about $40 - $50 per ha. To this can be added the effect of graduated weed and pest control, better quality. The effect may increase if different parts of a field have different cultivation records, e.g., application of farmyard manure. As already mentioned, all these estimates are based on experience gained in fields chosen because of their great variation. The effect is correspondingly lower in fields with lower variations.

In this experiment, the subdivision of the fields has been based on sub-fields which were mainly based on variation in soil type, yield, topography, aerial photographs and the farmers' and the advisers' experience. Other methods are available which are based on goal-oriented soil sampling combined with a statistical processing of the analysis results. This method is also to be tested and the possibility cannot be excluded that this will reveal an even larger field variation. In the future, we have to find out how the fields can be sub-divided in a rational and correct way.

REFERENCES

Danish Agricultural Advisory Centre (1995) *Fertilization According to the N-min method 1995* (in Danish).

Danish Plant Directorate (1995) *Green fields, Crop Rotation- and Fertiliser Planning and Fertiliser Bookkeeping,* Danish Plant Directorate, (in Danish).

Østergaard, H.S., *et al.* (1995) Nitrate Leaching Depending on Cropping System. *Nitrogen Leaching in Ecological Agriculture*, L. Christensen, (Ed.), AB Academic Press, pp 173-179.

IMPROVING A PLANTER MONITOR WITH A GPS RECEIVER

A.M. SARAIVA, S.M. PAZ, C.E. CUGNASCA

Agricultural Automation Laboratory, Computer Engineering Department, Polytechnic School, University of São Paulo, CP 61548 - São Paulo - SP - 05424-970 - Brazil

ABSTRACT

The basic function of a Planter Monitor is to check if all planting rows are working properly. If, in a particular row, the rate of seed planting is not within a pre-defined range, the Planter Monitor issues an alarm to the operator.

The Planter Monitor developed for research purposes by the Agricultural Automation Laboratory of the University of São Paulo (Brazil), has extra features. During the planting operation, it calculates statistical data - such as operating time under various abnormal conditions, seed population, planted area, average speed - and stores these data in its memory in order to transfer them to a computer, for further analysis and report generation, after the operation is completed.

Now another feature is being added to this Planter Monitor: a link to a GPS receiver. The positional information provided by the GPS receiver, when merged with the statistics already calculated, will generate a "planting map", reflecting how the planting operation was performed, and showing the spatial variability of seeds.

Although the final goal would be having a variable-rate planter, it is believed that the map obtained with a fixed-rate planter can be used as a site-specific management tool, and can help to explain the variability on yield maps.

INTRODUCTION

This paper presents a proposal for the inclusion of a GPS receiver in a Planter Monitor in order to generate a seed population map for use in Precision Farming systems.

Based on a Planter Monitor that already has some data-logging functions and an extra serial channel, alterations were proposed to accommodate a link to a GPS receiver and the logging of positional data.

The shift of the Monitor from an operational tool to a management tool is presented, in addition to the changes that were necessary to permit connection to the GPS receiver. Some considerations of the available options are made.

Planter monitoring: conventional versus management approach

Planter monitors have been on the market for decades in the USA and in some European countries and have changed dramatically as the electronic technology has evolved. These changes have focused on the information the monitors deliver to the operator, but in all cases the instrument remains basically a tool for the operator. The planter monitors available on the Brazilian market, which are North-American, Italian and German, do not provide facilities for recording information to enable the analysis of that information after the planting operation. A management approach to plant monitoring can help to identify frequent problems, not only in the machine, but also in the way the machine is used by the operator. Operators who are less skilled or trained, or even less concerned with the achievement of good results, can be monitored since the manager has a log of the important data of the operation. This is thought to be of special significance in Brazil where most of the operations in the field are made by employees with a limited cultural level, instead of being conducted by the farmers themselves.

To help solve this problem we developed a planter monitor which differs from the models available on the market, first, by having a modular design, but mainly by permitting the data-logging of parameters which are of interest to the farm manager (Saraiva *et al.*, 1993).

While the Monitor helps the operator by presenting seed rates, speed, planted area, etc., several other statistics are stored in the memory. These statistical data are not of interest to the operator during planting, but they can help future operation management and optimization. Examples of the logged data include percentage of time spent in operation; percentage of time spent in transport or in headlands; percentage of time in normal operation (i.e., speed and planting rate within a predefined range) and percentage of time under any fault condition (i.e., speed too high or too low, and/or planting rate too high or too low). After planting is completed, these data can be transferred from the Monitor to any personal microcomputer, and there, they can be stored on disk files and manipulated by a program developed by LAA. The program will generate reports and graphics that can be very useful to the manager or farmer.

PLANTING AND PRECISION FARMING

Although many agricultural operations and machines are already controlled by electronic equipment, improving the accuracy and quality of the operation, this has not been the case with planters (except in a few experiments) for many years. This was due in part to some technical problems concerning the mechanical concept of the machine, and also to the ignorance of the advantages of varying the application rate for the seeds.

As Precision Farming becomes a technological reality more data is necessary in order to address one of the main issues that remains unsolved, namely the ability to derive accurate recommendations based on the variability of the inputs. Initially soil and yield were the main parameters that were observed when trying to define site-specific practices, but since then many other variables have been considered. A better understanding of the complexity of the relationships in agricultural production systems is lacking, and this demands the integration of many data sources, and many variables, distributed throughout the production cycle. These "check points" can enhance our knowledge of the influence of each parameter on the final yield variability (Searcy, 1995).

As planting is the first operation that directly involves the plants (i.e., the seeds), it provides the first data layer related to the crop. It is clear that any problem occurring during planting will have a direct effect on the yield. If this variability in the stand due to planting is not taken into account, it is likely to be incorrectly attributed to some other cause.

The importance of considering the results of planting as one of the first crop-related layers of a Precision Farming system was pointed out by researchers who have been investigating systems for analyzing plant spacing and stand uniformity in the early stages of crop growth (Easton, 1996; Plattner and Hummel, 1996).

One advantage of using the Planter Monitor as a source for a first "seed population layer" is that the data can easily be obtained in parallel with the actual planting operation. The only additional cost is that of modifying the equipment once to add this new feature.

CHANGES IN THE PLANTER MONITOR

Based on the arguments above we decided to incorporate a GPS receiver in our Planter Monitor so that we could add positional information to the statistics currently being calculated, in order to generate a seed population map. The choice of GPS as the positioning system was based on its wide use in precision agriculture systems (Korte and Yule, 1996; Krüger et al., 1994; Stafford and Ambler, 1994), on its high accuracy when using the differential technique and on its decreasing price. This option demanded studies and changes to both the hardware and software of the equipment.

Modifications to the electronic circuit

Most of the Planter Monitors on the market have only one module, which is installed in the tractor cabin, and has to be connected to the sensors through a set of long cables, one for each sensor.

The LAA's Planter Monitor has two separate modules: the Planter Module and the Tractor Module. The Planter Module is installed in the planter, and it is connected to the seed flow sensors and to the speed sensor. Since it is located very close to the sensors, the many cables needed to connect them can be shorter than in other planter monitors. The Tractor Module is installed inside the tractor cabin and provides the interface with the operator, as well as performing most of the data processing. A single cable connects both modules, permitting data exchange and providing energy to the Planter Module. After planting, the same cable is used to connect the Monitor to the office computer in order to transfer all data. FIGURE 1 shows a scheme of LAA's Planter Monitor before and after the alterations to the circuit.

For the Monitor hardware there were two options: the installation of a "GPS card" on the Planter Monitor; or the connection of a GPS receiver to it.

The GPS card is a board with a common GPS receiver circuit, but without keyboard, display and memory. Cheaper and smaller than the conventional GPS receivers, this card can be incorporated into the equipment being developed or installed inside a computer, therefore allowing more compact units. In the case of the LAA's Planter Monitor, however, the Tractor Module, where this card would need to be installed, would have to be redesigned, to enable the card to be packed into its interior and to incorporate an adequate interface for the card.

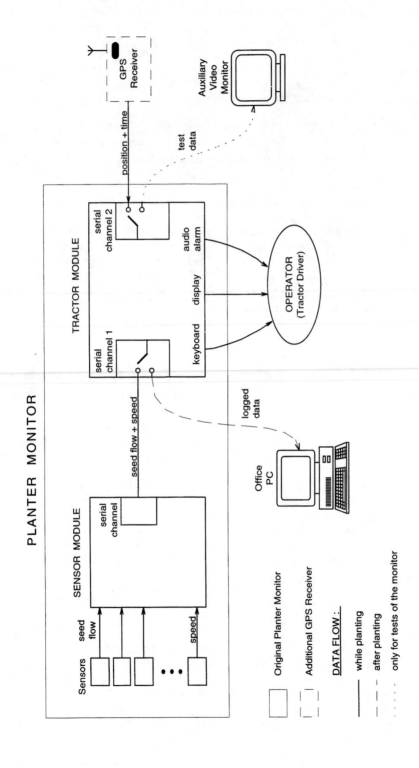

FIGURE 1. Schematic diagram of the Planter Monitor, including the GPS receiver.

At present, the second option seems to present some advantages. Firstly, the Tractor Module of the Planter Monitor has an extra serial channel, for testing purposes, in addition to its main serial channel used for communication with the Planter Module and for transfer of statistical data to the office microcomputer. The extra channel is fully compatible with the serial channel of the majority of the GPS receivers on the market. Secondly, the connection between the Tractor Module and the GPS receiver, through an external cable, offers a more modular solution. Depending on its application, the GPS receiver can be easily disconnected, allowing the Planter Monitor to keep operating in the way it was originally designed. Thirdly, the use of a conventional GPS receiver, with a display, permits the position verification, which can help during tests of the Monitor. Fourthly, the external GPS can be shared with other equipment more easily.

The main disadvantage of the second option, the cost, can be reduced when one considers that the same GPS receiver can be used with other field equipment and operations for precision agriculture.

With the adopted solution, no change in the electronic circuit of the Planter Monitor is required. It is only necessary to install a connector at the Tractor Module, so that the extra serial channel can be accessed from outside and connected to the GPS receiver.

Thanks to the system's modularity, another possibility is to implement the Tractor Module in a notebook computer or equivalent, provided it has two serial channels. The advantage of this option would be the possibility of substantially increasing the memory available, while working on a more friendly programming environment, thus allowing a more adequate graphical user interface.

Modifications to the internal program

Relative to the Monitor software, the Planter Module's program doesn't need any change since its interface to the sensors remains the same. However, the internal program of the Tractor Module has to be modified. Firstly, it has to support communication with the GPS receiver. The majority of receivers on the market use the "NMEA-0183" protocol. However, there is no uniformity of message selection. Each model may issue a different set of protocol messages. There are also some models that send messages that can be chosen by the operator. Some receivers do not follow the "NMEA-0183" protocol, but define their own standard. Therefore, it is necessary to be able to configure the set of messages accepted by the Planter Monitor, which means that different protocols can be selected.

The second modification to the program refers to the way the planting map is stored in the Tractor Module. Here, there are several options that must be evaluated given the constraints of the available memory. It is possible, for example, to store the GPS receiver position every time it is received, together with the statistics to be mapped (seed rate, tractor speed, etc.). This option, however, is memory consuming (the GPS receiver usually updates its position once a second), and this decreases the Planter Monitor autonomy. On the other hand, instead of storing the raw data, the Tractor Module processor could perform some data manipulation. For example, the field to be planted could be divided into a grid of rectangular cells, and the statistics obtained by the Monitor could be stored as an average for each cell. In this case, there is a considerable reduction in memory consumption, but some of the information originally sent to the office computer is lost. This option would also increase the complexity of the Monitor software and could possibly decrease its performance. A reasonable option seems to be to store the raw data, but at a lower rate than the GPS receiver update - for example, once every 10s.

The memory space available on the Tractor Module is 60 kbytes which is more than enough for its original purpose. Considering that map information would now be stored once every 10 s, this would give the Monitor an autonomy of about ten hours if we only store information of planting rate, speed, alarms and position. This suggests that for research purposes it is satisfactory to test the Monitor with its current hardware configuration. However, if we would like to store all the variables originally stored and do it once every 1s, this would require about 2 Mbytes of memory for the same autonomy of ten hours, which is impractical to have with this hardware. In this case we would need to implement the program of the Tractor Module in a notebook.

A last modification to the Tractor Module program is related to the transmission of data to the office computer. The new planting map, generated by the inclusion of the GPS receiver data, has to be added to the data currently being sent, so that this map can be manipulated by the office computer.

Modifications to the office computer program

Obviously, the office computer program that generates reports also had to be modified. The original program was written in Visual Basic and provides a user-friendly interface. The modifications being made will support both presentation of the maps and the generation of a GIS-compatible file, for software such as Idrisi or ArcView. The latter option will allow the integration of the map with a field information system based on a commercial GIS, which is also under development.

The modifications are expected to be ready in time for field tests at the beginning of 1997.

This project is part of a major Precision Agriculture project of the Research Group on Agricultural Automation, University of São Paulo, Brazil, which involves the Computer Engineering Department / Polytechnic School and the Rural Engineering Department / Escola Superior de Agricultura "Luiz de Queiroz". It is one of the first projects of the group on Precision Agriculture and it is providing knowledge and experience on related matters, such as GPS and mapping.

CONCLUSIONS

A Planter Monitor which has already been adapted for use as a mechanization management tool is being modified to allow connection to a GPS receiver. This will permit the storage of positional data along with the usual planting operation data: seed rate, speed, alarms, etc.

The necessary changes relate only to the software, since there is an extra serial channel to which an external GPS receiver can easily be connected. Software modifications are required to both to the Monitor data-logging software, and to the office computer software where report generation software will now handle map data. Additionally, the map data generated can be transferred to conventional GIS software.

For a greater autonomy and more frequent and detailed acquisition of data, a notebook computer could replace the proprietary Tractor Module hardware, maintaining the Planter Module and the current architecture. This would also provide better performance and user interface.

The seed population map obtained can be useful for Precision Agriculture information systems, as a first crop-related map, with no additional operational cost.

ACKNOWLEDGMENTS

The authors wish to express their gratitude to the Brazilian funding agency Financiadora de Estudos e Projetos - FINEP, for providing financial support for this project.

REFERENCES

Easton, D. (1996) Corn population and plant spacing variability: the next mapping layer. *Proceedings of the Third International Conference on Precision Agriculture.* Bloomington/Minneapolis, ASA-CSSA-SSSA.

Korte, H., Yule, I.J. (1996) Wide area network DGPS for use in Precision Farming. *1996 ASAE International Meeting*, St. Joseph, ASAE, paper 96-1.024.

Krüger, G., Springer, R., Lechner, W. (1994) Global Navigation Satellite Systems (GNSS). *Computers and Electronics in Agriculture*, 11, 3-21.

Plattner, C.E., Hummel, J.W. (1996) Corn plant population sensor for precision agriculture. *Proceedings of the Third International Conference on Precision Agriculture.* Bloomington/Minneapolis, ASA-CSSA-SSSA.

Saraiva, A.M., Cugnasca, C.E., Massola, A.M.A. (1993) Planter monitoring: a management approach. *1993 ASAE International Winter Meeting*, St. Joseph, ASAE, paper 93-1.552.

Searcy, S.W. (1995) Engineering systems for site-specific management: opportunities and limitations. *Site-Specific Management for Agricultural Systems*, Madison, ASA-CSSA-SSSA, pp 603-612.

Stafford, J.V., Ambler, B. (1994) In-field location using GPS for spatially variable field operations. *Computers and Electronics in Agriculture*, 11, 23-36.

DEVELOPMENT AND IMPLEMENTATION OF SITE SPECIFIC FERTILIZATION IN DENMARK. METHODS AND RESULTS

K. PERSSON

National Institute of Animal Science, Research Center Bygholm,DK-8700 Horsens, Denmark

L. MØLLER

KEMIRA DANMARK A/S, DK-7000 Fredericia, Denmark

ABSTRACT

Two different types of systems (Agri Matic and LORIS) are available in Denmark for site specific fertilization of farmland. The two systems consist of computer programmes and technical solutions for various commercial fertilizer spreaders. The systems differ from each other primarily in their way of determining position in the field. One system uses a set of fixed tramlines which must be traversed in a certain direction, while in the other system the position is determined by a DGPS-system. In addition, the methods by which the quantities to be applied at a single spot in the field are determined differ. The two systems have been tested and both show good results. The accuracy of the positioning systems and the settings on the machines have proved satisfactory when testing the systems at: 1) variable speed levels, 2) different spreading width, 3) changes in application levels, and 4) when trying to reproduce the spreading pattern whilst driving in the opposite direction. Only when making a sudden change in speed by stopping the tractor might there be a small time lag before the computer stops the spreader which can cause a small overdose with the LORIS system. Even when the expected quantity of fertilizer is applied, the overall distribution may be poor as the spreading pattern is often influenced by the flow rate. Calculations for different overall distributions have been carried out and show that the spreading performance is of great importance. Some three-point mounted spreaders need adjustment of the top link, fertilizer drop point or vane position on the disc according to the dose level and flow rate of the fertilizer. There is still no technical solution to this problem.

BACKGROUND

Generally both fertilizer and lime are applied by distributing a fixed pre calculated quantity evenly over the area to be treated. Seen from a cultivation as well as an environmental point of view, this is not an optimum solution. Knowledge of the variability in a cultivated area allows us to consider applying a graduated distribution of these plant growth factors. Varying the application of lime and fertilizer according to the requirements of each sub area probably makes it possible to increase utilization rate and to minimize the leaching of nutrients which is a global problem.

Locally varying the quantity of fertilizer applied is not a new idea - farmers have always been aware of the differing requirements of each sub area. However, the technical solutions to enable locally varying application have, however, not always been available.

With the increasing farm size, the need for technical auxiliary equipment increases, which could facilitate such a variation without the driver having to continuously estimate conditions.

METHODS

Basis for site specific fertilization

Site specific fertilization can be performed on the basis of estimates of potential/realized yield, soil type, topography, registered N-min content, humus content etc. A number of these parameters can be determined by measurements whereas others can only be determined from a visual estimate or the farmer's experience from previous cultivation.

Technical solutions

Two different systems for site specific fertilization are available in Denmark today (see Figure 1). The positioning of the spreader in the Agri Matic system is based on fixed tramlines in the field whereas in the LORIS system positioning is based on the use of DGPS signals.

The Agri Matic system

The Agri Matic system is based on a computer programme which maps the fields of a farm. When using the system, driving routines and tramlines are fixed for all relevant fields. This is achieved partly by determining the driving direction and partly by registering the position and length of the tramlines. Also obstacles in the field

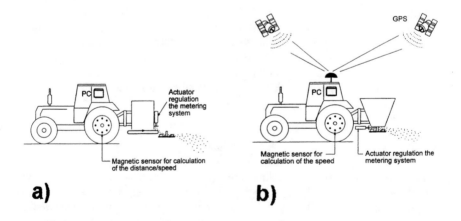

a)   **b)**

FIGURE 1. Principles of the two systems tested. (a) The Agri Matic system, where the position is located by the knowledge of the position of the tramline and the distance traveled. The actual fertilizer levels are adjusted by an actuator on the metering system. (b) The LORIS system, where the position is located by using DGPS signals. The speed is determined by a metering device on the tractor wheel. The actual fertilizer level is adjusted by an actuator on the metering system.

(e.g., poles or water holes) are included on the field maps. The tramlines are fixed by driving through the field.

The variation in the fertilizer application rate for each tramline is calculated on the basis of the factors mentioned previously, according to the decisions made by the user or the guidelines given by the consultant. In the few cases the Agri Matic system is used in practice the field maps are made on the basis of a few soil samples and the farmer's experience of soil texture and yields. The variable fertilizer level for each fixed tramline is recorded on the electronic field map. This map divides the field into quadrangles which are each given an individual level of fertilizer. The spreader will change the dose level every time it passes into a new area. This system is primarily used for the application of nitrogen, but is applicable to all types of fertilizers.

To use the Agri Matic system, portable PC is used for planning the fertilizer application and making the application maps. For spreading in the field, the tractor is equipped with a PC, a monitor, a joystick and an inductive sensor which is actuated by a number of magnets on a wheel rim. The sensor gives a signal which can be used to calculate the distance driven along each tramline. When spreading, the beginning and the end of each tramline must be given by the joystick to the computer and data from the tramline are recorded by the PC. The fertilizer spreader is a twin disc spreader, Bredal B2, equipped with an electric motor actuating the slider valve for the fertilizer application. When driving along each track, time lags are put into the controlling programme which are dependent on the spreading width to account for the actual backward spreading distance. The damper setting has a sliding adjustment in order to avoid distinct changes between two fertilizer levels.

The LORIS system

The LORIS system (LOcal Resource Information System) is a programme which creates fertilizer application maps for spreading in the field. The maps are created from a series of basic inputs such as the field boundary, average fertilizer requirements, and expected yield, plus a series of geo-reference data such as yield maps from previous years, soil texture and soil analyses. The input data can be imported as geo-coded point data or drawn in the programme as polygon data. All input data will automatically be converted to a net of grid points with a smooth transition between each point. The default grid point size is 10 m. The local fertilizer requirements are then calculated on the basis of fixed algorithms and illustrated as fertilizer maps comparable to topographical maps with contour lines. Data are transferred from the planning computer by means of a chip card which is placed in a controlling computer installed in the tractor. The fertilizer spreader (Bøgballe EXW - spinning disc type) is controlled according to the fertilizer map and the position in the field.

By using the DGPS system, the computer always 'knows' the position of the spreader and can thereby calculate the amount of fertilizer needed. The actual driving speed is measured by inductive sensors in the same way as in the Agri Matic system. The fertilizer map is expressed as a set of grid points and the computer calculates the actual fertilizer application level, by taking into account the level at nearby grid points. If the DGPS signals are lost the system is designed so that the spreader will automatically spread fertilizer at the average level for the whole field so that, in principle excessive, fertilization can be avoided. As in the Agri Matic system the spreader is equipped with a spindle motor actuating the damper slide of the spreader.

RESULTS

The two systems for site specific fertilizing have both been tested in practice and in on-going experimental trials. The results published here have been achieved by testing specially designed application maps under varying conditions. Figure 2 shows examples of tests of the Agri Matic System along a tramline in which there were 15 changes in the fertilizer level and also changes of speed. The test showed a good correlation between the setting predicted by the machine and the actual setting. In only a few spots was there a deviation of more than 10 kg N/ha at 20 m working width. This result is considered satisfactory.

Figure 3 shows the results of the LORIS system test. The ability of the system to regulate fertilizer application under various conditions was tested. The results shown derive from driving at a constant speed and regulating the fertilizer application from 50 to 150 kg N/ha. The system was tested by driving up and down the same tramline in order to examine the accuracy of the position determination. The results showed a high level of accuracy. The difference between to and fro is due to a minor time lag put into the programme. If the time lag is altered, the change will be increased. In the tests, the changes made by the system are correct in relation to the calculated ones. Unlike in the Agri Matic system, with the LORIS system the driver is free to decide the driving

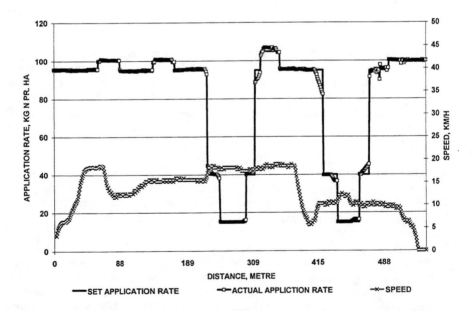

FIGURE 2. Results from tests using the Agri Matic system on a 514 metre tramline. The settings on the metering system and the actual speed were measured. The dose level was estimated by measuring the settings of the metering system and the actual speed. Working width 20 m.

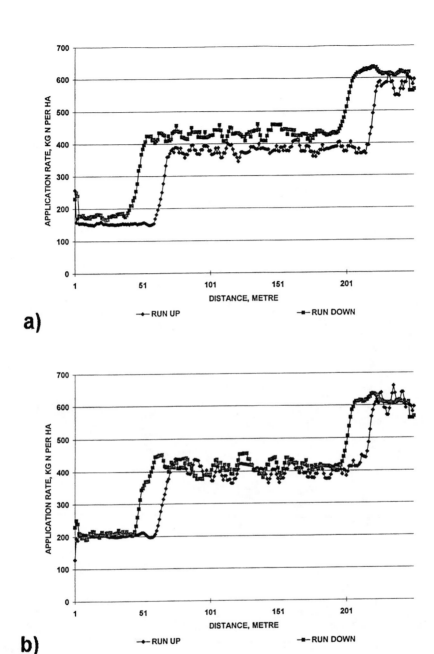

FIGURE 3.1. Results from tests of the LORIS system. The dose level was estimated by measuring the settings of the metering system and the actual speed. The test was carried out twice with the tractor traveling both up and down the field in the same track. a) 12 meters working width, 8 km/h and graduation from 200 to 400 to 600 kg/ha. b) 24 meters working width, 8 km/h and graduation from 200 to 400 to 600 kg/ha.

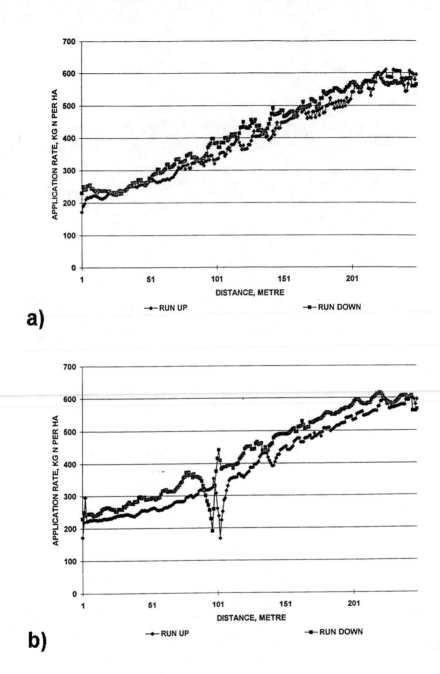

FIGURE 3.2. Results from tests of the LORIS system. The dose level was estimated by measuring the settings of the metering system and the actual speed. The test was carried out twice with the tractor traveling both up and down the field in the same track. a) 18 meters working width, 8 km/h and even graduation from 200 to 600 kg/ha. b) Sudden driving stop and measured delay in shutting off the metering system. Even graduation from 200 to 600 kg/ha.

direction within the field. For this reason, it is possible to drive in the direction where the changes in application level are minimized. The computer programme ensures a smooth transition between levels. However, the test showed that if the user draws an application map in the form of polygons where there is a clear line between two dose levels, and then drives along this line, the grid point principal may result in fluctuations in the dose level between the levels in the two adjoining areas. The significance of this depends on the difference in dose level and the grid size. A similar effect is observed when driving close to the field boundary because the boundary line will represent the average dose level. Research on this subject will continue.

In the Bredal spreader, changes in dose level when changing the speed are controlled mechanically and therefore a sudden stop will instantly stop the spreader. In the Bøgballe system, the dose level at varying driving speeds is controlled by inductive sensors. The test conducted here showed that when the tractor is stopped very abruptly there is a delay of 3.25 seconds before the spreader stops and this will cause a small overdose. In practice this is not very important.

Overall fertilizer distribution

In addition to measurements of the setting accuracy of the machines, the fertilizer distribution in the field was calculated on the basis of measured three dimensional distributions with different machines and quantity settings. The purpose of the calculations was to find the relationship between the distribution required and that actually obtained. In practice, we must assess not only the actual amount of fertilizer applied per metre, but also the ability of the spreader to distribute the fertilizer evenly.

From experience it is known, that the spreading pattern for some spreader types changes with the amount spread, and consequently, an even distribution which is a prerequisite in the case of site specific fertilizing, cannot be expected.

It has also been shown (Griepentrog, 1997) that the CV of the spreading distribution across the tramlines is influenced by the extent of overlapping on the transition between two doselevels. Griepentrog results shows 1) the lowest CV when spreading distribution is with single overlap, 2) a slightly higher CV with double overlapping and 3) the highest CV when there is no overlapping (full width spreader). This also has to be taken into account when testing spreading performance in precision farming.

CONCLUSIONS

From the tests conducted, it seems evident that there are technical solutions to the problems posed by site specific fertilization. In these tests, the Agri Matic system showed a high degree of accuracy in terms of the dosage at single spots in the field.

Likewise, the LORIS system showed a high level of accuracy when driving across areas of sharp changes in fertilizer level. When driving very close to the field boundary or when driving along a line marking an abrupt change in dose level, jumps in the application level may occur, depending on the dose level and distances from the grid points. Tests will be continued.

The total fertilizer distribution in the field depends not only on the quantity of fertilizer applied, but also on the spreading equipment. Examinations of the spreading properties of various spreaders show that the spreading distribution varies according to the dose level, in a manner which depends on the spreader. This is primarily due to the sensitivity of the spreaders to the setting of the feed point position or the vane position on the disc.

Further examination of this issue is required in order to make a final assessment of the possibilities for site specific fertilization.

REFERENCES

Gripentrog (1997) Teilflächenspezifische Düngung. *Landtechnik,* **52(1)**, 12-13.

A SITE-SPECIFIC IRRIGATION SYSTEM FOR THE SOUTHEASTERN USA COASTAL PLAIN

E.J. SADLER, C.R. CAMP, D.E. EVANS, L.J. USREY

Coastal Plains Soil, Water, and Plant Research Center, USDA-ARS, 2611 West Lucas Street, Florence, SC 29501-1241, USA

ABSTRACT

Yield maps from 1985 to present on a highly variable Coastal Plain field increasingly implicate soil water relations as the cause of spatial yield variation. Soils are sandy, often with dense horizons. Together, these factors limit water storage in the root zone, a conclusion reached through experience, observation, and process-level crop modeling. Management of water through irrigation in this region is complicated by the limited soil water storage and the significant chance of rain, which increases the risk of leaching if deficit irrigation is not practiced. However, spatially variable soils mean that even careful management of water, if done uniformly across the field, will still be improperly done on a significant portion of the area. It appears that the only method of addressing water stress on spatially variable fields while minimizing the potential of leaching under wetter areas is by site-specific water management. Therefore, in 1995, two site-specific center pivots were built by adding custom hardware to commercial center pivots. The first machine was used to control water and fertilizer application for a replicated experiment with 144 small plots (9 m radially x 7.5° [10-15 m]), which tested the hardware and software under controlled conditions. The second machine is undergoing modifications at the current time.

INTRODUCTION

The southeastern USA Coastal Plain is roughly the coastal one-third of Virginia to Georgia. It is comprised of nearly level, sandy surface soils and sandy clay subsoils (Pitts, 1974; USDA-SCS, 1986). The landscape contains numerous shallow (<3 m) depressions of varying size and unknown origin. Surface texture within the depressions is generally finer than that outside, where the soils are generally sandy loam or loamy sand, with extensive inclusions of sands. Many soils also have an eluviated E horizon of similar texture to the A, but with very little organic matter (<1%) and high bulk density (up to 1.8 g cm^{-3}). The sandy soils and root-restricting eluviated horizons combine to reduce available water holding capacity (commonly 20 to 40 mm) and thus make nonirrigated crop production a challenge in the area. To increase rooting depth, management practices commonly include subsoiling to a depth of about 0.4 m beneath the crop row to fracture the E horizon.

Coastal Plain climate is warm, humid, and cloudy. Average rainfall is >1000 mm/yr. Most summertime rain occurs during thunderstorms, causing June, July, and August to be the months with greatest rainfall, averaging from 100 to 150 mm/month. However, each month during the growing season has ranged from 20 to 250 mm during the past century. Such variability in rainfall, with the poor water relations described above, means that yield-reducing drought stress frequently occurs in an area that appears to have plentiful rain. Sheridan *et al.* (1979) reported a 50% probability of 22 day droughts during the growing season. Such drought dramatically reduces crop growth and yield.

Spatial patterns in crop growth, particularly during dry years, suggest water management may be critical for managing soil variability in the Coastal Plain (Karlen *et al.* 1990; Sadler *et al.*, 1995a; 1995b). Persistence of relative yield patterns for drought and non-drought years supports this assumption. Difficulties in scheduling irrigation for a center pivot on variable soils had illustrated the problems encountered when attempting to manage soil water under these circumstances (Camp *et al.*, 1988).

In 1991, a team designed a computer-controlled, variable-rate center pivot (see Camp and Sadler, 1994). Two commercial machines were acquired (description below), and modifications were made to achieve this objective. The first machine, used since 1995, was demonstrated under the controlled conditions of a replicated experiment on a reasonably uniform field. The second machine, which will be modified based on experiences with the first, will be the culmination of the project - variable-rate management of water, fertility, and pesticides on a highly variable Coastal Plain soil.

Literature and communication with independent researchers working toward similar goals contributed to the design of the machine. Lyle, W.L. (personal communication, 1992) described a multiple-orifice emitter design that could be individually switched to provide a series of stepwise incremental flow rates. This was part of the Low-Energy Precision Application (LEPA) system. Duke *et al.* (1992) and Fraisse *et al* (1992) switched sprinklers on and off for varying proportions of a base time period, usually 1 min. This design can provide a continuous range of application rates using a single nozzle, where other systems require additional nozzles, manifolds, and switches to achieve additional increments of rate. However, the on/off sprinkler action may be in or out of phase with the start-stop motion of the irrigation tower, impressing additional variability in application depth. This disadvantage is minimized when the wetted radius is larger, the alignment of the irrigation machine is controlled very closely, and the base time period of the sprinkler is small relative to the duration of tower stoppage. Stark *et al.* (1993) used a similar concept with a patented (McCann and Stark, 1993) control system for a variable-rate linear-move system, in which individual conventional sprinklers were controlled by computer. Three sprinkler sizes ($\frac{1}{4}$, $\frac{1}{4}$ and $\frac{1}{2}$ of full flow) provided $\frac{1}{4}$, $\frac{1}{2}$, $\frac{3}{4}$, and full irrigation. This system was installed on a field-scale center pivot, and uniformity of application was reported. Further developments on a linear move system were reported by King *et al.* (1995).

The objective of this presentation is to describe a variable-rate center pivot machine and to illustrate its capabilities to the European precision agriculture audience.

DESCRIPTION OF THE CENTER PIVOT

Commercial machine

The commercial system is described in Camp *et al.* (1996). It will be summarized here. Two small, 3-tower, 137 m commercial center pivots were purchased in 1993 (Valmont

Irrigation, Inc., Valley, NE[1]). In anticipation of increased load, truss design was heavier than normal; otherwise, the unit was conventional. A set of overhead sprinklers and a set of LEPA quad sprinkler heads on drop tubes were installed on both machines, to provide immediate ability to irrigate, albeit uniformly.

Modifications

PLC control system. All electrical output devices (solenoids, pumps, controllers, etc.) were controlled using a programmable logic controller (PLC: GE-Fanuc model 90-30, Charlottesville, VA) mounted on the mobile unit, about 5 m from the pivot point. Expansion units (3/pivot) with analog and digital cards were installed along the truss and connected by cable to the PLC. The PLC had an on-board 80386 PC with software written in Visual Basic (Microsoft Corp., Redmond, WA) to convert a map of control values to on-off settings in the directly-addressable solenoid control registers of the PLC. In order to determine location from the C:A:M:S® (Valmont Irrigation, Inc.) controller, the communication link required between the mobile PC and the stationary C:A:M:S was made with short-range radio-frequency modems (900 MHz, broad-band modems; Comrad Corp., Indianapolis, IN). The on-board PC repeatedly interrogated the C:A:M:S unit to determine the angle of the pivot and other parameters to provide assurance the system was functioning properly, and also exerted some control over the C:A:M:S unit, setting speed and shutting down in emergencies. The position in polar coordinates was found using the angle and the segment position on the truss. (The angle reported was found to be systematically in error, so a correction was determined with surveying techniques and built into the software.) When the location had been determined, the program checked whether a plot boundary had been crossed. If not, the interrogation cycle repeated. When a boundary was crossed, the expected application map was checked, the appropriate table lookup was performed, and the solenoid registers set accordingly.

Water delivery system. The design and modification of the manifolds and sprinklers for the first commercial pivot were done in cooperation with The University of Georgia Coastal Plain Experiment Station, Tifton, GA (Omary et al. 1996). The truss was segmented into 13 sections 9.1 m (30 ft) long (see Figure 1). Each section had three parallel, 9.1 m manifolds, each with six industrial spray nozzles at 1.5 m spacing. Water was supplied to each set of three manifolds directly from the boom via 5 cm (2 in) ports, drop pipes, a distribution manifold, and hoses. Each individual manifold had a solenoid valve, pressure regulator, low-pressure drain, and air entry port. The three manifolds and their nozzles were sized to provide 1x, 2x, and 4x a base depth at the position of the section, which depended on distance from the center to account for the greater area subtended per unit angle traveled. Octal combinations of the three manifolds provided 0x, 1x, 2x,...7x the base depth. The 7x depth was designed to be 12.5 mm (0.5 in) at 50% duty cycle on the outer tower. The small size of the unit, 120 m, meant that at 100% duty cycle, a full circle could be irrigated in less than 4 hr, and at a 17% setting, in less than 24 hr.

Distribution uniformity of the water application depth was examined for the worst-case

---

[1] Mention of tradenames is for information purposes only. No endorsement implied by USDA-ARS or any cooperator of preference over other equipment that may be suitable for the application.

scenario, in which one element (9.1 m square) was irrigated at a nominal depth of 12.5 mm and was surrounded by elements without irrigation. Distribution was measured using 50 cups spaced 0.3 m apart along a line in the radial direction. In the tangential direction, the 50 cups were staggered so that one line of 25 was beneath a nozzle, and the other line of 25 was between nozzles. Each test was repeated three times. As can be seen in Figure 2, spray carryover and drift caused an area about 3 m on either side of the nominal control zone to be irrigated at depths other than the target. This was expected from indivdual nozzle characteristics. Baffles are being considered to limit the carryover in the radial direction, but it appears that a buffer zone will be needed between elements in the tangential direction. All buffer areas are avoided for plot yield measurements.

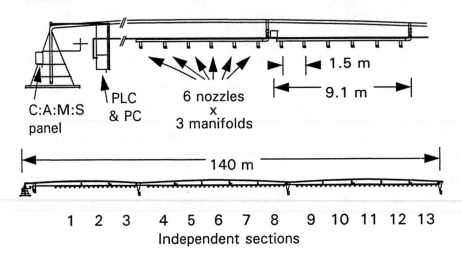

FIGURE 1. Side view of site-specific center pivot and closeup of tripod and example section.

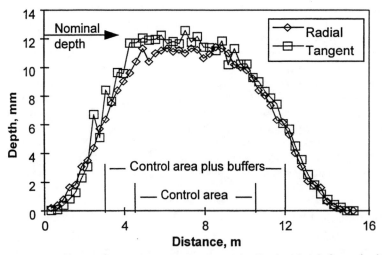

FIGURE 2. Radial and tangential distribution of application depth for a single element surrounded by unirrigated elements. Each point is the mean of three measurements.

Nutrient injection system. Injection of nutrients (Urea-ammonium nitrate, UAN) was accomplished using a 4-head, 24 V DC variable-rate pump (Ozawa Precision Metering Pump, model 40320), check valve, and nurse tank connected to the stationary vertical riser. Since the flow rate of water could vary depending on the spatial application schedule, the amount of fertilizer injected into the water supply pipe was varied proportionately in order to hold the concentration constant. This was done by the PC on board the PLC, which calculated the aggregate flow rate, the required injection rate, and the 0-5 V DC voltage required, and then reported that to the operator. Spatially-variable application of nutrients was done using a minimal, spatially-variable irrigation, but with uniform concentration.

Pesticide application system. A proprietary, ultra-low-volume (130 liter/hectare) pesticide application system was installed on the first pivot in summer 1996. The 13-segment organization and control system were used, although the pesticide system (pump, sprinklers, and nurse tank) was completely separate and used the pivot solely as a ground transport.

Canopy temperature system. The center pivots were rigged with aluminum booms and masts to hold small infrared thermometers for each of the 13 sections. The booms extended about 3 m in front of the leading edge of the manifold, and the masts were designed to adjust 1.5 m above or below the boom, which was at 3 m height. Pivot #1 had one IRT installed per section, with the footprint nominally centered within the 9.1 m section. Pivot #2 had two IRTs per section, with the footprints about 3 m inside the ends of the 9.1 m sections. The IRTs were Exergen Irt/c .3X with 3:1 field of view (~17°) and type K thermocouple leads. The IRTs were read using analog cards on the PLC, and the data were stored on the on-board PC. Figure 3 shows dry, bare soil surface temperature under pivot #2. The cooling trend as the pivot rotated clockwise from straight up (0 degrees) is evident, as are the grassed access roads at 0 and 175 degrees. Circular patterns are attributed to sensor differences from nominal calibrations.

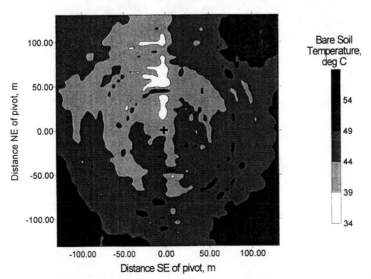

FIGURE 3. Map of dry, bare soil surface temperature, taken from 12:00 to 15:35 local standard time.

Use in replicated plot experiment. The pivot described above was sited on a relatively uniform soil area (USDA-SCS, 1986), chosen specifically for proving the technology under controlled conditions. The primary experimental objectives were to test rotation and irrigation effects on a corn-soybean rotation vs continuous corn under conservation tillage. A secondary objective was to test subsoiling against not doing so, in the possible trade-off of irrigation to manage water rather than subsoiling to increase the rooting depth.

Experimental design. There were 144 treatment plots in total: 4 replications x 3 rotations (corn-corn, corn-soybean, soybean-corn) x 2 tillage (subsoiled, non-subsoiled) x 3 water managements (rainfed, tensiometer, crop stress) x 2 nitrogen (single sidedress, incremental applications). In 1995, both the tensiometer and crop stress treatments were operated based on tensiometers. The individual plots were laid out in a regular 7.5° by 9.1 m (30 ft) pattern, which made the minimum plot length 10 m in section 7 and 15 m in section 13. As seen in Figure 4, the four replicates were sited in the outer annuli, which had the most uniform soil areas. The outer rings were used so that planting and other operations could be done without sharp turns. All operations were done on the circle rather than with straight rows to simplify operations in this experiment.

Fertilization during this experiment was achieved by injecting urea-ammonium-nitrate (UAN 24S) into the system. To prevent spray drift, 38-mm layflat hose was placed around the 2x nozzles and extended to the ground. The 2x nozzles provided 3.6 mm of irrigation at 50% duty cycle and 1.8 mm at 100%.

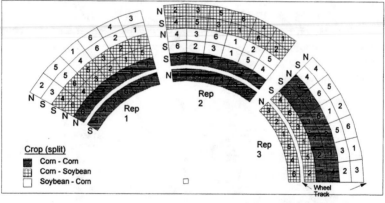

Grassed Waterway and Field Road

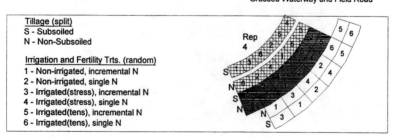

FIGURE 4. Plot plan for replicated field experiment used to test site-specific center pivot under controlled conditions. (After Sadler et al., 1996).

Conclusions

Since this description is of the design and testing of an apparatus, conclusions must be limited to the performance thereof. The site-specific center pivot evolved from the basic commercial machine in March 1995 to a functioning, proven technology by August. Control software was primitive and fragile initially, but similarly evolved through modification and experience such that operation was possible via the remote C:A:M:S unit by the end of the summer. The second year's experiences were acceptable. Prior measurements of system uniformity had demonstrated acceptable distribution within control elements as well as expected border effects between elements with contrasting application depths (Omary et al., 1996). Further tests confirmed this as fact. No evidence was seen that uniformity or border width had changed. Surface redistribution had been a concern during design, because of the small wetted radius of the sprinkler, but even the collection into layflat hose for fertilization did not cause excessive local ponding and runoff. Preliminary tests with infrared thermometers to map spatial variation suggest that methods must be developed to account for temporal skew, unit-specific calibration, and solar irradiance.

Plans are to complete outfitting the second pivot with variable rate irrigation, fertilization, and low-volume pesticide variable-rate application equipment based on experiences gained with the first pivot. The software will be modified to accommodate irregular soil unit boundaries, using map units as the primary control factor initially, but general enough to handle any spatial control factor.

REFERENCES

Camp, C.R., Christenbury, G.D., Doty, C.W. (1988) *Scheduling irrigation for corn in the Southeast*. C.R. Camp and R.B. Campbell (Ed.), Florence, SC, ARS-65, USDA-ARS, Chapter 5, pp 61-78.

Camp, C.R., Sadler, E.J. (1994) Center pivot irrigation system for site-specific water and nutrient management. ASAE Paper No. 94-1586, St. Joseph, MI, ASAE.

Camp, C.R., Sadler, E.J., Evans, D.E., Usrey, L.J., Omary, M. (1996) Modified center pivot irrigation system for precision management of water and nutrients. ASAE Paper No. 962077, ASAE Annual Meeting, Phoenix, AZ, July 14-18, 1996.

Duke, H.R., Heermann, D.F., Fraisse, C.W. (1992) Linear move irrigation system for fertilizer management research. *Proc. International Exposition and Technical Conference*, The Irrigation Association, pp 72-81.

Fraisse, C.W., Heerman, D.F., Duke, H.R. (1992) Modified linear move system for experimental water application. *Advances in planning, design, and management of irrigation systems as related to sustainable land use*, Belgium, Leuven, 1, pp 367-376.

Karlen, D.L., Sadler, E.J., Busscher, W.J. (1990) Crop yield variation associated with Coastal Plain soil map units. *Soil Sci. Soc. Am. J.*, **54**, 859-865.

King, B.A., Brady, R.A., McCann, I.R., Stark, J.C. (1995) Variable rate water application through sprinkler irrigation. Site-*specific management for agricultural systems. 2nd International Conference*, Madison, WI, ASA/CSSA/SSSA/ASAE, pp 485-493

McCann, I.R., Stark, J.C. (1993) Method and apparatus for variable application of irrigation water and chemicals. U.S. Patent No. 5,246,164.

Omary, M., Camp, C.R., Sadler, E.J. (1996) Center pivot irrigation system modification to provide variable water application depth. ASAE Paper No. 962075, ASAE Annual Meeting, Phoenix, AZ, July 14-18.

Pitts, J.J. (1974) Soil survey of Florence and Sumter Counties, South Carolina. USDA-SCS, U.S. Government Printing Office, Washington, DC.

Sadler, E.J., Bauer, P.J., Busscher, W.J. (1995a) Spatial corn yield during drought in the SE Coastal Plain. *Site-specific management for agricultural systems. 2nd International Conference*, Madison, WI, ASA/CSSA/SSSA/ASAE, pp 365-382.

Sadler, E.J., Busscher, W.J., Karlen, D.L. (1995b) Site-specific yield on a SE Coastal Plain field. *Site-specific management for agricultural systems. 2nd International Conference,* Madison, WI, ASA/CSSA/SSSA/ASAE, pp 153-166.

Sadler, E.J., Camp, C.R., Evans, D.E., Usrey, L.J., (1996) A site-specific center pivot irrigation system for highly-variable coastal plain soils. *Proceedings of the Third International Conference on Precision Farming.*

Sheridan, J.M., Knisel, W.G., Woody, T.K., Asmussen, L.E. (1979) Seasonal variation in rainfall and rainfall-deficit periods in the Southern Coastal plain and Flatwoods Regions of Georgia. *Georgia Agric. Exp. Sta. Res. Bull.* **243**, 73 pp.

Stark, J.C., McCann, I.R., King, B.A., Westermann, D.T. (1993) A two-dimensional irrigation control system for site-specific application of water and chemicals. *Agronomy Abstracts* **85**, 329.

USDA-SCS. (1986) Classification and correlation of the soils of Coastal Plains Research Center, ARS, Florence, South Carolina, USDA-SCS, South National Technical Center, Ft. Worth, TX.

CONSISTENCY IN YIELD VARIATION AND OPTIMAL NITROGEN RATE

L. THYLÉN

Swedish Institute of Agricultural Engineering, P.O. Box 7033, S-750 07, Uppsala, Sweden

ABSTRACT

In Sweden, the optimum nitrogen rate depends mainly on the yield potential. Therefore a yield map could be of significant help in determining optimum nitrogen rates. The objective of this study was to examine the consistency of variation in yield and optimum nitrogen rate.

Intensity trials of nitrogen were conducted during each of three consecutive years in a field containing zones of different yield potential. The nitrogen rates were 0, 50, 100, 150 and 200 kg N/ha, except during the first year when the 200 kg N/ha rate was excluded.

Over the study period, the spatial patterns of variation in both yield and optimum nitrogen rate varied consistently. However, as expected, variation in absolute values was inconsistent. The optimum nitrogen rate was higher in high-yielding areas than in lower-yielding areas.

INTRODUCTION

During recent years many yield-mapping systems have been commercialised in Europe and North America. Well known, are mapping systems from Case, Claas, John Deere, LH-Agro (AgLeader), Massey-Ferguson and RDS. The yield mapping system is an essential part of any precision farming system. Other components of high-precision farming include the use of variable rate technology (fertilizers, pesticides, lime, etc.) and consideration of the variation in the present crop.

For many farmers yield mapping is their first step towards precise farming. The detection of a variation in yield that is consistent over several years can help farmers pinpoint trouble spots. There is a great deal of literature discussing consistency in yield variation. For example, Auernhammer *et al.* (1995) reported a high degree of similarity in patterns of spatial variation in yield between years. Stafford *et al.* (1996) also reported similar spatial variation in crop yield between years.

At JTI, the first yield maps were made in 1992. To determine position, a dead reckoning system was used, and since then DGPS has been used for this purpose.

The aim of the present study was to investigate the consistency in yield variation and optimum nitrogen rate.

MATERIALS AND METHODS

Experimental field

The experimental field, located near Uppsala, was soil mapped in 1993. Samples were taken in a square grid (grid size 33 x 33 m). The soil is a heavy clay (35 - 60%) with a low organic matter content (1.7 - 2.7%).

Crops

During the study, the field was planted with summer barley and oats. In Sweden wheat is commonly planted after oats, but because of wet weather conditions this was not possible. The rotation order, starting in 1992 and ending in 1995, was oats, barley, oats, and barley. After the harvest in 1993, the size of the field was reduced since part of it became set-aside.

Yield mapping

The yield was mapped during harvesting. For positioning, we used a dead reckoning system in 1992 and a DGPS receiver in the following years. Yield values were obtained from a measurement system (flow-meter) mounted on the combine. Recorded yield data were managed according to Thylén and Murphy (1996).

Intensity trials with nitrogen

In each year from 1993 to 1995 three intensity trials were conducted in the field. Nitrogen rates in the parcels were 0, 50, 100, 150 and 200 kg N/ha except in 1993 when the 200 kg/ha rate was excluded (average nitrogen rate in Sweden for applications to summer crops is about 100 kg N/ha). All trials were performed in same two blocks throughout the three-year period, with each block containing five 10 x 2.4 m plots. The positions of the blocks were moved 10 metres after every year. The positions of the trials are shown in Figure 1.

A third-degree polynomial (equation 1) was fitted to the recorded yield data. By differentiating equation 1, the optimum nitrogen rate can be calculated (equation 2) by replacing y' with the price quotient (nitrogen/grain). In this study the price quotient was assumed to be 5:

$$y = a + bx + cx^2 + dx^3 \qquad (1)$$

$$y' = b + 2cx + 3dx^2 \qquad (2)$$

where:
y = yield
x = nitrogen rate
a, b, c and d = constants

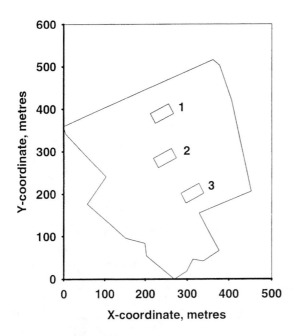

FIGURE 1. Lay-out of the nitrogen trials, which were placed about 150 m apart.

RESULTS

Crop yield

A short summary of the crop yield is presented in Table 1.

As can be seen from the yield maps (Figure 2) the pattern of variation in yield was fairly consistent. However, small areas did differ significantly from year to year. This variation could have been due to animal grazing, etc. The correlation table (Table 2) only includes areas that were yield mapped for four years.

TABLE 1. Summary of crop yield variation. In 1994 the oats crop suffered from a severe drought.

| | ANNUAL YIELD | | | |
	Minimum (ton/ha)	Average (ton/ha)	Maximum (ton/ha)	Standard deviation (ton/ha)
1992	3.66	5.77	7.05	0.60
1993	1.40	4.54	6.24	0.78
1994	1.73	3.40	5.20	0.73
1995	1.81	5.12	6.27	0.72

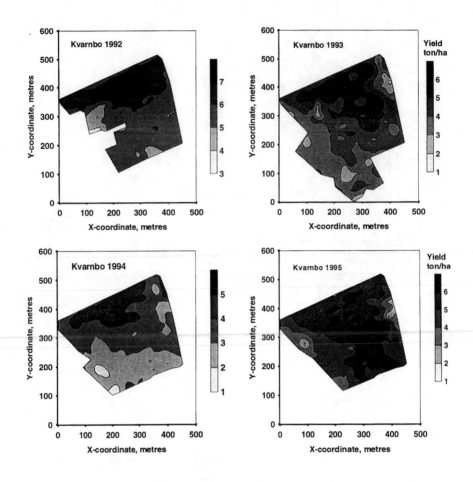

FIGURE 2. Yield maps from the experimental field.

TABLE 2. Correlation in yield between years. The correlation matrix was calculated based on four yield grids from four consecutive years of yield maps, each containing 981 observations. The area of every grid node was 100 m².

	1992	1993	1994	1995
1992	1	0.505	0.671	0.489
1993		1	0.566	0.706
1994			1	0.523
1995				1

Intensity trials with nitrogen

The optimum nitrogen rate varied significantly within the field (Table 3). No trials were conducted in very low-yielding areas such as field entries or headlands. Areas with higher yields had a higher optimum nitrogen rate. There was a good correlation (Table 4) between optimum nitrogen rates in different years.

Soil sampling

As the yield, the soil varied rather much. Many parameters showed a good spatial dependency. When looking at correlation between yield and soil chemistry the results were rather discouraging. There was a weak negative correlation between yield and phosphorus levels in the soil. However, there was a better correlation between yield and clay content (Figure 3).

DISCUSSION

At the site described the variation in yield has been fairly consistent. The yield pattern for other fields on the farm, yield mapped over several years, shows similar variation in crop yield. With a consistent variation in yield the information could be used for planning of variable rate application of nitrogen. However, of a higher economical interest are planning of set-aside and the possibility to locate areas with some kind of yield limiting factor.

TABLE 3. Variation in optimum nitrogen rate in the three test plots over a three-year period. The nitrogen/grain price quotient was estimated to be 5.

Year	Optimum N-rate (kg/ha)		
	Trial 1	Trial 2	Trial 3
1993	> 150	104	75
1994	57	35	0
1995	115	75	68

TABLE 4. Correlation matrix for the calculated optimum nitrogen rates for the three test plots over a three-year period. The values in Table 3 were used to calculate the correlation matrix.

	1993	1994	1995
1993	1	0.966	0.968
1994		1	0.870
1995			1

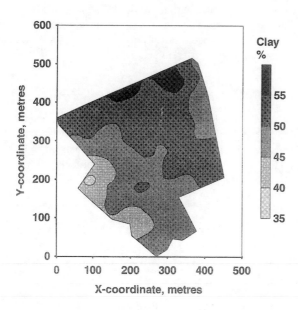

FIGURE 3. The variation in clay content was similar to variation in crop yield.

ACKNOWLEDGEMENTS

This study was funded by the Swedish Farmers' Foundation for Agricultural Research and the Royal Swedish Academy of Agriculture and Forestry.

REFERENCES

Auernhammer, H., Demmel, M., Pirro, P.J.M. (1995) Yield measurement on self propelled forage harvesters. *ASAE paper no. 951757.*

Stafford, J.V., Ambler, B., Lark, A.B., Catt, J. (1996) Mapping and interpreting the yield variation in cereal crops. *Computer and electronics in agriculture,* **14(2,3),** 101-119.

Thylén, L., Murphy, D.P.L. (1996) The Control of Errors in Momentary Yield Data from Combine Harvesters. *J. agric. Engng. Res.,* **64,** 271-278.

MULTIDISCIPLINARY APPROACH FOR PRECISION FARMING RESEARCH

G.W. BUCHLEITER, W.C. BAUSCH, H.R. DUKE, D.F.HEERMANN

USDA-Agricultural Research Service, Water Management Unit, Colorado State University, Fort Collins, CO 80523, USA

ABSTRACT

Research in precision farming requires a multidisciplinary approach because of the many complex interactions between various crop production factors that affect crop yield. A 5 year study involving 20 researchers was initiated in 1996 to quantify the variabilities of crop production factors and yield on two sprinkler irrigated corn fields. Current data collection activities are focused on characterizing spatial variabilities prior to implementing precision farming (PF) practices. Field data coupled with simulation models are used to produce GIS layers of information about variations in water, nutrients, vegetative material and pests. In the future, a statistical model will be developed to aid producers in analyzing their data and to make better PF management decisions. These PF practices will be implemented under center pivot irrigation and evaluated for economic and environmental benefits.

INTRODUCTION

Spatial variability of several yield limiting factors such as nitrogen, phosphorus, and organic matter, within a field has been recognized for a long time (Franzen *et al.*, 1996; Varvel *et al.* 1996). Producers have become much more interested in managing crop production inputs to compensate for the spatial variability with the introduction of new technologies such as yield monitors and intensive soil sampling and variable rate application equipment. To date, much of the technology that has been developed has focused on yield mapping and variable application of fertilizers. Intuitively, precision farming (PF) is an opportunity to optimize production inputs through variable rate applications which places materials where they can be used most efficiently. Precision farming is not just the use of high-tech equipment, but the acquisition and wise use of information obtained from that technology (Vanden Heuvel, 1996). The paradigm for improving management is to not only obtain more data from sub-areas within a field, but also have the ability to interpret these data to make better management decisions appropriate for that sub-area. Implementing PF requires significant investment and an increased level of management. Producers need the technological tools to be able to assess the variability and the potential for increased profitability, before making the necessary investments.

There are significant areas in the Great Plains and Pacific Northwest regions of the United States that are irrigated with self propelled irrigation systems (National Research Council, 1996). With careful water and nutrient management, these soils are very productive for both grain and root crops. However, ongoing environmental and economic pressures encourage producers to continually adopt more efficient management techniques such as PF.

The general long term thrust of our research effort is to evaluate the impacts of PF on water quality and the economic feasibility of PF under irrigated conditions. Initially, our main objective is to develop procedures that can: (1) quantify the scale and magnitude of variabilities of water, nutrients, weeds, insects, and diseases which affect yield; and (2) assess the causes of yield variability within a field under sprinkler irrigated conditions in the Great Plains region of the USA.

EXPERIMENTAL APPROACH

An integrated system that can economically quantify spatial variabilities, analyze the data and recommend appropriate management decisions, and reliably implement these decisions within the constraints of commercial farming practices, is crucial to the producer's long term adoption of PF. Such a system has three general components: (1) data collection (2) analysis and decision support, and (3) implementation of decisions. Much of the current PF research is focused on data collection and implementation using variable rate application of fertilizers, usually under rain-fed conditions. Typically research is focused on one or two variables to explain yield variations. usually within one or two scientific disciplines. The major gap in developing an integrated PF system is a thorough understanding of the complex interactions between the various aspects of crop production. Under irrigated conditions, water application is a management variable that has large impacts on yield as well as affecting other crop production factors.

The current practice on most commercial farms doing PF, is to develop variable rate application maps prior to planting and then apply fertilizer and chemicals accordingly. Documented economic and environmental benefits of PF are limited and are usually inferred or derived from crop yield response curves to fertilizers, or other relationships between various crop production variables (Lowenberg-DeBoer and Boehlje, 1996; Reetz and Fixen, 1995; Swinton and Ahmad, 1996). Many times the explanations for yield differences based on 1 or 2 years of data, are unable to account for yield differences in a year when growing conditions are different. This suggests that some pertinent factors affecting yield are not being properly included in the analyses.

There is temporal variability as well. Growing seasons are rarely the same so there are significant management decisions to be made through the growing season that can affect yield. Several large companies either have or are developing the capabilities to provide images of entire fields to farmers on a regular and timely basis throughout the growing season. Correct interpretation of these images along with necessary equipment to accomplish variable rate applications, can enable farmers to make better PF management decisions throughout the growing season.

We feel our research must be conducted within the constraints of commercial farms to evaluate PF as a complete system similar to the way producers might view it when considering adopting it. Our approach is to involve all of the applicable scientific disciplines in a common research project to more fully explore and understand the interactions of the various crop production factors at the field scale. Since many disciplines are involved and multiple components are required for a fully integrated system, a systems approach is needed to define the system boundaries and how input from the various disciplines can be synthesized in a common framework. A working group of about 20 researchers organized and agreed to conduct research in support of this overall objective on PF. Areas of technical expertise include soil fertility, plant pathology, weed science, entomology, plant pathology, remote sensing, ET, irrigation system evaluation and modeling, variable rate application of water and chemicals, hydrological and water quality modelling, statistics, and agricultural economics. These researchers from different agencies and different university departments agreed to contribute resources where possible from their individual research programs toward the necessary data collection. Initially only a modest amount of 'seed' money was available to initiate planning and help coordinate data collection activities. Two farmers who are very interested in PF, agreed to allow this multidisciplinary research on their commercial farms.

We recognized that good communication and delegation of some of the coordination work to smaller teams were necessary to efficiently develop sound research plans. A five-member multidisciplinary team was selected to manage the project. Its primary functions are to facilitate communication between researchers and farm cooperators and to oversee coordination of data collection and analytical activities among the researchers. A project coordinator was hired to carry out these duties on a daily basis. A second five-member multidisciplinary team was organized to develop a comprehensive written project plan which identified the data to be collected and how it will be analyzed to quantify the causes of yield variability. Our intent is to encourage cooperation and dialogue among researchers so sampling strategies are consistent and common data sets are collected which will strengthen analytical procedures and facilitate future model development as needed.

Identification of field sites and project planning began in 1996. The first two years of this anticipated 5 year study will focus on determining the magnitude and spatial variability of the various factors affecting yield, collecting necessary data for computer modeling, and collecting baseline information that documents current production practices and their impact on groundwater quality. Several researchers will be pursuing alternative and less expensive ways to quantify spatial as well as temporal variability. This information will be used to evaluate and refine future data collection procedures and to improve our understanding of how interactions of various factors affect yield. The remaining years of the project will focus on developing PF management recommendations, implementing PF practices under center pivot irrigation systems, and evaluating the economic and environmental benefits of the PF practices. Our goal is to characterize variabilities down to a 10 m x 10 m size, although management decisions would be made at a larger scale of approximately 0.5 ha.

ANALYTICAL APPROACH

Multivariate regression models, physically based computer simulation models, and neural network are several general approaches for relating yield to the various factors affecting it. The disadvantage of regression models is that although they relate cause with effect, they can only be used reliably within the range of input data that were used to develop the regression equation and they are limited in explaining or furthering the understanding of the physiological causes for yield variation. Many of the computer simulation models reflect the interests and biases of the developers and consequently treat different subsystems within the larger soil-water-plant-atmosphere system at different levels of detail.

A review of the papers and abstracts from the 3rd International Conference on Precision Agriculture indicated that statistical approaches are the prevalent general approach (Mallarino, et al., 1996; Han et al., 1996), although simulation models are being used in several instances to help quantify the effects of PF (Booltink et al., 1996; Pachepsky et al., 1996). The ability of either of these approaches to quantify the causes of yield variability over a range of conditions is unknown. Based on the literature and the combined experience of our research team, both approaches have merit. With this multidisciplinary approach there are a large number of parameters with many interactions that may not be understood well enough to model physically. A literature review of crop simulation models for corn indicated there were no physiologically based models that were validated and robust enough to predict yield based on water and nutrient statuses while simultaneously accounting for the impacts of pests, weeds and diseases on yield. Significant modifications or additions to a model are required to include the effects of weeds, insects, and diseases on yield.

Even though there is not a single simulation model that considers all the factors affecting yield, models can be used to quantify variability of parts of the system. Our initial efforts are to develop a general multivariate model for yield prediction and use physical computer models to economically quantify variability of certain layers of information. From a simplistic, crop physiological point of view and in the absence of yield reducing disease, insects, or weeds, four general factors - light, water, nutrients, and the amount of active photosynthetic material - determine maximum yield. Light is temporally variable but typically unaffected by management. On the other hand, vegetative material (plant density), water, and nutrients are greatly affected by management decisions. If there is a deficiency in any one of these factors, yield is reduced. Since the deficiency's impact on yield is cumulative, vegetative material, water, and nutrients are monitored through the growing season and quantified.

The analysis component combines the various pieces of raw data into layers of information for each of the general factors identified as affecting yield. Recognizing that the amount of data that can be collected to properly characterize a field will be voluminous and dynamic, a Geographical Information System (GIS) is an excellent tool for analyzing and visualizing the spatially related data. Variability of a specific factor or a combination of factors can be easily represented and visualized as a GIS layer of information. Figure 1 is a conceptual representation of the various layers of information that are used for a multivariate regression analysis of the factors affecting yield. Determining the impact of each of these factors requires use of deterministic models representing physical processes to combine several elementary layers. For example, layers with field slopes, water application, and infiltration rates are combined using an infiltration-hydrological routing model to generate two additional layers - available water and percolated water below the root zone. A deterministic ET-nutrient model updates the vegetative cover layer using information from the environment, water, and nutrient layers. For the insect and disease layers, the map delineations are the grid cell boundaries and reflect whether or not specific diseases or insects were present at the sample sites. Multiple layers may be included to reflect temporal variability.

The complex interactions between various factors are probably not possible to model in sufficient detail with GIS alone, so a multivariate regression model will be developed to predict yields. Data is obtained by querying the GIS database for the parameter value of each layer at numerous locations around the field represented by the vertical arrows in Figure 1. These values are the dependent variables in the multivariate regression analysis.

Actual yields throughout the field are measured at harvest with a commercial yield monitor mounted on the farmer's combine. Determining the full effect of weeds, insects, and diseases on yield is problematical in farmer's fields where the usual management practice is to aim for nearly complete control. Weed densities are predicted from weed seed bank counts taken

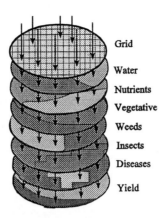

Figure 1. Conceptual representation of GIS layers.

from soil samples and weed seedling counts after crop emergence. Existing predictive models that assume no control measures, use the predicted weed densities to estimate yield losses. A GIS layer showing predicted yields will be derived using the multivariate model results and adjusted for yield depressions caused by pests.

A GIS layer indicative of the difference between actual yield and predicted yield, indicates the range in unexplained yield variability. To determine the adequacy of the model to properly estimate yield reductions because of suboptimal growing conditions, appropriate sampling points are selected across the field. Estimated yield is plotted against actual yield for each sampling point. A linear regression of these data is useful in assessing the accuracy of the model over a range of conditions. A regression equation with a slope of 1 and zero intercept indicates perfect correspondence with no bias. Since the model does not respond perfectly to all of the variations, the regression equation reflects both inadequacies in the model as well as measurement errors. If necessary, the model is adjusted so predicted yield more closely matches the actual yield.

SAMPLING APPROACH

Initially, research is focused on evaluating existing variabilities and their impact on yield, with no attempt to incorporate PF in the crop production practices. An essential project requirement was to determine the variabilities of factors affecting yield for two fields totaling 125 ha. Although the scales of spatial dependencies for the various factors were unknown during the planning stage, a maximum grid size of 76.2 m x 76.2 m was set. This sampling density is comparable other research reported in the literature as well as the highest densities being done commercially. Researchers responsible for each of the factors of interest agreed to develop layers of information at that scale as a minimum requirement.

The cost of obtaining the necessary data for each cell in a smaller grid was a big concern. Data collection approaches differ for the various factors of nutrient, water, vegetative material, weeds, insects, and diseases. Logistically and financially it is impossible to physically measure all factors at a higher density so alternative methods of describing variability are being used. Some researchers felt that the range of spatial dependency for some of the factors exceeded the 76.2 m spacing, making it possible to use geostatistics to estimate intermediate values. Depending on the available resources, additional samples were taken at closer spacings, usually in a 15.2 m x 15.2 m grid at several random locations in the field. These data were used to develop the variograms needed for kriging. Soil properties such as nutrient levels, organic matter, pH levels are estimated over the entire field using standard point sampling procedures and kriging.

In situ sorptivity and conductivity tests were taken at spacings ranging from 0.76 m to 76 m. Since soil water movement processes are highly nonlinear and soil parameters vary randomly over an area, adjustments to parameter values are necessary to correctly estimate soil properties and water movement on an aggregate scale. Geostatistical models were developed to predict values for soil properties at field locations that are intermediate to field sampling sites. These models will be used in designing sampling schemes that measure variations at a scale that is important to crop response, and is commensurate with the size of the subarea to be managed.

Since the stage of growth when water deficiency occurs has a big impact on yield, characterization of water availability throughout the growing season is crucial. Water applied during an irrigation is a controllable factor having both spatial and temporal variability. Pressurized self propelled irrigation systems are probably the best method for being able to control water and apply where desired, but they do not guarantee perfect uniformity. Poor design of sprinkler packages, incorrect installation of sprinklers and nozzles, fluctuating hydraulic conditions, and wind drift, can all cause problems in water application. Problems in obtaining accurate measurements of water application under sometimes difficult and varying field conditions and the large amount of labor involved in sampling at a small enough scale to capture variability, make exclusive use of field measurements unfeasible. An alternate approach is to use computer simulation models that use a limited amount of calibration data, to predict outcomes over a much wider area and/or range of conditions. A computer model has been developed to design and evaluate center pivot sprinklers. (Fraisse *et al.*, 1996; Heermann, 1990). Water application depths are calculated using field measurements of pumping level, operating pressure, sprinkler nozzle types and sizes, and sprinkler elevations. Sets of these measurements are taken for various topographical conditions, so the necessary simulations can be made to characterize water application over the entire field. A limited number of can measurements are taken at close intervals to verify model predictions. Rainfall is measured with recording raingauges at seven locations in the center and on the periphery of each field. Variability of precipitation can be very large for the thunderstorm activity that is typically experienced during the summer months.

Another approach to reduce data collection costs, is to characterize crop status with remote sensing. Variations in vegetative matter are determined primarily from remote sensing of reflectance at various frequencies in the visible and near infrared wavebands. Measurements of crop leaf area and plant densities are used for verification and calibration of the remotely sensed data. These remote sensed data will be used in conjunction with deterministic water and nutrient balance models to delineate similar areas on the GIS water and nutrient layers.

RESULTS

This project provides insight about multidisciplinary research on two levels. One level is an examination of how a systematic process can integrate different perspectives and types of data into a unified analysis that is useful in developing an integrated PF system. The second level is an evaluation of the performance of the PF system for economic savings and impacts on the environment.

One of the initial tasks was to identify the best field sites to collect the necessary data for assessing the causes of yield variability. Approximately sixty fields irrigated with center pivot sprinklers and growing corn, sugar beets, and onion, were available as potential field sites. Available resources of money and labor dictated that only two fields could be selected. Yield monitoring and the resultant yield maps were deemed essential in documenting and analyzing causes of yield variability. Although yield monitoring of root crops such as potatoes, onions, and sugar beets is being done, the prevalent type of yield monitors is for grain crops. Since much of the team's experience and expertise is working with corn, it was decided that initially work should be focused on corn only, which is the predominate grain crop under irrigation in the Great Plains.

It was desirable that the sites include significant variations in vegetation, topography, soil characteristics, weed, insect, and disease pressures, to cover as big a range as possible of the factors affecting yield. Other significant factors that could aid in data collection and interpretation were the availability of historical yield data, digitized soil maps, site accessibility for data collection activities, and the proximity of the two fields chosen because of the likelihood of both sustaining damage if hail occurred during the growing season. Since several years of data are necessary to evaluate yield variability under different climatic conditions, it was highly desirable that any site selected be planted to corn for several years in order to obtain a thorough and more complete understanding of the causes of yield variability. Only a few of the fields had yield maps in the previous 2 years. Since the normal crop rotation scheme included a crop other than corn every 3 or 4 years, it generally was not possible to identify any fields with a yield map history and still have two years of corn production.

With all of the factors to consider, a systematic approach which ranked the desirability of the identified factors and criteria for scoring each factor was set up. Based on the farmer's sense of variability and 1996 cropping plans, the sixty potential fields were narrowed down to twelve. Aerial color photographs were taken twice during the 1996 growing season, as well as interviews with crop consultants and site visits, to obtain a qualitative assessment of expected crop variability. Changes in cropping plans, lack of observed variability and atypical irrigation practices made all of the fields unsuitable and the systematic approach almost useless. Based on their revised 1997 cropping plans, farmers identified six fields as potential sites where corn will be grown for the next two years, and having the most variabilities in soil types and topographical relief. A site visit by members of the management team made the final selection of two field sites comprising 125 ha. The benefit of a systematic approach for site selection was that it required considerable discussion about what the essential requirements are and encouraged creative thinking across various disciplines about how various sampling tasks might be approached. It was very difficult to come up with completely acceptable weights, scoring functions, and scores in order to do a rigorous ranking of the various sites. The final decision was based primarily on meeting the required criteria.

Base maps with accurate horizontal and vertical control were prepared for both field sites using differential GPS equipment. Since topography has a significant effect on several of the factors affecting yield, a topographical map with an accuracy of > 6 cm was prepared for all researchers to use. Since most of the 1997 data collection activities are focused on characterizing spatial variabilities prior to implementing PF practices, we expect to present maps and statistics which reflect the measured variabilities of the various factors affecting yield at the two field sites.

SUMMARY AND CONCLUSIONS

A review of current research in PF indicates that at most, only a couple of factors affecting crop yield are considered in most studies. There are numerous factors affecting yield variability which interact with each other. A multidisciplinary team has been formed to investigate the extent and causes of yield variability for corn under sprinkler irrigated conditions over a 5 year time span. Both field sampling and several kinds of models will be used to develop multiple GIS layers that will combined in a multivariate model to predict yield. Actual yields will be compared with predicted yields for various locations within several fields to evaluate the suitability of the predictive model.

REFERENCES:

Booltink, H.W.G.,Verhagen, J., Bouma, J., Thornton, P.K. (1996) Application of Simulation Models and Weather Generators to Optimize Farm Management Strategies. In: *Proc. of 3rd Int'l Conference on Precision Agric.,* Robert, P.C., Rust, R.H., Larson, W.E. (Eds.)., Madison, WI., AS, COSA, SSSA. pp 343-360.

Fraisse, C.W., Heermann, D.F., Duke, H.R. (1995) Simulation of Variable Water Application with Linear-move Irrigation Systems. *Trans. of ASAE.* **38**(5):1371-1376.

Franzen, D.W., Cihacek, L.J., Hofman, V.L. (1996) Variability of Soil Nitrate and Phosphate Under Different Landscapes. In: *Proc. of 3rd Int'l Conference on Precision Agric.,* Robert, P.C., Rust, R.H., Larson, W.E. (Eds.)., Madison, WI., ASA, COSA, SSSA. pp 521-529.

Han, S., Schneider, S.M., Evans, R.G., Rawlins, S.L. (1996) Spatial Variability of Soil Properties on Two Center Pivot Irrigated Fields. In: *Proc. of 3rd Int'l Conference on Precision Agric.,* Robert, P.C., Rust, R.H., Larson, W.E. (Eds.)., Madison, WI, ASA, COSA, SSSA. pp 97-106.

Heermann, D.F. (1990) Center Pivot Design and Evaluation. In: *Proc. of Third Nat. Irrig. Symp.,* ASAE, St. Joseph, MI. pp 564-569.

Lowenberg-DeBoer, J. and Boehlje, M. (1996) Revolution, Evolution or Dead-end. Economic Perspectives on Precision Agriculture. In: *Proc. of 3rd Int'l Conference on Precision Agric.,* Robert, P.C., Rust, R.H., Larson, W.E. (Eds.)., Madison, WI , ASA, CSSA, SSSA. pp 923-944.

Mallarino. A.P., Hinz, P.N., Oyarzabel, E.S. (1996) Multivariate Analysis as a Tool for

358

Interpreting Relationships Between Site Variables and Crop Yields. In: *Proc. of 3rd Int'l Conference on Precision Agric.*, Robert, P.C., Rust, R.H., Larson, W.E. (Eds.)., Madison, WI, ASA, COSA, SSSA. pp 151-158.

National Research Council (1996) *A New Era for Irrigation : Report prepared by Committe on the Future of Irrigation in the Face of Competing Demands.* National Academy Press. Washington D.C. pp 49, 128-133.

Pachepsky, Y., Trent, A., Acock, B. (1996) Apparent Spatial Variability of Crop Model Parameters as Estimated from Yield Mapping Data. In: *Proc. of 3rd Int'l Conference on Precision Agric.*, Robert, P.C., Rust, R.H., Larson, W.E. (Eds.)., Madison, WI , ASA, CSSA, SSSA. p 277.

Reetz, H.R., Fixen, P.E. (1995) Economical Analysis of Site-Specific Nutrient Management Systems. In *Proc. Site Specific Management for Agricultural Systems.*, Robert, P.C., Rust, R.H., Larson, W.E. (Eds.)., Madison, WI., ASA, COSA, SSSA. pp. 743-752.

Swinton, S. M. and Ahmad, M. (1996) Returns to Farmer Investments in Precision Agriculture Equipment and Services. In: *Proc. of 3rd Int'l Conference on Precision Agric.*, Robert, P.C., Rust, R.H., Larson, W.E. (Eds.)., Madison, WI., ASA, CSSA, SSSA. pp 1009-1018.

Vanden Heuvel, R.M. (1996) The Promise of Precision Agriculture. *Jour. of Soil and Water Cons.*, **51,** 38-40.

Varvel, G.E., T.M.Blackmer, D.D. Francis, J.S.Schepers (1996). Similitarities Between Organic Matter Content and Phosphorus Levels When Grid Sampling for Site-Specific Management.In: *Proc. of 3rd Int'l Conference on Precision Agric.*, Robert, P.C., Rust, R.H., Larson, W.E. (Eds.)., Madison, WI ., ASA, CSSA, SSSA. p 577.

USE OF NITROGEN SIMULATION MODELS FOR SITE-SPECIFIC NITROGEN FERTILIZATION

T. ENGEL

Technische Universität München, Lehreinheit für Ackerbau und Informatik im Pflanzenbau, Lange Point 51, 85350 Freising, Germany

ABSTRACT

Site-specific nitrogen fertilization is necessary from an economical and an ecological point of view. The necessary technical equipment is available, but there are many open questions concerning the correct way to calculate the fertilizer amounts. The first step is to identify and localize homogeneous areas. This paper presents a method whereby the fertilizer amount for each area can be calculated using simulation models for crop growth and nitrogen dynamics. The building block system 'Expert-N' can be used to figure out which simulation model or combination of simulation approaches fits best for a specific site. Resulting strategies for the actual and further development of simulation models are discussed.

INTRODUCTION

Most fields have a high spatial variability of the soil characteristics which causes considerable yield variations of more than 100% (Birrel *et al.*, 1995). The reason is usually the origin of the soils, but it can also be caused by different management practices. Traditionally the farmers manage their fields uniformly and tend to ignore the inherent spatial variability which is unfavourable from an ecological and an economical point of view. The nitrogen fertilization is calculated using the average yield. Therefore there are spots in the field where the economic potential yield is not reached and other spots with lower yields where the environment is polluted because the surpluses of nitrate can be leached during the winter. In future, site-specific management can make an important contribution to realizing an economically successful and ecologically acceptable agriculture (Schnug and Holst, 1994).

PROBLEM

Requirements for site-specific fertilizer applications are the identification and localization of homogeneous areas in the field. For all homogeneous areas the fertilizer demand has to be calculated separately. The basis for the calculation can be the yield of the previous year, the yield potential or the long-term average yield of the homogeneous area. The necessary technical equipment to realize this concept comprises the localization of the tractor using GPS (global positioning system) and the electronic steering of the fertilizer spreader by the on-board computer. This technical equipment has reached a quality which allows its use for site-specific farming (Schueller and Wang, 1994). For agricultural purposes GPS has to be used in the differential mode where the receiver needs an error signal from a stationary mounted receiver. An obstacle is still the availability of this error signal which could be sent by radio stations (Beuche and Hellebrand, 1997). As soon as the error signal is available everywhere and without any interruptions the system can be used in the agricultural practice. While the technical realization of precision farming is nearly solved, there are still problems in how to support the farmer when making decisions concerning appropriate treatments and levels of treatment to be spatially applied (Blackmore *et al.*, 1995).

SITE-SPECIFIC FERTILIZATION OF PHOSPHORUS AND POTASSIUM

The concept of site-specific fertilization with phosphorus and potassium is not very difficult because it is based on simple nutrient balances. Usually agricultural soils with normal soil fertility have stored enough nutrients to fertilize the amount which is taken up by the plants so that the soils do not lose their fertility in the long-term. A yield map which can be created by modern combines shows the yield of the previous year. There is a strong correlation between the yield and the uptake of the plant so that the yield map can be used to calculate the site-specific P- and K-uptake of the previous year. By consideration of the site-specific soil supply which can be determined by soil analysis a site-specific application map for the fertilization of phosphorus and potassium can be calculated.

SITE-SPECIFIC FERTILIZATION OF NITROGEN

The concept of site-specific nitrogen fertilization is much more difficult. Some of the reasons are shown in the following sections.

Special problems of the nitrogen fertilization

Due to the complexity of the nitrogen cycle (see Figure 1) a simple nutrient balance is totally insufficient to calculate the amount of nitrogen fertilization. Therefore the sole use of a yield map as basis for the calculation is not possible. In contrast to P and K some of the different components of the nitrogen cycle can be leached and transformed to other components. The basis for the calculation must be the yield potential or the expected yield of a homogeneous area which is strongly influenced by the water balance, the weather of the specific year and the phytosanitary situation. The use of simulation models to model the nitrogen dynamics in soil, plant and atmosphere can contribute to making progress in the area of site-specific nitrogen fertilization.

Use of nitrogen simulation models to calculate fertilizer recommendations

The integration of nitrogen simulation models in a computer-aided site-specific plant production is a big challenge for the agricultural sciences. The objective is the forecast of the deficit between the nitrogen demand of the plants and the nitrogen supply of the soil to apply the appropriate amount of nitrogen fertilizer at the right time and place. The principal problem is that the fertilization has to make good a deficit which does not exist at the time of application. Therefore, it is necessary to estimate the future course of nitrogen demand and supply. This prognosis shall cover the period between the actual application and the time of the next fertilizer application.

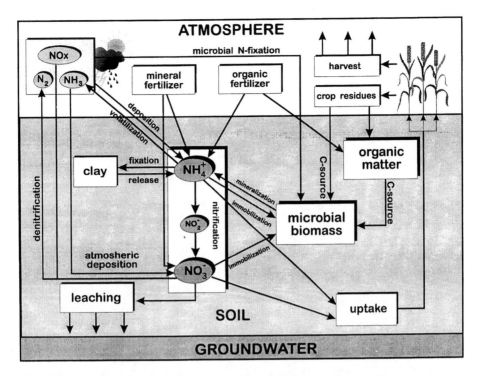

FIGURE 1. Components of the nitrogen cycle in soil, plant and atmosphere.

Due to their high time resolution (time step 1 day or less) nitrogen simulation models are adequate and absolutely necessary tools to consider the nitrogen dynamics between two application dates. First of all it is necessary to define appropriate times for nitrogen applications depending on the development stages of the plants. Since long-term weather forecasts are not available, the simulation models have to use weather data which are based on the site-specific long-term average weather. The resulting simulation run shows the expected course of the nitrogen demand (if no deficit occurs) and the actual nitrogen supply of the soil. The difference between the two nitrogen pools can be used as the fertilizer recommendation.

If the real weather differs very much from the long-term weather, it is necessary to consider the effect of the actual weather for the calculation of the next fertilizer application. If the nitrogen supply is lower or the nitrogen uptake is higher the next fertilizer recommendation has to be increased and must be applied earlier to avoid nitrogen deficiency. If the conditions are more positive than expected (higher N supply or lower uptake) the next fertilizer recommendation can be reduced. This example shows that it is necessary to observe the weather conditions during the season and to run the simulations some days before the next expected fertilizer application date.

From the point to the field

Simulation models usually simulate the vertical water and nutrient flow at a specific point which should be representative for an area with homogeneous soil conditions. So the first step should be the identification and localization of homogeneous areas. Then the simulation of crop growth and nitrogen cycle can be used as the basis for the calculation of the fertilizer demand. The resulting fertilization concept uses the difference between the simulated nitrogen demand of the plants and the simulated supply of the soil for each homogeneous area.

IDENTIFICATION OF HOMOGENEOUS AREAS

Different methods can be used or should be combined for the identification of homogeneous areas (Long *et al.*, 1995).

From the technical point of view there is no problem creating yield maps using DGPS and yield sensors on harvesters (Birrel *et al.*, 1995). But yield maps cannot be the only source for the identification of homogeneous areas, because the distribution of the yield is different every year depending on the weather and the crop (Werner, 1995).

Additional information can be obtained by the analysis of satellite images but the resolution is often not good enough for the use in precision farming. Aerial photographs have a higher resolution and help especially in the diagnosis of the stone fraction of the soils (Amon, 1995). Tomer *et al.* (1995) showed that infrared photography is a valuable tool for assessing variability of agricultural soils and associated crop growth. According to Blackmer *et al.* (1995) aerial photography and photometric sensors show promise as techniques for identifying N deficit areas.

Soil sampling together with geostatistical analysis is suitable for the creation of soil maps on the scientific level, but it is much too expensive for practical purposes (Murphy *et al.*, 1995).

The overlay of available digital maps using geographical information systems (GIS) is another possibility to get homogeneous areas. The problem is that digital maps of soil properties or field borders are usually not available in Germany.

This shows that there is no general recipe on how to define homogeneous areas. Therefore, the author cannot give a general recommendation. In every case it is necessary to use the available tools and to combine them as efficiently as possible. One solution could be to perform a rough classification using yield maps and remote sensing techniques. Problem zones of this first approach could be refined by soil or crop sampling.

PROBLEMS OF NITROGEN SIMULATION MODELS

The potential and necessity of simulation models are well accepted among scientists. Therefore many international teams of scientists are working on the development of simulation models of nitrogen dynamics. But the following topics prevent the use of simulation models in the calculation of fertilizer applications:

- Simulation models require a large amount of input data. These input data consist of extensive information about the field management, daily weather data and a detailed description of the soil profile (soil texture, water content at saturation, field capacity and permanent wilting point, organic matter content, organic nitrogen, initial values for water and nitrogen content, etc.). These data are not available for most of the farmers' fields. They can only be determined through considerable financial efforts.

- Crop growth models are important modules of nitrogen simulation models. But the existing crop models do not cover the whole spectrum of important plants.

- Simulation models are often developed and validated for specific regions and specific soils. It is unknown whether the simulation models give valid results for other regions or other weather conditions.

- Models usually provide a huge amount of output data, but no site-specific fertilizer recommendations. The interpretation of the simulation results is too difficult for other users.

THE BUILDING BLOCK SYSTEM 'EXPERT-N'

To reduce or eliminate the existing problems, the research project 'Expert-N' was initiated (Engel, 1995). The objective was to develop a building block system of model approaches which can be used for the integration and comparison of model approaches and which represents a flexible and modular simulation system itself. The building block system can be used for the simulation of the water, heat and nitrogen dynamics in soil and plant. 'Expert-N' has an easy-to-use user interface based on MS-Windows. All model functions which describe the single processes in an agroecosystem are programmed according to the ANSI-C standard so that they can be used on different computer platforms. The actual release 'Expert-N 1.0' is available for MS-Windows 3.11. A 32-bit version or Windows 95 or Windows NT is in preparation.

'Expert-N' is delivered in three different versions which are adapted to the knowledge of different user groups:

- A model developer has access to all routines of the system so that he can integrate his own model approaches and extend the building block system step by step.

- Scientists can create 'new models' by the new combination of existing modules. By comparison with measured values the optimal model approach combination for a specific site can be determined.

- A model user like a consultant or farmer only has access to module combinations which cannot be changed and which are well tested and validated.

This procedure supports the different user groups in an optimal way and avoids the misuse of the system.

The simulation of water, heat and nitrogen dynamics is executed depending on the chosen model configuration. A comfortable user interface with buttons and combo boxes allows the user to define which processes he wants to simulate and which model approaches he wants to use. This choice can be done in the configuration window which is shown in Figure 2. The configuration window contains information about the desired time control and the modules which are actually chosen for water balance, heat flow, nitrogen dynamics and crop growth. Each process in the four function groups has at least two different modules the user can choose.

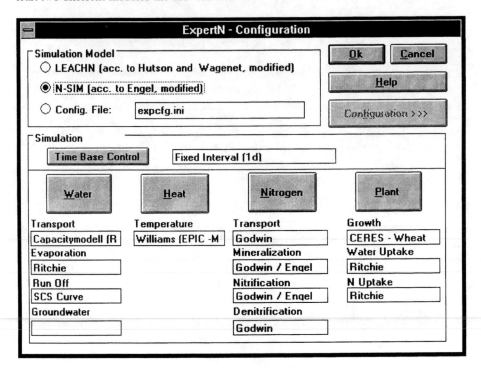

FIGURE 2. Configuration window in 'Expert-N'.

The actual release of 'Expert-N' contains the model approaches from the models LEACHM (Hutson and Wagenet, 1992), N-SIM (Engel, 1991) and the CERES crop growth models for wheat (Ritchie, 1991), barley and maize. Additionally several hydraulic functions and mathematical algorithms are implemented to solve the partial differential equations. The user also has the possibility of linking his own functions and model approaches using the 'dynamic-link-library' concept of Windows. This helps to extend the function libraries of 'Expert-N' step by step.

So far nitrogen simulation models are seldom used by farmers or crop consultants. A few farmers and consultants have started using the system 'Expert-N' for the calculation of fertilizer recommendations. The feedback is encouraging and hopefully supports the use and further development of simulation approaches.

NECESSARY FURTHER DEVELOPMENTS

The use of simulation models is especially suited to fine-tune fertilizer amounts and application dates. Booltink *et al.* (1997) achieved a reduction of 100 kg N-fertilizer and more on sandy and clayey soil profiles simulating site-specific management scenarios for winter wheat. But there are still many requirements for the further development of models so that they can be used more on a broader scale.

The available simulation systems have to be extended especially with more crop model approaches to cover all important agricultural crops.

Jones *et al.* (1994) suggest a standardization of the data formats for the existing models. This would facilitate the data exchange between modelers and force the model development.

To run simulations for whole regions or water catchments it is necessary to couple simulation models with geographic information systems. First steps have been made (Engel *et al.*, 1995; Franko and Schenk, 1995), but there are many open questions. First of all it is difficult to choose suitable model approaches because most of the existing approaches were developed for the one-dimensional vertical water and nutrient flow. But it is also necessary to simulate lateral flows especially in hilly regions, e.g. because the run off from higher fields has to be considered as an additional water input in lower fields. According to Addiscott (1995) it is necessary to find and use the best suitable mechanistic model which was validated for the specific purpose and scale. Addiscott supposes that on the larger scale functional models are more suitable than complex mechanistic models.

ECONOMIC BENEFITS

The economic benefits of spatially variable nitrogen applications are difficult to estimate. The existing research results are very heterogeneous. In many cases it causes only a redistribution of the costs.

According to Kitchen and Hughes (1995) the savings of fertilizers are compensated by higher costs for data collection and technical equipment (GPS receiver etc.). Other authors (Vetsch *et al.*, 1995; Reetz and Fixen, 1995; Booltink and Verhagen, 1996) found considerable money returns.

Better distribution of nitrogen fertilizers can also reduce nitrate leaching (Booltink *et al.*, 1997) but the positive effects are usually not expressed in economic terms.

CONCLUSION

There is a lack of decision support tools for site-specific fertilizer recommendations. Simulation models can be important tools for the calculation of site-specific fertilizer recommendations. Future research should focus on the further development of flexible, easy-to-use simulation tools like 'Expert-N' to support the special needs of precision farming.

REFERENCES

Addiscott, T.M. (1995) Modelling the fate of crop nutrients in the environment: problems of scale and complexity. *European Journal for Agronomy*, **4(4),** 413-417.

Amon, H. (1995) Abgrenzung und Bewirtschaftung von Teilschlägen mit Hilfe von Fernerkundung und Elektronik. *Dissertation*, Technische Universität München-Weihenstephan.

Beuche, H., Hellebrand, H.J. (1997) DGPS-Ortung mit öffentlichen Referenzdiensten - Bewertung für den landwirtschaftlichen Einsatz. *Zeitschrift für Agrarinformatik*, 2(5), 31-38.

Birrel, J.S., Borgelt, S.C., Sudduth, K.A. (1995) Crop yield mapping: comparison of yield monitors and mapping techniques. *Site-Specific Management for Agricultural Systems*, American Society of Agronomy, pp. 15-31.

Blackmer, T.M., Schepers, J.S., Meyer, G.E. (1995) Remote Sensing to Detect Nitrogen Deficiency in Corn. *Site-Specific Management for Agricultural Systems*, American Society of Agronomy, pp. 505-512.

Blackmore, B.S., Wheeler, P.N., Morris, J., Morris, R.M., R.J.A. Jones (1995) The role of precision farming in sustainable agriculture: a European perspective. *Site-Specific Management for Agricultural Systems*, American Society of Agronomy, pp. 777-793.

Booltink, H.W.G., Thornton, P.K., Verhagen, J., Bouma, J. (1997) Application of simulation models and weather generators to optimize farm management strategies. *Precision Agriculture*, American Society of Agronomy, in press.

Booltink, H.W.G., Verhagen, J. (1996) Using decision support systems to optimize barley management on spatial variable soil. *Systems Approaches for Agricultural Development (SAAD II)*, Kluwer Academic Publishers

Engel, T. (1991) Entwicklung und Validierung eines Simulationsmodells zur Stick-stoffdynamik in Boden und Pflanze mit Hilfe objektorientierter Programmierung. *Dissertation*, Technische Universität München-Weihenstephan.

Engel, T. (1995) Expert-N - Ein Baukastensystem für Stickstoffmodelle als Hilfsmittel für Beratung, Forschung, Wasserwirtschaft und politische Entscheidungsträger. *Abschlußbericht zu einem Forschungsvorhaben des BMBF*. 30 pp.

Engel, T., Jones, J.W., Hoogenboom, G. (1995) AEGIS/WIN - A Windows Interface Combining GIS and Crop Simulation Models. *Paper No. 953244 of the Society for Engineering in Agricultural, Food and Biological Systems (ASAE)*, 12 pp.

Franko, U., Schenk, S. (1995) Gebietsbezogene Modellierung der Humusakkumulation in Abhängigkeit von Standort und Bewirtschaftung. *Referate der 16. GIL - Jahrestagung in Kiel 1995, Berichte der GIL*, C. Noell and J.M. Pohlmann (Eds), Gesellschaft für Informatik in der Land-, Forst- und Ernährungswirtschaft, **7**, pp. 84-91.

Hutson, J.L., Wagenet, R.J. (1992) LEACHM: Leaching Estimation And Chemistry Model: A process-based model of water and solute movement, transformations, plant uptake and chemical reactions in the unsaturated zone. Version 3.0. *Department of Soil, Crop and Atmospheric Sciences, Research Series No. 93-3*, Ithaca, New York, Cornell University.

Jones, J.W., Hunt, L.A., Hoogenboom, G., Godwin, D.C., Singh, U., Tsuji, G.Y., Pickering, N.B., Thornton, P.K., Bowen, W.T., Boote, K.J., Ritchie, J.T., Balas, S. (1994) Input and Output Files. *DSSAT version 3*. G.Y. Tsuji, G. Uehara and S. Balas, (Eds), Honolulu, Hawaii, IBSNAT, University of Hawaii,. pp 1-94.

Kitchen, N.R., Hughes, D.F. (1995) Comparison of variable rate to single rate nitrogen fertilizer application: corn production and residual soil NO_3-N. *Site-Specific Management for Agricultural Systems*, American Society of Agronomy, pp. 427-441.

Long, D.S., Carlson, G.R., DeGloria, S.D. (1995) Quality of field management maps. *Site-Specific Management for Agricultural Systems*, American Society of Agronomy, pp. 251-271.

Murphy, D., Oestergaard, H., Schnug, E. (1995) Lokales Ressourcen Management - Ergebnisse und Ausblick. *Technik für die kleinräumige Bestandesführung*, KTBL-Arbeitspapier No. **214**, pp 90-101.

Reetz, F.R., Fixen, P.E. (1995) Economic analyses of site-specific nutrient management systems. *Site-Specific Management for Agricultural Systems*, American Society of Agronomy, pp. 742-752.

Ritchie, J.T. (1991) Wheat phasic development. *Modelling plant and soil systems*. J. Hanks, and J.T. Ritchie (Eds), American Society of Agronomy. Madison, WI, USA, pp 31-54.

Schnug, E., Holst, P. (1994) CAF - Realisierung einer ökologischen und ökonomischen Landwirtschaft. *Elektronikeinsatz in der Außenwirtschaft*, KTBL-Arbeitspapier No. **175**, pp 175-178.

Schueller, K.J., Wang, M.W. (1994) Spatially-variable fertilizer and pesticide application with GPS and DGPS. *Computers and Electronics in Agriculture*, **11**, 69-83.

Tomer, M.D., Anderson, J.L., Lamb, J.A. (1995) Landscape analysis of soil and crop data using regression. *Site-Specific Management for Agricultural Systems*, American Society of Agronomy, pp. 273-284.

Vetsch, J.A., Malzer, G.L., Robert, P.C., Huggins, D.R. (1995) Nitrogen specific management by soil condition: managing fertilizer nitrogen in corn. *Site-Specific Management for Agricultural Systems*, American Society of Agronomy, pp. 465-473.

Werner, A. (1995) Ziele und Möglichkeiten der kleinräumigen Bestandesführung. *Technik für die kleinräumige Bestandesführung*, KTBL-Arbeitspapier No. **214**, pp 7-26.

THE ADEQUACY OF CURRENT FERTILIZER RECOMMENDATIONS FOR SITE SPECIFIC MANAGEMENT IN THE USA

G.W. HERGERT

University of Nebraska WCREC, RT. 4 Box 46A, North Platte, NE, 69101, USA

W.L. PAN

Washington State University, Department Crop and Soil Science, Pullman, WA, 99164, USA

D.R. HUGGINS

University of Minnesota, Lamberton, MN 56152, USA

J.H. GROVE

University of Kentucky, Lexington, KY, 40546, USA

T.R. PECK

University of Illinois, Urbana, IL, 61801, USA

ABSTRACT

Many fertilizer recommendations used in the USA were developed 30 to 40 years ago from data averaged over many soils. Past data provides a good starting point for site specific fertilizer recommendations but bulk soil sampling has complicated interpretations. Soil test critical levels were developed from small homogeneous areas in fields, but the information was extrapolated to the whole field which encompassed much wider variation. Many soil test parameters are log normally distributed so a few high testing cores can skew the average. Past data can be reviewed and used to develop new recommendations/algorithms if sufficient sites are available. To make major improvements in site specific recommendations many soils/landscape sites that were not a part of the original database will require research. New partnerships between researchers, producers, fertilizer dealers and agricultural consultants will be required to develop new databases in a time of declining funding for agriculture.

INTRODUCTION

Variable rate fertilizer application (VRA) and site specific management (SSM) provide one of the greatest challenges and opportunities for improving fertilizer use efficiency since the wide scale adoption of soil testing in the 1950s and 1960s. The assumption of VRA is that it will more closely match productivity, input efficiency and profitability compared to uniform application (Sawyer, 1994). However, SSM and VRA raise many questions about current fertilizer recommendations. Are they adequate for this new technology?

The objective of this paper is to review past soil test correlation and calibration work and the resulting fertilizer recommendations in the USA and to determine what, if any, changes are required to improve fertilizer recommendations for SSM/VRA.

HISTORICAL OVERVIEW

Soil testing has been available for many years (Bray, 1929) and is generally understood to include chemical or physical measurements made on a soil. The process for developing the actual fertilizer recommendations, however, is often criticized because it relies on judgment as well as science. The process has been likened to a black box where information is fed in and an answer (fertilizer recommendation) comes out.

Many of the fertilizer recommendations still in use today were developed in the 1950s and 1960s at a time when fertilizer was inexpensive and responses to fertilizer were usually dramatic. The 'cost' of not having adequate fertility was to limit yield whereas over-application simply increased soil fertility more rapidly. This masked errors in correlation, calibration and fertilizer application. Fertilizer application (band vs. broadcast, manure) and changing tillage practices have increased heterogeneity and complicated both the sampling and interpretation processes (Peck and Soltanpour, 1990).

A major problem with soil test interpretation on a field basis has been the assumption of normality of the soil test parameter. In many cases data in a field are not distributed normally, but log normally (Hergert et al., 1995b; Parkin et al., 1988). The complication this presents with composite sampling is obvious (Figure 1). This has led to confusion and questioning of the credibility of soil testing. The critical levels were not the problem because plots were small and assumed to be homogeneous. The problem was one of scale when the small plot information was extended to the field scale which encompassed much wider variability.

Due to differing views on interpreting soil test values, however, numerous articles and reviews on the philosophy of soil test interpretation have been written (McLean, 1977; Olson et al., 1987). The science is not in question but the interpretation is open to debate. Much of the past controversy needs to be resolved as we approach a new era of reviewing, validating and revising soil test correlation and calibration for VRA.

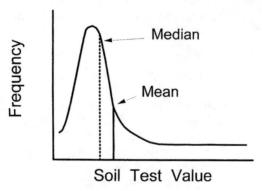

FIGURE 1. A log-normal frequency distribution.

Previous soil test correlation, calibration and interpretation will be the basis for SSM/VRA fertilizer recommendations as additional research and interpretation are conducted. The assumption is that the recommendations may not be totally correct, but are a good starting point. This assumption, however, raises the question of what must be done to improve them.

ADEQUACY OF PAST DATA AND FUTURE RESEARCH NEEDS

For N, P, K and lime three questions will be reviewed: (1) What is the adequacy of current fertilizer recommendations for VRA/SSM? (2) Can past correlation and calibration data be recovered and reanalyzed to provide improved site specific recommendations? and (3) What new data will be required to provide site specific recommendations if the old data is inadequate?

Nitrogen

Adequacy of current recommendations. Crop and soil N budgeting remains the foundation of current N recommendations for grains in the USA (Bock and Hergert, 1991). Since most N recommendations in the humid region of the USA have been related to a type of unit N requirement (UNR), yield potential is the primary determinant of N rate. UNR is defined as the amount of N fertilizer required per unit of yield if no N were supplied from non-fertilizer sources. Due to this fact, determining appropriate yields and N response on a site specific basis becomes imperative for improving N recommendations for SSM/VRA.

Soil nitrate-N testing is a standard management tool in low rainfall areas of the USA where dryland cropping and irrigation are practiced (Hergert, 1987) and has been incorporated into revised N recommendations (Gerwing and Gelderman, 1996; Hergert et al., 1995a). N recommendations have been improved for maize in many humid regions of the USA because of recent research with a pre-sidedress N test (PSNT) (Fox et al., 1989; Magdoff, 1991; Schmitt and Randall, 1994).

For states that have not had a concerted effort to improve N recommendations in recent years, older data may be sufficient if there were sufficient soils or sites represented in the initial database to develop site-specific N recommendations (Hergert, 1987). If not, new research will be required.

Reanalysis of past data. As researchers look for past data, many may find that it is not available or easily accessible. Older hand-written data can be computerized but data on punch cards may be difficult to retrieve. Many of the assumptions that went into past recommendations and algorithms should be reviewed. An alternative rational to basing maize N recommendations on yield has been suggested (Vanotti and Bundy, 1994). Little difference in the economic N rate for high and low yielding years was shown and is similar to optimum N rate and economics considering N/crop price ratios (Bock and Hergert, 1991).

Another challenge will be to determine if enough varied sites (eroded areas, hill tops, wet spots, saline spots) were used to develop the initial recommendations. In most instances a minimum of 20 to 25 site years may be required to develop different

373

response functions that are truly site specific (Nelson and Anderson, 1977).

New data required for improved SSM/VRA recommendations. Future research will require more experiments like those of Vetch *et al.* (1995) to provide improved site-specific data. In most instances new data can enhance the existing database. The question of 'yield stability' across years (temporal variability) also must be addressed if yield goal is to be used as the basis for fertilizer N recommendations. Economic analysis (Bock and Hergert, 1991) will be required to provide economically and environmentally sound N recommendations. This is imperative since most producers will want to somehow use information from yield monitors.

Phosphorus and potassium

Adequacy of current recommendations. Soil tests for P currently used in the USA include the Bray-1 developed in the 1940s (Bray, 1948), the Mehlich-1 and Olsen methods developed in the 1950s (Mehlich, 1953; Olsen *et al.*, 1954) and the Mehlich-3 developed in the 1980s (Mehlich, 1984). A review of Bray's original work and recent research indicates little change in the critical concentration of Bray-1 P (Figure 2) or potassium is required to produce optimum yield (Bray, 1948; Mallarino and Blackmer, 1992).

Reanalysis of past data. Most USA university laboratories use the sufficiency concept for P and K although there is disagreement where fertilization should cease. Agronomists must agree on critical concentrations that are economically and environmentally sound as a first step in providing P and K recommendations that are site specific.

A second step for improving P recommendations for SSM/VRA will be reanalyzing past data to reestablish some of the specificity lost as data were averaged over sites and years. In Illinois, different correlation curves were initially developed for different subsoil P levels (Figure 3), but are not used extensively today. A similar division based on subsoil P has been included in recently revised Iowa recommendations (Voss, 1995).

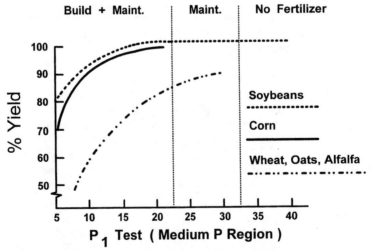

FIGURE 2. Bray's original correlation of relative yield and soil test P.

374

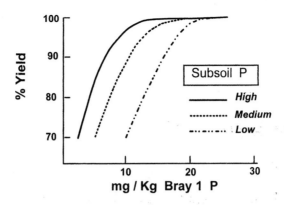

FIGURE 3. Correlation of relative corn yield to subsoil P and soil test P in Illinois.

A third step in improving P and K recommendations for SSM/VRA will be developing more landscape/field position-based recommendations. Severe P deficiencies on eroded hilltops and side slopes of the Palouse region of eastern Washington (Guettinger and Koehler, 1967) were used for side-by-side strips with different fertilizer treatments to encompass variation which was built into the recommendations.

New data required for improved SSM/VRA recommendations. If data are not available to be reanalyzed by soil type or landscape position, additional sites will be required. A new approach to soil test correlation-calibration will be required in many areas as base funding for this type of research is no longer available at most USA Land Grant Universities. Partnerships between researchers, producers, fertilizer dealers and agricultural consultants will be required to do field-scale testing to develop data for new response curves. This will be a challenge, but should provide improved data for SSM/VRA P and K recommendations.

Lime

Adequacy of current recommendations. Field studies to establish crop response to liming and to determine the 'biological lime requirement' (Black, 1993) are much less numerous than those for N, P, or K. Researchers have long recognized that extensive sampling could provide the basis for spot lime application (Linsley and Bauer, 1929).

The lack of field data may be particularly acute for fields under conservation tillage soil management. It has been well documented that the 'pH profile' of the surface 30 cm is substantially different than that found in the moldboard plowed soils on which most past liming data were obtained. Near-surface pH values are often significantly reduced and this has resulted in recommendations to reduce the depth of sampling in such fields when characterizing soil acidity (James and Wells, 1990).

Illinois research shows that current lime recommendations work well in spite of the limitations discussed above. Two fields have been grid sampled for several years to characterize the spatial variation of pH and several other soil parameters in the top 20 cm (Franzen and Peck, 1995). This research points out the importance of developing detailed pH and lime recommendation maps then following with later soil analysis to determine where lime had the most effect. With yield maps, correlation of yield, soil type, pH and liming can be evaluated to determine the most economic approach for liming in the future.

Reanalysis of past data. The lack of sufficient SSM/VRA data sets from field studies with lime may not be a great concern. The usual basis for lime recommendations is a 'target pH'. Though it is not precisely clear why an acid soil must be thoroughly neutralized (van Lierop, 1990), target pH is often somewhere between 6.4 and 7.0. This is considerably greater than that required by many crops to produce optimal yields (Woodruff, 1967). Laboratory data on changes in soil pH as a function of lime amendment may provide much of the needed information.

New data required for improved SSM/VRA recommendations. Although considerable data regarding acid soil neutralization exists, further effort will be required to translate it into information-models-algorithms useful to SSM/VRA. Optimistically, if field soils are mapped once for their colloid quantity (texture, organic matter content) and quality (clay mineralogy), and the relationships between these and buffer capacity are well described, then real-time pH sensing would be all that is required to drive lime delivery in the field. Different tillage systems might result in different operating depths for the sensor and a different algorithm for the rate controller.

CONCLUSIONS

In many states in the USA, lime and fertilizer recommendations were developed by averaging over many soils. A lack of SSM crop response information is not likely to hinder fertilizer or lime rate recommendations if current models/algorithms are used. Soil fertility research approached from SSM offers the researcher, and indeed the consultant, fertilizer dealer and/or individual producer, the opportunity to correlate and calibrate new and old extractants/procedures. Perhaps it would be more honest to admit the frailties in the soil test knowledge framework and create teams or partnerships with those (producers, agricultural consultants, fertilizer dealers) who will soon have part of the data to improve that base of knowledge for all of us as we develop truly site specific lime and fertilizer recommendations.

REFERENCES

Black, C.A. (1993) Soil fertility evaluation and control. Boca Raton, FL, Lewis Publishers.

Bock, B.R., Hergert, G.W. (1991) Fertilizer nitrogen management. *Managing Nitrogen for Ground Water Quality and Farm Profitability,* R.F. Follett (Ed.), Madison, WI, SSSA, pp 139-164.

Bray, R.H. (1929) A field test for available phosphorus is soils. *Ill. Agric. Exp. Stn. Bull.* **337,** 589-602.

Bray, R.H. (1948) Correlation of soil tests with crop response to added fertilizers and with fertilizer requirements. *Diagnostic Techniques for Soils and Crops,*

H.B. Kitchen (Ed.), Washington, DC, The American Potash Institute, pp. 53-86.

Fox, R.H., Roth, G.W., Iversen, K.V., Piekielek, W.P. (1989) Soil and tissue nitrate tests compared for predicting soil nitrogen availability to corn. *Agron. J.,* **81,** 971-974.

Franzen, D.W., Peck, T.R. (1995) Sampling for site-specific application. pp. 535-551. *In* P.C. Robert, R.H. Rust, and W.E. Larson (ed.) *Site-Specific Management for Agricultural Systems.* ASA, CSSA, SSSA. Madison, WI.

Gerwing, J. Gelderman, R. (1996) Fertilizer Recommendation Guide. South Dakota State Univ. Coop. Ext. Serv. EC750.

Guettinger, G. Koehler, R. (1967) Phosphorus for winter wheat. Circ. 423, Inst. of Agric. Sci. State College of Washington.

Hergert, G.W. (1987) Status of residual nitrate-nitrogen soil tests in the United States of America. *Soil Testing: Sampling, Correlation, Calibration, and Interpretation,* J.R. Brown (Ed.), SSSA Spec. Pub. 21, Madison, WI, SSSA, pp 73-88.

Hergert, G.W., Ferguson, R.B., Shapiro, C.A. (1995a) Fertilizer Suggestions for Corn. Univ. Nebraska NebGuide G74-174 *(Revised).*

Hergert, G.W., Ferguson, R.B., Shapiro, C.A., Penas, E.J., Anderson, F.B. (1995b) Classical statistical and geostatistical analysis of soil nitrate-N spatial variability. *Site-Specific Management for Agricultural Systems,* P.C. Robert, R.H. Rust, and W.E. Larson (Eds), Madison, WI, ASA, CSSA, SSSA, pp 175-186.

James, D.W., Wells, K.L. (1990) Soil Sample collection and handling: Technique based on source and degree of field variability. *Soil testing and plant analysis,* R.L. Westerman (Ed.), 3rd ed., Madison, WI., SSSA, pp 25-44.

Linsley, C.M., Bauer, F.C. (1929) Test your soil for acidity. Univ. of IL, Col. of Agric. and Agric. Exp. Stn. Circ. 346.

Magdoff, F.R. (1991) Understanding the Magdoff pre-sidedress nitrate test for corn. *J. Prod. Agric.,* **4,** 297-305.

Mallarino, A.P., Blackmer, A.M. (1992) Comparison of methods for determining critical concentrations of soil test phosphorus for corn. *Agron. J.,* **84,** 850-856.

McLean, E.O. (1977) Contrasting concepts in soil test interpretation: sufficiency levels of available nutrients versus basic cation saturation ratios. *Soil Testing: Correlating and Interpreting the Analytical Results,* M. Shelly (Ed.), ASA Spec. Pub. 29, Madison, WI, ASA, CSSA, and SSSA, pp 55-74.

Mehlich, A. (1953) Determination of P, Ca, Mg, K, Na and NH_4. North Carolina Soil Testing Div. Mimeo, Raleigh, NC.

Mehlich, A. (1984) Mehlich-3 soil test extractant: A modification of Mehlich-2 extractant. *Commun. Soil Sci. Plant Anal.*, **15**, 1409-1416.

Nelson, L.A., Anderson, R.L. (1977) Partitioning of soil test-crop response probability. *Soil Testing: Correlating and interpreting the analytical results*, M. Stelly (Ed.), ASA Spec. Pub. 29, Madison, WI, ASA, CSSA, and SSSA, pp 19-38.

Olson, R.A., Anderson, F.N., Frank, K.D., Grabouski, P.H., Rehm, G.W., Shapiro, C.A. (1987) Soil Testing Interpretations: Sufficiency vs. Build-Up and Maintenance. *Soil Testing: Sampling, Correlation, Calibration, and Interpretation*, J.R. Brown (Ed.), SSSA Spec. Publ. 21, Madison, WI, SSSA, pp 41-52.

Olsen, S.R., Cole, C.V., Watanabe, F.S., Dean, L.A. (1954) Estimation of available phosphorus in soils by extraction with sodium bicarbonate. USDA Circ. 939, Washington, DC., U.S. Gov. Print. Office.

Parkin, T.B., Meisinger, J.J., Chester, S.T., Starr, J.L., Robinson, J.A. (1988) Evaluation of statistical estimation methods for log normally distributed variables. *Soil Sci. Soc. Am. J.*, **52**, 323-329.

Peck, T.R., Soltanpour, P.M. (1990) Principles of Soil Testing. *Soil Testing and Plant Analysis - Third Edition*, R.L. Westerman (Ed.), Madison, WI, SSSA, pp 3-9.

Sawyer, J.E. (1994) Concepts of variable rate technology with considerations for fertilizer application. *J. Prod. Agric.*, **7**, 195-201.

Schmitt, M.A., Randall, G.W. (1994) Developing a soil nitrogen test for improved recommendations for corn. *J. Prod. Agric.*, **7**, 328-334.

van Lierop, W. (1990) Soil pH and lime requirement. *Soil testing and plant analysis 3rd ed.*, R.L. Westerman (Ed.), Madison, WI, SSSA, pp 73-126.

Vanotti, M.B., Bundy, L.G. (1994) An alternative rationale for corn nitrogen fertilizer recommendations. *J. Prod. Agric.*, **7**, 243-249.

Vetch, J.A., Malzer, G.L., Robert, P.C., Huggins, D.R. (1995) Nitrogen specific management by soil condition: managing fertilizer nitrogen in corn. *Site-Specific Management for Agricultural Systems*, P.C. Robert et al. (Eds), Madison, WI, ASA, CSSA, SSSA, pp 465-474.

Voss, R.D. (1995) ISU revised nutrient (fertilizer) recommendations. *Proc. Seventh Annual Integrated Crop Management Conf.*, Iowa State Univ., Ames, IA.

Woodruff, C.M. (1967) Crop response to lime in the midwestern United States and Puerto Rico. *Soil acidity and liming*, R.W. Pearson and F. Adams (Eds), Agronomy Monogr. 12, Madison, WI, ASA, pp 207-231.

ENVIRONMENTAL AND ECONOMIC EFFECTS OF SITE SPECIFIC AND WEATHER ADAPTED NITROGEN FERTILIZATION FOR A DUTCH FIELD CROP ROTATION

T.J. DE KOEIJER, G.J.M. OOMEN

Department of Ecological Agriculture, Wageningen Agricultural University, Haarweg 333, 6709 RZ, Wageningen, The Netherlands

G.A.A. WOSSINK

Department of Farm Management, Wageningen Agricultural University, Hollandseweg 1, 6706 KN Wageningen. The Netherlands

ABSTRACT

Dutch within-field differences in soil fertility are significant. Therefore the expectations regarding economic and environmental benefits of site specific management are high. However, research into the effects of site specific management on nitrogen fertilisation practice in Dutch arable farming showed no improvements in economic or environmental impacts. In case of environmental restrictions however, site specific fertilising can limit negative income effects. Combining site specific and weather adjusted nutrient management based on split applications promises a significant reduction of input use while maintaining economic results. In this precision agriculture, knowledge of the organic matter turnover seems to be important.

INTRODUCTION

Due to the increasingly stringent environmental legislation, site specific management is receiving more and more attention. For a sustainable agriculture, optimal efficiency of input use is crucial. The nitrogen input for a certain cropping period can be divided into an internal input and an external input. The internal input is the result of nitrogen mineralisation and nitrogen leaching in the preceding period. The external input just before and during the growing season can be matched to the requirements of a crop by taking into account the measured or expected internal input. The nitrogen mineralisation in a given period is related to crop rotation and previous manuring. Leaching and, to a lesser extent, also mineralisation depend on physical and morphological soil properties. Since variability in those properties in Dutch cropping fileds is significant, site specific management is expected to result in improved efficiency and therefore economic and environmental benifits. Due to the influence of the internal input supplied by the dynamics of the organic matter, site specific management has to be studied at the level of crop rotations.

Most research on site specific management originates from North America. Schueller (1992) gives an overview and states that, although field variability is very important, it is not studied as much as would be expected. The Dutch situation is characterised by very small fields compared to the avarage field size in the USA, and so the economic and environmental effects of site specific fertilising have been tested on hypothetical soils (Van Noordwijk and Wadman, 1992). They found that with increasing spatial variability in a field, the optimal economic nitrogen supply increases too while, in case of environmental restrictions, the allowed fertiliser rate decreased with increasing spatial variability. Booltink and Verhagen (1997) researched the effects of spatially variable fertilisation for one crop. However the did not analyse the effects of site specific management separately. Instead the researched the total effect of site specific fertilisation, optimum time windows and optimal amounts for nitrogen.

The objective of this paper is to evaluate the economic and environmental effects of site specific management for nitrogen fertilisation for a typical Dutch arable crop rotation. Two technologies are considered: a) application of fertiliser and timing on each site in line with the current fertilisation recommendations and b) split fertilisation, i.e. to adjust the nitrogen supply to the crop uptake.

MATERIALS AND METHODS

To analyse the relationships between physical yield, nitrogen input, heterogeneity and nitrogen leaching, a model is used to simulate the nitrogen dynamics in a crop rotation (Habets and Oomen, 1994). The N-Dicea model uses relatively simple data which can be either provided by farmers or estimatid form historical data. The model calculates the amount of nitrogen leaching during a growing season in a particular crop rotation.

Crop growth is not simulated in the N-Dicea model. If the relationship between available nitrogen and crop yield is known form another source, however, the effects of lower nitrogen fertilisation can be calculated. If the amount of nitrogen uptake needed for optimal growth is not available, the resulting yield loss can be calculated. A detailed overview of the functions and parameters used in the model has been described in De Koeijer *et al.* (1996).

Calculations were performed for a plot situated in the North-West of the Netherlands. The plot is 300 by 200 metres and part of the 'Van Bemmelen' experimental farm. The soil is marine sediment, and the polder was reclaimed in around 1930. In the plot, four different soil profiles can be distinguished. Three of these react similarly for nitrogen, so for this research it was sufficient to divide the plot into only two different zones. The field consists of approximately one third sand and two thirds clay.

In the simulations for the different fertilisation strategies, weather data from one year, 1985, were used. This year can be characterised as average rainfall (so water was not a limiting factor) with average temperatures. The four year crop rotation in the case study consisted of potato, winter wheat, sugarbeet and spring barley. The model was validated by performing simulations for 1994 and 1995 for which the yields and N-input were recorded in the experimental field on the Van Bemmelenhoeve. The yields measured differed significantly for the two zones. In 1995, the yield of winter wheat amounted to 8000 and 10000 kg/ha on sand and clay, respectively. In 1994 for potato, the yield varied even more from 25000 to 45000 on sand and clay respectively (Verhagen *et al.* 1995). So the variation in crop yield was considerable, especially for potato, where the ratio of the yield level found for the two zones was almost two.

For the calculations regarding the economic viability of site specific management, the yield ratio between the sand and clay zone for sugarbeet and spring barley is assumed to be equal to the yield ratio between zone 1 and 2 for winter wheat. In contrast to potato, sugarbeet and spring barley have an intensive root system similar to that of winter wheat. This assumption was necessary because appropriate information for sugarbeet and spring barley was lacking.

The prices used in the model calculations were based on KWIN (PAGV, 1994) which gives the average price for 1989-1993 as DFL 0.1955 for potato, DFL 0.3768 for wheat, DFL 0.1067 for sugarbeet and DFL 0.4322 for barley per kg product.

SCENARIOS

The following nitrogen fertilising strategies were simulated:
a) A uniform application rate according to the nitrogen recommendations in the Netherlands (IKC-at, 1993). In this strategy, the field is regarded as having a homogeneous clay soil although in fact one third of the field consists of sand, the nitrogen residue in the soil was considered in terms of the average value for the whole field.
b) Site specific, also according to the nitrogen fertiliser recommendations (IKC-at, 1993). In this strategy clay and sandy soils are fertilised differently and the nitrogen residue in the soil was evaluated separately for both zones.
c) Even distribution of the maximum amount of fertiliser permitted by the environmental restriction that the drinking water norm should not exceed 50 mg nitrate per litre of ground water.
d) Site specific accounting for the mentioned environmental restriction.
e) Site specific and weather adjusted application without yield reduction. In the last strategy there should be enough nitrogen available for the plant throughout the growing season while minimising the residue of nitrogen. To assess nitrogen demand during the growing season for each crop at four time points, the amount of fertiliser necessary to reach the non nitrogen limited yield level is calculated. The calculated mineralisation and leaching of nitrogen during the preceding periods is taken in account in assessing the nitrogen application in the latter strategy.

For the simulations of the environmental scenarios (c and d), the environmental restriction of 50 mg nitrate per litre of ground water was converted into the amount of nitrogen loss allowed per ha. The amount of nitrogen loss allowed depends on the soil characteristics and the crop. The amount of nitrogen loss which can be tolerated is taken as the starting point for the simulations of the environmental scenarios. With this knowledge the maximum nitrogen input is sought and the possible yield reductions calculated. In the scenarios where there is an even distribution of fertiliser the average nitrogen loss allowed per crop was used. In the site specific scenarios the amount of allowed nitrogen loss was calculated per zone.

RESULTS

In Table 1, quantities of N-fertiliser applied, N-leaching, and the income effects per year are presented as the annual average for the complete crop rotation of four years. In the traditional strategy, i.e. even fertilising according to the fertiliser recommendations,

TABLE 1. The average N-fertilising (kg/ha), N-leaching (kg/ha) and income effects (DFL per ha) per year.

Fertilising strategy		Zone[*]	N-input kg/ha	N-leaching kg/ha	income-effect DFL/ha
a.	Uniform application according to recommendations	S C	166 166 166	112 92 99	0[**] 0 0
b.	Site specific application according to recommendations	S C	183 157 166	131 83 99	-18 9 0
c.	Uniform application within environmental norms	S C	88 88 88	59 27 38	-164 -81 -109
d.	Site specific application within environmental norms	S C	49 111 90	43 36 38	-365 42 -93
e.	Site specific and weather adjusted	S C	79 71 74	78 30 46	90 99 96

[*] Zone S is characterised by a sand soil type.
 Zone C is characterised by a clay soil type.
[**] Growth was not limited by shortage of N

for clay there is no shortage of nitrogen although the amount of nitrogen residue, especially in the fourth year is very low. Nitrogen leaching on sand is higher than nitrogen leaching on clay. In the site specific fertilising strategy, the quantity of applied fertiliser is higher for sand than for clay. This is caused by greater leaching of the nitrogen residue in sand than in clay, so that in spring the internal nitrogen supply in sand is lower and therefore the fertiliser recommendation is higher.

There are also differences in the leaching of nitrogen within the crop rotation. Leaching of nitrogen is significantly higher for potato than for the other crops. This is caused by higher fertiliser applications for potato which has a relatively inefficient rooting system and by the easy decomposition of the nitrogen-rich crop residues.

Insights into the effect of site specific fertilisation can be gained by comparing the strategies according to the fertiliser recommendations for the total field with site specific fertilisation. According to the model results, site specific fertilisation does not lead to a decrease in average nitrogen input or nitrogen leaching, if the current recommendations are followed. The distribution of N-fertiliser between the clay and sand zones differ, in spring, hardly any nitrogen is left on sand and so more nitrogen has to be administered. On clay, less nitrogen has been leached, more nitrogen is available in spring, and hence the fertiliser input is lower. The distribution of nitrogen leaching under a site specific regime varied more than in the traditional scenario. The leaching of nitrogen on sand is higher because of the higher nitrogen input but this effect is compensated by lower nitrogen leaching on clay. The economic effect is zero.

To meet the drinking water norm of 50 mg nitrate per litre of ground water, the nitrogen input must be reduced by 47% for the case study. This reduces nitrogen leaching by 62%. In potato and sugarbeet this restriction results in considerable yield reductions on both soils. For spring barley, yield reduction only occurs in the sand zone. The yield reduction for potato is larger on clay than on sand while for sugarbeet the largest yield reduction occurs on sand. The yield reduction is 191 DFL/ha/year. The yield loss is not compensated for by the reduction in input costs of 74 DFL/ha/year.

Sugar beets are especially sensitive to lower nitrogen input. According to the current recommendations for this crop, the nitrogen has to be applied early in the season and therefore a large part leaches, especially in sand. In winter wheat, this problem is less important because the nitrogen is applied in three stages.

In the strategies where nitrogen leaching is not allowed to exceed the drinking water norm in any year or zone, growth inhibition occurs due to nitrogen deficiency. In particular this happened on sand for potato, sugarbeet and to a lesser extent for barley. On clay, only potato showed a small yield reduction. Comparison of the two environmentally restricted strategies provides insight in the economic effects of site specific fertilising when nitrogen input has to comply with environmental restrictions. Location specific nitrogen application has a small positive economic effect of about 20 DFL per ha per year in the total rotation.

The last strategy is an attempt to find the most efficient fertilising strategy, utilising the N-Dicea model. For each application, the nitrogen input was determined according to the already known weather of the following period. On sand, more nitrogen was applied despite the lower needs of the crop in order to compensate for the higher leaching. A reduction in nitrogen input to 45% was realised with this strategy. The nitrogen leaching is reduced to 46% of that occurring in the traditional strategy. The strategy does not comply with the drinking water norm.

DISCUSSION

Under the current farming practice of fertilising according to the fertiliser recommendations, site specific fertilising alone does not result in positive economic effects when compared to the traditional strategy. Therofere it is not economically advantageous to adapt the amount of mineral nitrogen to the soil zones, if the current recommendations are followed. The spatial distribution of nitrogen Under site-specific fertilising is different to that in the traditional strategy. Nitrogen leaching during the winter is higher on sand than on clay. Therefore, the nitrogen input should be higher on sand. It is clear that the fertiliser recommendations are risk averse. The fertiliser recommendations are not based on the maximum yield level and nitrogen uptake but on the soil and weather characteristics in which the expected losses are already taken into account. The current nitrogen recommendations are evaluated on the basis of a fixed amount of nitrogen fertiliser minus an estimate of the quantity of mineral nitrogen in the soil (Noordwijk and Wadman, 1992). Therefore, the site specific distribution of nitrogen does not result in higher yields. The risk of too much nitrogen leaching during bad weather conditions for optimal crop growth is smaller under site specific management. This means that site specific management reduces the risk that there may not be enough nitrogen available on sand. Even though in this research the calculations were performed for normal wet weather conditions, no positive yield effect was found. However, the calculations showed that the nitrogen available would not have been sufficient for an optimal crop growth if the weather conditions had been slightly worse.

The effect of site-specific application on ground water quality was also not significant. As explained above site specific management alone simply leads to more leaching of nitrogen in very bad weather conditions.

The scenarios with environmental restrictions can have smaller negative economic effects than those presented here if higher leaching of nitrogen in the crops with high economic returns can be compensated for by lower leaching of nitrogen in winter wheat and barley. The economic effect of site specific fertilisation could also be improved by compensating for the high level of leaching on sand by a lower level of leaching on clay.

The last scenario illustrates how much nitrogen is applied to avoid risk. The simulations show that the amount of nitrogen can be reduced to 45% without yield losses. This scenario indicates the importance of weather adjusted as well as site specific management. This scenario could be put into practice by applying the repeatedly calculated fertiliser input in several doses spread over the growing season.

An important role in the results is played by the very high net mineralization of organic matter. In the chosen crop rotation (based on the actual farming practices of the Van Bemmelenhoeve), the net annual mineralization on the clay and sand soil amounts to 50 respectively 75 kg N/ha/year. A considerable part of this nitrogen is released in autumn and lost by leaching. This implies that, for precision agriculture, not only time and site specific management is important but also that the organic matter balance needs a lot of attention as well. Application of catch crops will improve the organic matter balance while in the mean time the leaching of nitrogen in the winter season will be reduced.

In this research, it is assumed that the two different soil types in the experimental field are more or less homogeneous. In practice, however, the exact locations of these differing soil types are not known and more zones may exist. Furthermore, the border between the two soil types is gradual. This means that the simulation results give the maximum differences because in practice location specific fertilising is not as accurate as we have assumed.

In this research, the effects of site specific fertilising of nitrogen have been calculated. If site specific management was not only applied for nitrogen but also for phosphate and potassium the economic effects under environmental restrictions might be more significant. Furthermore, the use and emissions of pesticides need to be added for a total environmental and economic assessment. As there are interactions between fertilising and chemical crop protection, the calculated economic and environmental effects of site specific management might be more significant when applied to both types of inputs.

CONCLUSIONS

In-field differences in soil fertility are significant. Therefore site specific management should result in a more efficient use of fertiliser which would have economic and environmental benefits.

Using current fertiliser recommendations for nitrogen in the Netherlands, our case study showed no reduction in nitrogen input or leaching under site-specific management.

The fact that no increase in yield occurred is due to the fertiliser recommendations for field crops used in the Netherlands. These recommendations are based on soil characteristics and averaged weather characteristics in which possible risks of input losses are already taken into account, rather than on uptake by the plant.

Site specific fertilising can limit the negative income effects of environmental restrictions of nitrogen emission.

Site specific and weather adjusted fertiliser management can reduce fertiliser input significantly and lead to a modest increase in income.

Precision agriculture should not only pay attention to the site specific fertilisation, optimum time windows and optimal amounts for nitrogen but to organic matter dynamics as well.

ACKNOWLEDGEMENT

The authors would like to thank F.J.M. Verhees who did most of the calculations as part of his MSc-research.

REFERENCES

Booltink, H.W.G and Verhagen, J. (1997) Using decision support systems to optimize barley management on spatial variable soil. In: Systems approaches for Agricultural Development. Volume 2. Applications of systems approaches at the field level, proceedings of the second international symposium on systems approaches for agricultural development, held at IRRI, Los Ba~nos, Phillipines, 6-8 December 1995. Kropff, M. J., Teng,P.S., Aggarwal, P.K. , Bouma, J., Baouman ,B.A.M., Jones, J.W. and Van Laar H.H. (editors). Kluwer Academic Publishers. Dordrecht/Boston/London, pp 219-233.

Habets, A.S.J., Oomen, G.J.M. (1994) Modelling nitrogen dynamics in crop rotations in ecological agriculture. *Nitrogen mineralization in agricultural soils: proceedings symposium held at the institute for soil fertility research*, J.J. Neeteson and J. Hassink (Eds), Haren, AB-DLO, pp 255-268.

IKC-at (1993) Stikstofbemestingsrichtlijnen voor de akkerbouw en de groenteteelt in de volle grond, Lelystad, Informatie en Kenniscentrum Akker- en Tuinbouw, The Netherlands.

Koeijer, T.J. de, Wossink, G.A.A., Verhees, F.J.M. (1996) Environmental and economic effects of spatial variability in cropping fields: nitrogen fertilization and site specific management. *Proceedings Interconference Symposium International Association of Agricultural Economists: Economics of Agro-Chemicals*, Wageningen April 24-28, 1996.

Noordwijk, M. van, Wadman, W.P. (1992) Effects of spatial variability of nitrogen supply on environmentally acceptable nitrogen fertilizer application rates to arable crops. *Netherlands Journal of Agricultural Science,* **40**, 51-72.

PAGV (1994) Kwantitatieve informatie voor de akkerbouw en groenteteelt in de vollegrond 1995, Lelystad, PAGV.

Schueller, J.K. (1992) A review and integrating analysis of spatially-variable control of crop production, *Fertilizer Research,* **33**, 1-34.

Verhagen, A., Booltink, H.W.G., Bouma, J. (1995) Site-specific management: balancing production and environmental requirements at farm level, *Agricultural Systems,* **49**, 369-384.

STRATEGIES FOR SITE-SPECIFIC NITROGEN MANAGEMENT

R.B. FERGUSON, G.W. HERGERT, J.S. SCHEPERS

University of Nebraska, Department of Agronomy, Lincoln, NE, USA

ABSTRACT

Site-specific nitrogen (N) management has been approached with a variety of methods for controlling the N application rate. The accuracy of the N rate map derived from these approaches determines the economic and environmental effectiveness of site-specific N management. Four approaches are considered here for development of N rate maps for irrigated maize - site-specific N management according to grid soil sampling, soil series/topography, yield maps and remote sensing. All four methods have advantages and disadvantages. Grid soil sampling can generate accurate maps with annual, high density sampling, which is not cost-effective in most situations. Rate maps based on soil series vary considerably in their accuracy according to the site, but take advantage of existing spatial information. Yield maps and remotely sensed images contain data at high spatial densities, but many factors other than N can influence yield and remotely sensed images. Since economic and environmental benefits of site-specific N management for maize are likely often to be small, relatively inexpensive approaches to site-specific N management will be required to be economically feasible. It is likely that an integrated approach, using spatially dense information from yield maps and remotely sensed images, along with directed soil sampling, baseline grid soil sampling and new, more detailed soil surveys will be used to manage N site-specifically.

APPROACHES TO APPLICATION MAP DEVELOPMENT

An effective evaluation of variable rate technology (VRT) N application requires an accurate application map. In most research to date, existing uniform N recommendations based on expected yield and measures of soil N availability are applied spatially (Gotway *et al.*, 1996; Snyder *et al.*, 1996; Kitchen *et al.*, 1995). Various approaches have been used to attempt to quantify the N supply potential of fields spatially. These can be generally divided into four categories: grid soil sampling, soil series - topography, yield mapping, and remote sensing. These categories can be further described in terms of their information extent and spatial density. Grid soil sampling can provide extensive information about the chemical and physical characteristics of the soil at selected points, but the spatial density of this information is likely to be limited because of the expense of collecting and analyzing soil samples. Yield maps and remotely sensed images using current technology can provide spatially dense, but not very extensive, information about the field.

GRID SOIL SAMPLING

This method of generating rate maps is currently the most commonly used commercial method for phosphorus (P) and potassium (K), and has been used for N rate maps as well. When commercially practiced, the sampling density is most often one sample for every 1.2 - 1.6 ha, usually representing a composite of several soil cores collected around a grid point. Grid density can have a major influence on the accuracy of the N application map. Figure 1, from Ferguson *et al.* (1996), illustrates this point. At this site, in Lincoln County, Nebraska, two sampling densities, both much denser than commonly used commercially, provide substantially different N maps. The coarser density misses a systematic pattern present in soil residual nitrate levels, most likely resulting from patterns of livestock fencing in the past. Forty five percent of the field received a different N recommendation with the coarser grid compared to the fine grid. However, coarser grid densities can provide acceptably accurate maps, as illustrated in Figure 2. At this site, a grid density similar to that used commercially provides an acceptable map at a density many times less than actually collected in the study. The area of the field which received a different recommendation at the lower density grid was only 17.6%. The dilemma is that it is currently difficult to predict the minimum grid size necessary to provide an acceptably accurate map. Further, grid sampling annually to detect nitrate levels in soil is not likely to be a realistic management option, due to cost and labor constraints.

SOIL SERIES - TOPOGRAPHY

Nitrogen rate maps based on soil series or field topography take advantage of spatial information which is already known, and in some areas of the U.S. is becoming available electronically. Several researchers have found topography-based N rate maps to be effective (Solohub *et al.*, 1996). Others have found topography or soil series based application maps to provide similar results to grid sample based maps (Thompson and Robert, 1994), while still others have found relatively low spatial correlation between crop response to N and soil map units (Everett and Pierce, 1996). Figures 3 and 4

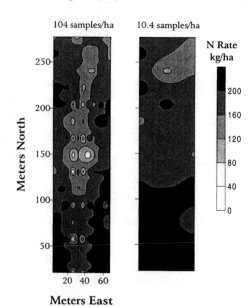

FIGURE 1. Recommended N rate (1994) at two grid sampling densities - 104 and 10.4 samples per ha, Lincoln County, NE, USA.

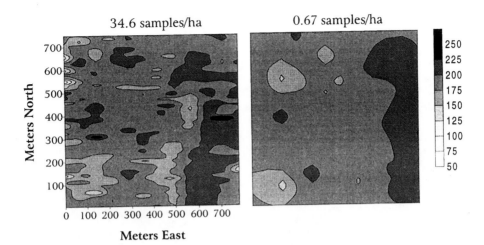

34.6 samples/ha · 0.67 samples/ha

Meters North

700
600
500
400
300
200
100

0 100 200 300 400 500 600 700

Meters East

250
225
200
175
150
125
100
75
50

FIGURE 2. Recommended N rate (1994) at two grid sampling densities - 34.6 and 0.67 samples per ha, Buffalo County, NE, USA.

illustrate both situations. Figure 3, from a field in Adams County, Nebraska, illustrates a situation where the soil residual nitrate map, here developed from grid soil sampling, is closely correlated with relative elevation and soil series in the field. At this site, residual nitrate-N is highest in the lowest area of the field, which also corresponds to the less productive Butler and eroded Holder soils. In this case, residual nitrate-N is higher where crop yield and consequently crop removal was less the preceding year. Figure 4 illustrates a situation, from Clay County, Nebraska, where the entire field is mapped as one series, and yet substantial variation in N rate is recommended from grid sampling - the recommended N rate in most years ranges from 134 to 212 kg N ha^{-1}, within one soil mapping unit.

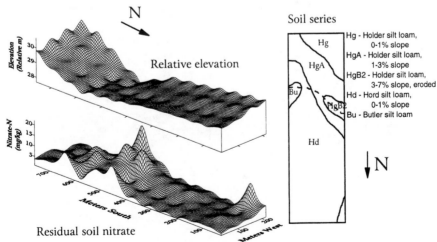

FIGURE 3. Relative elevation, residual soil nitrate-N (1992) and soil series, Adams County, NE, USA.

389

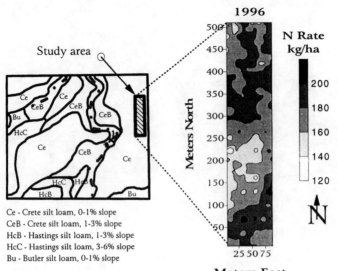

FIGURE 4. Soil series and 1996 N rate map, Clay County, NE, USA.

One criticism of grid soil sampling is that it ignores what we already know about fields. Soil surveys contain spatial information which is ignored when we arbitrarily impose fixed grid patterns for soil sampling on a field. On the other hand, arbitrary grid sampling may often expose things we did not know about fields which have resulted from prior management and are unrelated to soil series or topography. Figure 5 illustrates the location of the residual effects on soil P from a farmstead animal lot described by grid soil sampling the same field in Clay County, Nebraska. The existence of this area of high phosphorus concentration was unknown prior to grid soil sampling.

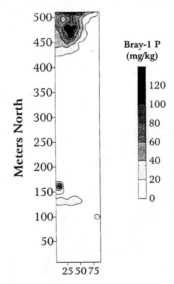

FIGURE 5. Soil Bray-1 phosphorus distribution (1993), Clay County, NE, USA.

This pattern of P in the field has likely been present for over 30 years - the length of time it has been farmed by the current owner, who has never applied manure or pastured livestock in the field - and likely much longer. The systematic pattern of residual N and resulting N fertilizer recommendation in Figure 1 is a similar example.

YIELD MAPPING

The most rapidly growing area of site-specific management is in yield monitoring and mapping. There is considerable interest in using yield maps as management tools. One potential application of yield maps is providing the basis for VRT application of fertilizers. A yield map, assuming it accurately represents the yield from the field, can be quite useful. It will reflect crop removal of nutrients on a spatial basis, and may use the plant as an integrator of soil parameters which might more reasonably reflect nutrient availability than a soil test. Kitchen *et al.* (1995) found productivity trends in fields which persisted over years and crops, but suggested that the use of yield maps to identify areas of relative productivity in fields should be based on an accumulation of as many years of mapped yields as possible. Davis *et al.* (1996), found yield levels from unfertilized plots were more accurate predictors of spatial patterns of N requirement than were yield levels from well-fertilized plots (168 kg N ha^{-1}). These results suggest that producers may have difficulty using yield maps produced from adequately fertilized fields to determine areas within fields having different N requirements. Many factors will influence yield other than N availability - often N may not be the yield limiting factor. Consequently, yield maps may not very accurately portray N availability to the crop, or predict N requirement for the coming season. Figure 6, from Clay County, Nebraska, illustrates spatial patterns of maize grain yield in four subsequent years, and the relationship of yield to organic matter. In 1993 and 1994, high wind damage to the crop in July for the most part masked the influence of soil organic matter on yield patterns. In 1995 and 1996, the yield map was significantly correlated to organic matter in the field. In two years of four, the pattern of soil organic matter was a primary factor influencing the yield map, and appears to represent productivity potential stable over time assuming other factors are not limiting yield.

REMOTE SENSING

Like the yield map, remotely sensed images can use the crop as an integrator of soil nutrient availability. Remote sensing appears to have considerable potential to monitor crop N status and also provide an application map if N deficiency is detected. Two primary remote sensing approaches are under investigation - real-time, ground based sensing and aerial or satellite sensing. Ground-based sensors, located on high clearance equipment or center-pivot sprinkler systems, may have the capability to detect crop N stress and apply N as necessary at the same time. Aerial or satellite sensors have the potential to detect developing crop N stress, and provide a N rate map for sidedressing or high clearance application. Remotely sensed images can correlate well with other spatial sources of information, as shown in Figure 7. At this site, a remotely sensed image collected in July, 1995 correlates visually with a map of organic matter (Figure 6). However, without the prior existence of the organic matter map, the cause of patterns observed in remote-sensed images may be a matter of speculation. Considerable research is needed in ground-truthing sensor technology - relating observations at

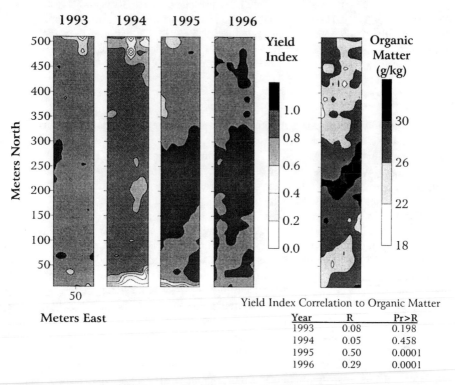

Yield Index Correlation to Organic Matter

Year	R	Pr>R
1993	0.08	0.198
1994	0.05	0.458
1995	0.50	0.0001
1996	0.29	0.0001

FIGURE 6. Yield index (observed/maximum yield), organic matter, and Pearson Correlation Coefficients, 1993-1996, Clay County, NE, USA.

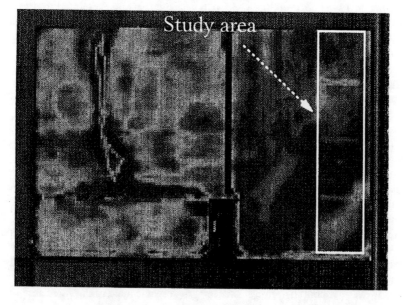

FIGURE 7. Resource21® Enhanced Vegetation Image, July 1995, Clay County, NE, USA.

specific spectral bands to specific stresses in crops. Currently, commercial availability of remote-sensing data is relatively limited. This is likely to change rapidly, as several partnerships have plans to install satellite remote sensing systems within the next few years.

AN INTEGRATED APPROACH

It is likely that future, successful implementation of VRT N application will rely on a combination of the above methods for generating the application map. The spatial density available from yield maps, remotely sensed images, and in some cases soil series/topography sources will be coupled with the intensive information available from soil sampling. In many cases, yield maps and remotely sensed images will be calibrated by directed soil sampling, in which areas of fields most likely to contain different organic matter or other nutrient levels are sampled to quantify the nutrient supply potential predicted by the map or image. It may be likely, in cases where prior management has the potential to significantly influence non-mobile nutrient levels in the soil, that an initial grid sampling at some determined optimum density will be collected, followed by N application based on a combination of crop removal, N stress detection, and soil supplying capability.

ECONOMIC AND ENVIRONMENTAL BENEFITS

The economic benefit of site-specific N management is still unknown. Some investigators show that VRT N application can be economically feasible in some years (Snyder *et al.*, 1996; Malzer, 1996), while others show no economic gain (Kitchen *et al.*, 1995). Summarizing field comparisons of variable to uniform N application in Nebraska, we find generally no significant reductions in the total amount of N applied with variable application. Averaged over 19 site/years, variable rate application resulted in the application of 3% less total N than uniform application, with little or no effect on grain yield. Malzer (1996), noted that factors currently used in making N recommendations, including expected yield, cropping history, organic matter and residual nitrate-N only predicted a portion of the potential benefit of site-specific N management, suggesting that additional factors will need to be considered to maximize profitability of site-specific N management.

The environmental benefit of site-specific N management is more difficult to quantify, and consequently few studies to date have addressed this question. Kitchen *et al.*, (1995) documented a decrease in root zone NO_3-N with variable N application. Redulla *et al.*, (1996), found similar residual NO_3-N levels remaining in soil with uniform or variable N application. In Nebraska, we have measured significant reductions in residual NO_3-N with variable application compared to uniform application in two site/years (Hergert *et al.*, 1996). Generally, reductions in leachable nitrate with site-specific management noted to date are relatively small.

SUMMARY

Theoretical calculations using existing datasets of spatial information, indicate that, at least part of the time, site-specific N management is both profitable and can reduce N loss to the environment. However, when site-specific N management has been actually compared to uniform application in the field, real differences economically or environmentally are often quite small or non-existent. It will be critical to identify soils and cropping systems with the most potential benefit from site-specific N management. For such sites, it will be essential that procedures be developed to generate accurate N application maps inexpensively. If variable rate N application from a map is to be practical, it must not cost much more to produce the application map than it costs to produce a recommendation for uniform N application. From this standpoint, real-time remote sensing on a fertilizer applicator or irrigation system has an advantage, since no rate map is needed - the applicator simply responds to the crop status at the time of application. However, real-time application of N according to crop N status, if it ultimately becomes practical, will not be an option for some cropping situations. Additional work is needed on identifying those soils and cropping systems which are most likely to reflect positive benefits, either economically, environmentally, or both, to site-specific N management.

REFERENCES

Davis, J.G., Malzer, G.L., Copeland, P.J., Lamb, J.A., Robert, P.C., Bruulsema, T.W. (1996) Using Yield Variability to Characterize Spatial Crop Response to Applied N. *Precision Agriculture, Proceedings of the 3rd International Conference, June 23-26, 1996,* P.C. Robert, R.H. Rust and W.E. Larson (Eds), Minneapolis, MN, pp 513-519.

Everett, M.W., Pierce, F.J. (1996) Variability of Corn Yield and Soil Profile Nitrates in Relation to Site-Specific Management. *Precision Agriculture Proceedings of the 3rd International Conference, June 23-26, 1996,* P.C. Robert, R.H. Rust and W.E. Larson (Eds), Minneapolis, MN, pp 43-53.

Ferguson, R.B., Hergert, G.W., Gotway, C.A., Fleming, K., Peterson, T.A. (1996) Soil Sampling for Site-Specific Nutrient Management. *Proceedings of the Great Plains Soil Fertility Conference, March 4-6,* Denver, CO.

Gotway, C.A., Ferguson, R.B., Hergert, G.W., Peterson, T.A. (1996) Comparison of Kriging and Inverse-Distance Methods for Mapping Soil Parameters. *Soil Sci. Soc. Am. J.,* **60**, 1237-1247.

Hergert, G.W., Ferguson, R.B., Gotway, C.A., Peterson, T.A. (1996) The Impacts of Variable Rate N Application on N Use Efficiency of Furrow-Irrigated Corn. *Precision Agriculture, Proceedings of the 3rd International Conference, June 23-26, 1996,* P.C. Robert, R.H. Rust and W.E. Larson (Eds), Minneapolis, MN, pp 389-397.

Kitchen, N.R., Hughes, D.F., Sudduth, K.A., Birrell, S.J. (1995) Comparison of Variable Rate to Single Rate Nitrogen Fertilizer Application: Corn Production and Residual Soil NO_3-N. *Site-Specific Management for Agricultural Systems, Proceedings of the 2nd International Conference, March 27-30, 1994,* P.C. Robert, R.H. Rust and W.E. Larson (Eds), Minneapolis, MN, pp 427-441.

Malzer, G.L. (1996) Corn Yield Response Variability and Potential Profitability of Site-Specific Nitrogen Management. *Better Crops,* **80(3)**, 6-8.

Redulla, C.A., Havlin, J.L., Kluitenberg, G.J., Zhang, N., Schrock, M.D. (1996) Variable Nitrogen Management for Improving Groundwater Quality. *Precision Agriculture, Proceedings of the 3rd International Conference, June 23-26, 1996*, P.C. Robert, R.H. Rust and W.E. Larson (Eds), Minneapolis, MN, pp 1101-1110.

Solohub, M.P., van Kessel, C., Pennock, D.J. (1996) The Feasibility of Variable Rate N Fertilization in Saskatchewan. *Precision Agriculture, Proceedings of the 3rd International Conference, June 23-26, 1996*, P.C. Robert, R.H. Rust and W.E. Larson (Eds), Minneapolis, MN, pp 65-73.

Snyder, C., Havlin, J., Kluitenberg, G., Schroeder, T. (1996) An Economic Analysis of Variable Rate Nitrogen Application. *Precision Agriculture, Proceedings of the 3rd International Conference, June 23-26, 1996*, P.C. Robert, R.H. Rust and W.E. Larson, (Eds), Minneapolis, MN, pp 989-998.

Thompson, W.H., Robert, P.C. (1994) Evaluation of Mapping Strategies for Variable Rate Applications. *Site-Specific Management for Agricultural Systems, Proceedings of the 2nd International Conference, March 27-30, 1994*, P.C. Robert, R.H. Rust and W.E. Larson (Eds), Minneapolis, MN, pp 303-323.

HOLES IN PRECISION FARMING: MECHANISTIC CROP MODELS

B. ACOCK

USDA, ARS Remote Sensing and Modeling Lab., Beltsville Agricultural Research Center-West, Beltsville, MD 20705, USA

Ya. PACHEPSKY

Duke University Phytotron, Department of Botany, Duke University, Durham, NC 27708, USA

ABSTRACT

Precision farming employs many advances in engineering to determine location within a field, record soil and plant properties including yield, and to apply spatially-variable amounts of various agricultural inputs. Most researchers in this area are engaged in measuring soil nutrients in a grid pattern over the field, sometimes overlaying nutrient treatments, assuming that nutrient imbalance is responsible for the yield variation. They are failing to find any consistent correlation between nutrients and yield. This practice is so prevalent that for some the principal question in Precision Farming has become, "How many holes per acre?" This is the first hole in Precision Farming - a lack of understanding of the reasons for yield variation. We will be able to prescribe treatments for yield variability only when we understand the reasons for that variability, and we will understand only when we consider all the major factors limiting plant growth. The only practical way of considering all these factors is to use mechanistic crop models: models that attempt to simulate the principal mechanisms in the soil/plant/atmosphere system occupied by the crop. Empirical models that make no attempt to represent mechanism are unlikely to explain yield variation or provide reliable prescriptions for Precision Farming. Because of limitations in our knowledge of plant behaviour, mechanistic crop models are still imperfect tools. It will be necessary to validate and possibly calibrate each crop model for various points in the field before using it in an iterative procedure to simulate and understand the pattern of yield variability observed. Even when we do understand all the reasons for yield variability, any prescription will depend on weather pattern. Precision farming will only work when we can prescribe management action during the course of the growing season. For these reasons, mechanistic crop models are essential to precision agriculture. Rather than discarding them because of their imperfections, we should continue to improve them.

INTRODUCTION

In the 1960s, engineers developed a range of sophisticated controllers for greenhouses. They could control greenhouse heating and ventilation as complex functions of light, windspeed, humidity, etc. The greenhouse growers brought these controllers to the plant physiologists and asked how they could use them to best advantage. The plant physiologists were unable to tell them. This started the first exercise in crop modelling in the UK, at what was then the Glasshouse Crops Research Institute.

We now have a similar situation in precision agriculture. The engineers have given us Global Position Sensors (GPS) to determine where we are in a field, yield monitors to measure yields on the move, computers to prepare maps of spatial variability, and precision applicators to place agricultural inputs where we want them in the field. Now we must learn how to use these new tools, and the new information they provide, to enhance farm profitability and reduce agricultural pollution. Having measured the variability in yield, we must develop prescriptions to apply agricultural inputs in spatially variable patterns. We have yet to formulate a coherent plan for doing this.

If we discover a low-yielding spot in the field one year, will it still be there the next year? Should we increase inputs to increase yield in that spot, or reduce inputs to reduce waste and pollution? These questions can only be answered if we understand why the yield is low in that place. The majority of scientists working in precision agriculture are digging holes in a pattern over the field, analyzing the soil nutrients present and attempting to correlate these concentrations with yield. This activity is so dominant that many farmers and consultants, and even some scientists, believe that the only remaining question in precision agriculture is "How many holes per acre?". In fact, as we shall demonstrate in this pair of articles, there are more important questions, but the answer to that particular question is, "More than you can afford". The principal hole in precision agriculture is the hole in our understanding of the reasons for yield variations. To look for simple correlations with single soil variables is to ignore the progress we have made in the last 160 years. It takes us all the way back to 1834 when J. B. Boussingault (cited by Russell, 1961) started the first fertilizer experiments in field plots.

In this paper, we briefly review the demands placed on crop scientists by precision agriculture, and our past progress in understanding crop behaviour. We examine the tools available to us for explaining yield variability and discuss how those tools might be used to develop prescriptions that respond dynamically to weather and management events during the season.

THE SIZE OF THE HOLE IN PRECISION AGRICULTURE

Farmers, aided by crop scientists, have learned how to optimize the major environmental factors, especially soil nutrients, which are so easily and cheaply amended. However, since no single factor controls yield on a farm, yield does not correlate well with any single factor. We need to consider all environmental factors, and their interactions, and the changes that occur over time, before we can understand yield variation.

Consider a (hypothetical) field that has a slope and a depression at the base of the slope. In a dry year, the slope is too dry to support a good crop but the depression has adequate water. In a wet year, the slope has enough water but the depression develops anoxic conditions and the crop suffers. Rain tends to leach the soil nutrients downslope. In our field, the depression also acts as a frost pocket and occasionally crops there are nipped by frost. In developing a prescription for this field, we must take all these factors into account. Presumably, the fertilizer will be needed where the crop growth is strongest, but at the start of the season we can have no idea where that will be. The weather for the season is unknown. Now, some decisions will have to be made, based on the weather of

398

past years. However, we will never be able to manage yield variability until we can account for and respond to events that occur during the season. That includes both weather and management events.

In other words, a prescription is not something static that is developed at the start of the season and then applied. It has to be dynamic, and we need tools that enable farmers or consultants to make management decisions quickly during the season. The best way of doing that is to use mechanistic crop models.

PROGRESS IN DEVELOPING CROP MODELS

In the first half of the 19th century, the factors most limiting crop growth were the major plant nutrients: nitrogen, phosphorus and potassium. Determining the optimum levels of these and other soil factors for various crop, soil and weather combinations occupied crop scientists intensively for the next 100 years, and still continues today. In the course of this work, several models were proposed to explain the responses of crop yield to soil factors. The additive model became enshrined in statistics in the Analysis of Variance, Liebig (1855, cited in Russell, 1961) proposed a multiplicative model in his Law of the Minimum, and Blackman (1905) proposed a limiting factor model in his Law of Limiting Factors.

In 1959, the first controlled-environment chambers enabled scientists to start examining aerial environmental factors (Hughes, 1959). This led to a rapid advance in our understanding of how plants respond to their aerial environment. In a few years, the plant physiologists seemed to have plotted every plant process as a function of all the environmental variables affecting it. In 1965, computers became available to crop scientists, and immediately several groups started trying to synthesize our knowledge into crop models: mathematical equations describing the relationships between environmental factors, and crop growth and development. Many of the early models were simple correlative models. For example, yield might be expressed as a function of the solar radiation integral for the season, April mean temperature, August rainfall, and a few other variables. Then more mechanistic models were developed to describe the principal processes involved in the plant soil and atmosphere. Today there are mechanistic or semi-mechanistic models extant for most crops, and some are being used by farmers to make management decisions.

THE TOOLS AT OUR DISPOSAL

As a result of research over the last 160 years, we now have a variety of tools to use in explaining yield variability. They all have strengths and weaknesses.

Empirical models

Empirical models, sometimes called regression, statistical or correlative models, describe relationships between variables without referring to the processes connecting those variables. They usually consist of a single equation. Neural networks produce models of this type. They are frequently used to summarize experimental data and they can reproduce those data closely.

The use of empirical models for making predictions is problematic. If all major factors affecting the crop are included in a model, then it will generally make accurate predictions within the range of the database used to develop it. If some major factors are excluded or predictions are being made for conditions outside the range of the database, then its predictions will usually be wrong (Chanter, 1981). The reasons for this are intuitively obvious. If factors A and B interact, then crop response to factor A will depend on the level of factor B. If factor B is not included in the model, the predictions of the model cannot be accurate. Most datasets are collected in the field, and all datasets on yield variability are collected there. The range of environmental conditions is necessarily limited. Equations fitted over these limited ranges will not predict what happens outside the ranges.

Finally, empirical models are not very useful for understanding yield variability because they have none of our understanding of crop behaviour built into them. Often the parameters are nonsense from the physiologist's point of view (Laird and Cady, 1969). They may be qualitatively wrong, e.g., a negative slope on the response of yield to solar radiation, or quantitatively wrong, e.g., a response slope that is too steep or too flat.

Mechanistic models

Mechanistic models, sometimes called explanatory or process models, attempt to represent processes occurring in the crop, soil and atmosphere. Typically, mechanistic crop models that predict yield, represent processes at the plant and organ level (Acock and Acock, 1991). For example, they might represent photosynthesis, transpiration, leaf expansion, abscission, etc. At the organ level, the model is empirical, e.g., the equations used for leaf photosynthesis rarely represent causality between variables. So the term mechanistic merely indicates that some attempt has been made to represent processes in the model. There are no completely mechanistic crop models because (a) our knowledge of the processes is incomplete, (b) they would contain more parameters than we could evaluate, and (c) our computers probably could not handle the calculations of processes at the most fundamental, molecular level and give us an answer in a reasonable time. Obviously, there are many degrees of mechanism in these models and we must distinguish between them.

The primary advantage of mechanistic models is that they can incorporate our collective knowledge of crop, soil and atmospheric processes. However, we still need to concern ourselves with the databases used to develop them, the ways in which they handle interactions of factors, and their ability to simulate crop responses to the environment as it changes during the growing season.

The range of environmental conditions in the database

A model based on our understanding of processes is more likely to make correct predictions beyond the range of the database used to develop it. This is because the form of the equations (i.e., logistic, hyperbolic, etc.) is more likely to be correct. However, it is highly desirable that even mechanistic models be developed from as wide a range of environmental data as possible. Since field data are always limited, this implies that data from controlled-environments must be used. Even though plant growth in controlled-environments is not always exactly the same as in the field, controlled-environments give us the correct form of the equation over a wide range of conditions, and parameters can be adjusted as needed from field data.

400

Interactions of factors

Crops are affected by soil temperature, air temperature, solar radiation, rain, humidity, wind, soil nutrients, soil depth, soil bulk density, soil hydraulic conductivity, weeds, pests, diseases, etc. Doubtless the reader can add to this list. The principal models for describing how these factors interact are the additive, multiplicative, and limiting factor models. The additive and multiplicative models do not work well when a large number of factors are considered (Acock and Acock, 1991). These models typically contain separate functions for each factor, e.g., A = f(air temperature), where A has a value between zero and one. In additive models, these functions are summed and in multiplicative models the product is calculated. Clearly, in additive models that account for many factors, each factor will contribute very little to the overall result, and a deficiency in one factor will have little effect. Predictions from additive models will tend to be too high. In multiplicative models, as the number of functions less than unity increases, the predictions become progressively lower (Swartzman, 1979).

The limiting factor model is more intellectually satisfying and comes closer to representing the interactions found in many datasets. In order to grow, a leaf needs carbohydrate, various elements supplied by the soil, adequate turgor pressure, and a temperature high enough for the chemistry to occur. Deficiencies in any of these may limit leaf growth. The proportion in which these factors are required is not fixed. So, for instance, the nitrogen content of leaf tissue can vary over a wide range. The carbon content per unit leaf area is also quite variable. However, this variability too can be incorporated in a limiting factor model (e.g., Acock and Trent, 1991).

Size of time step

Most mechanistic models run in daily time steps, i.e., the state of the crop and soil is updated once each simulated day. This makes them fairly dynamic, but there is much to be gained by going to an hourly time step. Plant water stress, in particular, is something that is usually zero at dawn, develops during the morning and is relieved in late afternoon. It is difficult to simulate the processes involved without using a time step of less than a day. Also, with an hourly time step it is possible to have the model report what factor was most limiting for each hour. GLYCIM does this by listing a string of 24 letters for each day, e.g., CCCCCCTTWWWWWTNNNCCCCCCC, where T = temperature, W = water, C = carbon, and N = nitrogen (Trent et al., 1996). This gives farmers a realistic picture of what is happening to their crops. They cannot respond on an hourly basis, and they cannot take any action to correct most of the deficiencies. However, they can see the full complexity of the crop they are managing, and realize that, even in the course of a single day, no one factor limits crop growth.

The weaknesses of mechanistic models

From the arguments above, the best tools for understanding crop yield variability would seem to be mechanistic crop models with a time step of about an hour, with a limiting factor model for interacting factors affecting crop growth, and developed from a controlled-environment database covering a wide range of environmental factors. Such models can incorporate our knowledge about the processes in soil, plant and atmosphere. However, their yield predictions are not as accurate as we would like them to be.

There are several reasons for this. Firstly, it is difficult to represent in a computer what we know about plants. Plant organs grow simultaneously and continuously under the influence of their environment. That environment is determined in part externally and in part internally through the movement of materials between organs. Modellers have to represent these simultaneous and continuous processes on a single CPU that performs discrete calculations in sequence. This is a minor problem that has largely been solved either by using small time steps and carefully sequencing the calculations of processes, or by using iterative procedures.

Secondly, and much more seriously, our collective knowledge of plant behaviour is limited. Nothing uncovers knowledge-gaps more rapidly than trying to develop a mechanistic model of a crop. For example, we still do not understand stomatal control well enough to model it. We cannot predict how many branches will form on a plant. We cannot predict when old leaves will abscise. Worse, we understand very little about the feedback processes that operate in crops. Plants are remarkably resilient, resisting extreme environmental conditions and still producing seed. They evolved this ability over the millennia because they are rooted to one spot, and must cope with whatever happens there. Unlike animals, they cannot move off in search of a more favorable environment; so they evolved adaptive mechanisms instead. At present we do not know how these operate. It is fairly easy to develop a model that simulates the death of crops, or very high yields under the right circumstances. It is less easy to model a crop that produces a modest yield in difficult circumstances. Crop models are a success to the extent that we understand plant behaviour, and a failure to the extent that we do not.

WHAT DO FARMERS NEED?

GLYCIM is a soybean crop model that is mechanistic, based on controlled-environment data, uses an hourly time step and incorporates a limiting factor model. It has been used on farms in the Mississippi valley since 1991. It has the weaknesses mentioned above, yet the farmers continue to use it. Their criteria for judging the model differ from those used by scientists. To scientists, the yields are not accurate enough. (As the scientists responsible for the model, we are troubled by how large the errors can sometimes be.) To the farmers, yield estimates are incidental. If the model gives them insights into the behaviour of their crop and if the yield changes logically in response to management actions, the model is useful. In other words, the credibility of the mechanism in the model is more important to farmers than the accuracy of the yield estimate. The farmers view the model as another source of advice, not the unassailable truth.

We do not suggest that we should be satisfied with our current crop models, but we do suggest that, imperfect as they are, the current models are useful tools. We feel very strongly that returning to techniques developed 100 years ago and abandoning all that we have learned in the meantime is not an appropriate response to the weaknesses of mechanistic crop models. This is not the time to turn back; it is the time to press on forward and complete our understanding of crop processes. The information gained from soil analyses can and should be used within the broader context of an examination of all relevant factors affecting yield variability.

EVOLUTION NOT REVOLUTION

Yield maps typically look like a pointillist painting in close-up. Managing the small scale variability that they show is completely beyond us at present. Just because the engineers can measure variability on this scale does not oblige us to manage it on the same scale. Rather we should see precision agriculture as an exercise in managing gradually smaller and smaller areas of land. This is an exercise in evolution, not revolution.

USING MECHANISTIC MODELS IN PRECISION AGRICULTURE

If mechanistic models were perfect tools, we could input soil, weather, crop genetic and management data, and predict yield. Because they are still imperfect tools, we must validate and, if necessary, calibrate them before use. For this purpose, we must collect complete datasets for various parts of the field with different characteristics. If prior years' yield maps are available, we choose one or more areas of unusually high and low yield, and one area of normal yield. Soil cores are taken in these areas to characterize the soil. Management actions and weather are recorded. Samples of crop row are harvested weekly and used to determine the dry weights and dimensions of various organs on the plants. Predictions of the model are compared with observed plant data for these areas and model parameters are adjusted if necessary. (The adjustment of model parameters is a separate topic, but we should note here that no parameter in a mechanistic model should ever be adjusted without a compelling scientific reason. A mechanistic model in which a large number of parameters is fitted to the data, quickly becomes an empirical model.)

The adjusted model can now be used to explain yields in the rest of the field. Where information on the spatial variability of management actions, soils and other environmental factors are available, these are used as input to the model. Where the model predictions do not agree with the observations, various hypotheses to explain the discrepancies are tested one by one. This leads to further environmental measurements and a gradual discovery of the salient characteristics of a field. The companion paper by Pachepsky and Acock (1997) justifies the use of models in this way.

This process of discovery is likely to take place over several years. Meanwhile, each year, the known characteristics can be used in a crop model with several weather scenarios to predict likely yield outcomes for various management actions. The choice of action will depend on the relative likelihood of the scenario's occurring, and the amount of risk that the farmer can stand. These ideas are already in use with crop models to decide on management actions field by field. With precision agriculture, they will be used to decide on management actions for each management zone within the field.

These techniques are currently being put into practice on Hess Farms Inc., Ashton Idaho, in a project called Site Specific Technologies for Agriculture (SST4Ag). The Idaho National Engineering Laboratory is providing engineering expertise, the USDA, ARS Remote Sensing and Modeling Laboratory, Beltsville, MD., and the Irrigated Agriculture Research and Extension Center, Prosser, WA., are providing expertise in soils, weather and crop science.

CONCLUSIONS

Many different factors contribute to yield variation, and all major factors must be taken into account in explaining and predicting it. Prescriptions must be dynamic to respond to factors such as weather and management actions, that change during the growing season. Mechanistic crop models can be used to consider all relevant factors and their changes during the season, to test hypotheses about reasons for yield variation, and to develop prescriptions. The mechanistic content of these models is more helpful to farmers than their ability to predict yield. For these reasons, mechanistic crop models are essential to precision agriculture. Rather than discarding them because of their imperfections, we should continue to improve them.

REFERENCES

Acock, B., Acock, M.C. (1991) Potential for using long-term field research data to develop and validate crop simulators. *Agronomy Journal*, **83**, 56-61.

Acock, B., Trent, A. (1991) The Soybean crop Simulator GLYCIM: Documentation for the modular version 91. Miscellaneous Series Bulletin 145. University of Idaho, Agricultural Experiment Station, Moscow, Idaho, 241pp.

Blackman, F.F. (1905) Optima and limiting Factors. *Annals of Botany*, **19**, 281-295.

Chanter, D.O. (1981) The use and misuse of linear regression methods in crop modelling. *Mathematics and Plant Physiology*, D.A. Rose and D.A. Charles-Edwards (Eds), London, Academic Press, pp 253-266.

Hughes, A.P. (1959) Plant growth in controlled environments as an adjunct to field studies: experimental applications and results. *Journal of Agricultural Science*, **53**, 247-259.

Laird, R.J., Cady, F.B. (1969) Combined analysis of yield data from fertilizer experiments. *Agronomy Journal*, **61**, 829-834.

Pachepsky, Ya., Acock, B. (1997) Holes in Precision Farming: spatial variability of essential soil properties. *Proceedings of the First European Conference on Precision Agriculture*, J.V. Stafford (Ed.), London, SCI.

Russell, E.W. (1961) *Soil Conditions and Plant growth*, 9th edition, Longmans, London, 688pp.

Swartzman, G.L. (1979) Evaluation of ecological simulation models. *Contemporary Quantitative Ecology and Related Ecometrics*, G. P. Patil and M. Rosenzweig (Eds), Fairland, MD., International Cooperative Publishing House, pp 295-318.

Trent, A., Acock, B., Reddy, V. R. (1996) *Wingly, a Soybean Simulation for Windows: User's Manual.* USDA, ARS, Remote Sensing and Modeling Lab., Beltsville, MD, 64pp.

ASSESSMENT OF A METHOD FOR ESTIMATING THE NITROGEN REQUIREMENTS OF A WHEAT CROP BASED ON AN EARLY ESTIMATE OF COVER FRACTION

N. AKKAL, M.H. JEUFFROY, J.M. MEYNARD

Unité d'Agronomie, Institut National de la Recherche Agronomique, 78850 Thiverval Grignon, France

P. BOISSARD, J. HELBERT P. VALERY

Unité de Bioclimatologie, Institut National de la Recherche Agronomique, 78850 Thiverval Grignon, France

P. LEWIS

Department of Geography, University College London, United Kingdom

ABSTRACT

For decision making in precision farming, we tried to control nitrogen supply. For this purpose, we proposed to model the relationship between the soil cover fraction (CF) or the reflectance factor (RF) of the crop and the leaf area index (LAI). Then LAI is used as an input to a crop growth model to predict the biomass. The estimate of nitrogen requirements over time is based upon the critical nitrogen content, and upon the potential biomass. To obtain a robust relationship between either the cover fraction or the reflectance factor of the crop and the LAI, a particular attention was paid to the study of three-dimensional (3-D) structure of the crop which explains the leaves angular behaviour over the tillering period. This approach allowed us to take into account the effects of significant agronomic factors such as plant density, variety and sowing dates upon plant structure.

INTRODUCTION

In some regions of North and East France, farmers have frequent problems with the death of individual plants in wheat crops during winter. Low values of leaf area index (LAI < 1.5), which can be observed at the end of winter, cause decreases in light interception over the whole season. Consequently, the potential yield is reduced and it is important to quantify this reduction in order to optimize the inputs, especially nitrogen supply.

For decision making, the problem is to estimate the potential yield and then nitrogen requirements early in winter. For this purpose, we worked on characteristics of the canopy such as cover fraction and reflectance, which are easy to acquire. Therefore it was necessary to model the relationship between either the cover fraction or the reflectance factor of the crop and the LAI, in order to estimate the LAI_0 (leaf area index at time t_0). Existing methods based on vegetation indices (Baret and Guyot, 1991), can also be applied in order to retrieve the LAI but this empirical approach is rather limited since it is not possible to take into account precise geometrical parameters. Using the

LAI_0, we applied the scheme proposed by Monteith (1972; 1977) to predict the potential growth biomass. The estimate of nitrogen requirements over time can be based upon the critical nitrogen content according to the potential biomass production over time, as shown by Justes *et al.* (1994).

To study these different relationships, we used models to obtain a robust method: nitrogen requirements and yield of the winter wheat must be correctly estimated whatever variety, sowing date, density, observation date before closing canopy and conditions of nitrogen nutrition may be. Consequently, models must take into account these covariables.

LEAF AREA INDEX ESTIMATE LAI_0 FROM COVER FRACTION OR REFLECTANCE

Design of the experiment and data acquisition

All measurements were done at the Institut National de la Recherche Agronomique (INRA), on three varieties (Soissons, Thésée and Pernel) three densities (70, 150 and 350 plants m^{-2}) and two sowing dates (early on mid October and late at mid November) with three replications. The experiment was conducted over 2 years, the first year being used for model construction and the other for validation purposes.

Cover fraction measurements (CF)

The cover fraction was considered as a relevant variable because it can be assessed during crop walking either by farmers or their advisers. As far as we were concerned, CF was extracted from digitised colour photographs taken at 2 m by a camera fitted with a 50 or 100 mm lens under vertical view angle. The measured plots were 0.75 m x 0.50 m and the sampling rate was 3 plots per treatment. Measurements were taken every 200 degree days up to the beginning of stem elongation (Soissons variety, 150 plants m^{-2}). On the other treatments, only one measurement was made during the tillering period.

Radiometric measurements (RF)

Radiometric and cover fraction measurements were made at the same plots and on the same dates. Radiometric data were acquired in the green, red and near infrared bands using a CIMEL radiometer (SPOT simulation) placed at a height of 2 m.

Leaf area index estimate before the closing of canopy

We can see in Figure 1 that for a given CF, LAI_0 depends upon variety. This is obvious mainly during the latter part of the tillering period because during this period, we observed that tillers compete together and change their angular direction. This has some consequence on the canopy structure. In fact, estimation of total leaf area (LAI_0) from leaf projections required assumptions concerning the 3-D structure i.e., the leaves' angular distribution and the leaves' aggregation (Ross, 1981).

Therefore it is of great importance for estimating the LAI_0 from CF or RL, to characterise and to model the 3-D structure of the plant inside the canopy. 3-D structure

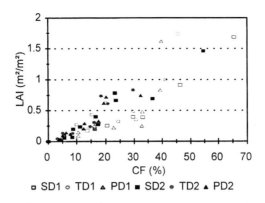

FIGURE 1. The relationships between LAI_0 and CF, for several treatments. S = Soissons, T = Thesee, P = Pernel, D1 = early sowing date, D2 = late sowing date.

models should allow us to carry out simulations of the soil cover fraction and the reflectance factor and to analyse the dependency of 3-D structure upon agronomic conditions at tillering. *A priori*, the structure parameters depend strongly not only upon agronomic practices but also upon genotype. In a first time a tool was developed to precisely analyse the 3-D geometry of wheat at tillering.

3-D structure study of the wheat canopy. In order to acquire the 3-D measurements, an original methodology adapted from close range photogrammetry was applied. This method, first developed to study the adult maize canopy by INRA and ENSEA (Ivanov *et al.*, 1995) was then adapted to wheat (Boissard *et al.*, 1995; 1996; Lewis and Boissard, 1997). We used the plants stereophotographs (stereo digital photography), to extract main parameters such as leaf inclination and twist angles, leaf azimuth at insertion as functions of the curvilinear abscissa. All measurements were made on individual plant extracted from field according to a sampling procedure.

Approach for wheat canopy simulation. A simulation tool was also developed in order to simulate canopies from plant measurements and calculate the cover fraction and the reflectance factor.

The 3-D canopy simulation was carried out at date t_0 by using the Botanical Plant Modelling System (BPMS) (Lewis, 1990; 1996), which was interfaced with the 3-D stereodata acquisition method.

Validation. A first validation of the canopy simulation was carried out on given variety and density.

Initially, canopies were simulated at different CF for only Soissons at 150 plants m^{-2}. Simulated plots (0.75 m x 0.50 m) have the same dimension as real ones. Cloning was performed by using the BPMS on a randomised set of plants among 10 types of plants. Then LAI_0 and CF simulated values were calculated from the projected set of elementary triangles onto the horizontal plane. Simultaneously, the corresponding real LAI_0 and CF values were obtained by planimetry technique and from the digital photo processing. Figure 2 shows simulated from 3-D model and measured values of CF and LAI_0.

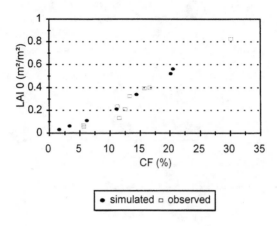

FIGURE 2. Comparison between simulated from 3-D model and observed data of LAI_0 and CF for Soissons 150 plants m^{-2}, at early sowing date.

Study of the sensitivity (future work). A sensitivity study of the geometrical model to the geometrical parameters (leaf angles behaviour) should provide us an arranged list of factors depending upon agronomic conditions (variety, density and sowing date). From that, we expect to be able to build robust relationships between the cover fraction (CF) or the reflectance factor (RF) of the crop and LAI_0.

Towards an operational use. The 3-D model will allow us to construct abacus relating LAI_0 to either CF or RF for different varieties. The abacus will make possible to estimate the LAI_0 corresponding to the CF or RF measured in the field. Since we will be supposed to know which are the relevant structural parameters, a minimum number of plant measurements will be necessary to build a new abacus.

MODELLING THE NITROGEN REQUIREMENTS UNTIL ANTHESIS

Modelling the dry biomass as a function of time until anthesis (DMt)

From LAI_0 estimated from either CF or RF early in the winter, we modeled the cinetic evolution of the LAI until anthesis. Then, the LAI evolution was used to model the intercepted radiation over time. The total accumulated biomass was computed according to the Monteith model (Monteith, 1972; 1977) which requires the LAI curve until anthesis as it is shown in Figure 3.

Modelling the leaf area index (LAI_t) as a function of time

The purpose of the LAI model is to predict the potential LAI_t at each time until anthesis, from a single measurement of LAI_0. The LAI_t value must be estimated whatever the variety, the density, the sowing date and the LAI_0 observation date may be.

The data acquired during the first year showed that, until the occurrence of LAI maximum, the evolution of LAI_t over thermal time could be fitted by a four parameters symmetrical logistic curve (Figure 4).

$$f = n + \frac{d - n}{1 + \exp(-b(dj - m))} \tag{1}$$

f: expected value;
n: low asymptote;
d: high asymptote;
m: point inflexion abscissa;
b: slope parameter;
dj: independent variable (sum of degree days).

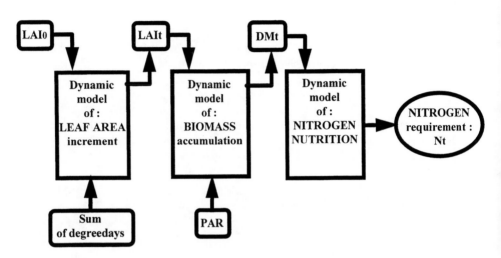

FIGURE 3. Modelling of the nitrogen requirements until anthesis.

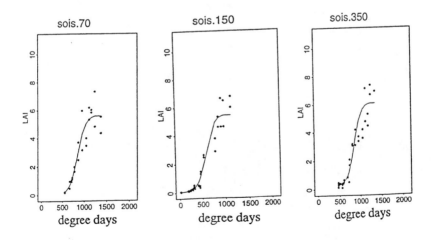

FIGURE 4. Fitting the logistic curve to the LAI data (Soissons variety at 70, 150 and 350 plants m^{-2}).

At the ear emergence the LAI reaches its maximum value and the LAI function can be represented by a plateau between ear emergence and anthesis.

Model parameters estimation. From the variations of the four parameters according to the treatments, linear models linking the parameters n, d, m and b to the independent variables genotype, plant density, sowing date and LAI_0 were established on the second year dataset.

Validation model. The model validation was carried out from the first year dataset.

Estimate of nitrogen requirements

The total amount of the nitrogen required in order to obtain the potential biomass can be calculated from the amount of the maximum accumulated biomass and the critical nitrogen content (Justes *et al.*, 1994), as is shown in Figure 3.

$$Nt = \alpha \, (DM)^{-\beta} \qquad (2)$$

where DM is the amount of biomass accumulated in the shoots and expressed in t.ha^{-1}, and Nt is the total N concentration in shoots expressed in % DM. α and β are the parameters.

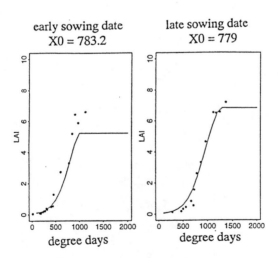

FIGURE 5. Example of validation of the LAI model for two sowing dates (Soissons variety at 150 plants m^{-2}). X0 : observation date (degree days).

CONCLUSION

The general aim of this work was to develop a new tool for decision making in the context of precision farming and especially for estimation of nitrogen requirements.

In this paper, we used the photogrammetric approach in order to analyse the 3-D structure of the wheat crop. The 3-D structure simulations allowed us to obtain a relationship between the cover fraction and the LAI_0 specially designed to a given variety and density we also proposed:

i) a LAI model, where LAI_0 is considered as an input variable;
ii) a nitrogen nutrition model.

The quality of all the models mainly depends on the fitting of parameters. In order to test the robustness of these models, it is necessary to test them in various pedoclimatic conditions. In the future we will pay attention to analysing the sensitivity of the different models to the variety and then to identifying relevant varietal parameters. Up to now, the data processing is not completed. The whole dataset will constitute a database especially designed to take into account the 3-D structure of wheat.

ACKNOWLEDGEMENTS

The authors would like to thank C Chabanet (previously at INRA laboratoire de Biométrie, Jouy en Josas, Fr) and C Hennequet (INRA laboratoire de Biométrie, Jouy en Josas, Fr), for their contributions.

Financial supports have been provided by Institut Technique des Ceréales et Fourrage (I.T.C.F, Fr), under conventions INRA / ITCF 1996 / 1997 and by INRA under the AIP Avenir 1995/1997.

REFERENCES

Baret, F., Guyot, G. (1991) Potentials and limits of vegetation indices for LAI and PAR assessment. *Remote Sensing of Environment*, **35**, 161-173.

Boissard, P., Akkal, N., Valery, P., Lewis, P., Meynard, J.M. (1995) 3-D plant characterization and modelling aimed at the remote control of winter wheat growth. *Proceeding of the International Colloq on Photosynthesis and Remote Sensing*, 28-30/08/95, Montpellier, Earsel INRA, pp 287-292.

Boissard, P., Akkal, N., Valery, P., Helbert, J., Lewis, P. (1996) Dynamic 3-D plant modelling: a morphological and structural approach based upon stereo data acquisition. *Modelling in Applied Biology: Spatial Aspects, Aspect of Applied Biology,* **46,** 125-129.

Ivanov, N., Boissard, P., Chapron, M., Andrieu, B. (1995) Computer stereo plotting for 3-D reconstuction of a maize canopy. *Agric. and Forest Meteorology,* **75**, 85-102.

Justes, E., Mary, B., Meynard, J.-M., Machet, J.-M., Thelier-Huchel, L. (1994) Determination of critical nitrogen dilution curve for wheat crops. *Annals of Botany*, **74**, 397-407.

Monteith, J.-L. (1972) Solar radiation and productivity in tropical ecosystèms. *J. App. Ecol.*, **IX**, 747-766.

Monteith, J.-L. (1977) Climate and the efficiency of crop production in Britain. *Phil. Trans. R Soc. Lond. B.*, **281**, 277-294.

Lewis, P. (1990) Botanical plant modelling for remote sensing simulation studies. M. Phil. PhD thesis, Transfer, University College, London, 137 pp.

Lewis, P. (1996) A Botanical plant modelling system for remote sensing simulation studies. M. Phil. PhD thesis, Transfer, University College, London, 349 pp.

Lewis, P., Boissard, P. (1997) The use of 3-D plant modelling and measurement in remote sensing. *Seventh International Symposium on Physical Measurements and Signature in Remote Sensing*, 7-11 April 1997, Courchevel, Fr, Balkema (in press)

Ross, J. (1981) The radiation regime and architecture of plants stands. *Junk Pub,* The Netherlands, 391 pp.

USING CROP SIMULATION MODELS TO DETERMINE OPTIMUM MANAGEMENT PRACTICES IN PRECISION AGRICULTURE

R. MATTHEWS, S. BLACKMORE

School of Agriculture, Food and Environment, Cranfield University, Silsoe, Bedfordshire MK45 4DT, UK

ABSTRACT

The CERES-Wheat model was used to derive response curves of grain yield to nitrogen applied for a range of incident solar radiation intensities representing the effects of slope and aspect. Distribution of applied nitrogen was then optimised by reiteratively maximising the marginal responses to incremental units of fertiliser, with two different goals; (a) to maximise the yield obtained from a given amount of fertiliser, and (b) to minimise the amount of fertiliser required to obtain a specified yield. A uniform application of 150 kg N ha^{-1} was used as the baseline in each case.

Since the uniform application rate is close to the plateau of the response curves for most of the solar radiation regimes used in the example, there was only a small effect ($\sim$ +1%) of optimised variable rate applications of N fertiliser. However, the same yield as that obtained with the uniform application could be obtained with 15% less N fertiliser when the optimal variable rate distribution was determined.

The approach demonstrates the possibility of using crop simulation models to predict yield response curves for different regions of a variable field to aid in the development of treatment maps for variable rate applications of specific inputs.

INTRODUCTION

Precision farming involves the use of spatial information at a sub-field level to help farmers make better informed management decisions to achieve their goals. However, although significant technical advances have been made in measuring and displaying variation in crop yields across a field, it is not always clear how to determine the best management practice for each part of a field in order to achieve these goals. For example, does a farmer apply more fertiliser to the lower yielding areas to try and raise yields to the average, or are these low-yielding areas at their potential yield already, and would he therefore be advised to apply more to the high yielding areas, believing they are able to make better use of it?

Answers to these questions are not clear-cut, but can be found using optimisation techniques based on a knowledge of the marginal yield response to a unit of the input in question for all the different regions of a field. The marginal yield response can be obtained from the curve describing yield response to the level of a particular input for each homogeneous region of the field. However, in most cases the response curves of each of these regions are unlikely to be known. While mini-experiments on each homogeneous region with a range of application levels of the input will provide this information, this approach is time-consuming, labour-intensive, and results are likely to be specific to that field only. An alternative, and perhaps complementary, approach is to use crop simulation models to predict the likely yield response to different levels of a particular input. Such models offer a cost-effective way in which agronomic knowledge accumulated from numerous previous experiments, usually with treatments of uniformly applied inputs on small and relatively homogeneous plots, can be extended to larger spatially-variable fields.

As an example of using crop models in this way, this paper describes an approach in which a wheat simulation model is used to determine the yield-fertiliser response curves for a range of incident solar radiation levels that might be encountered due to slope and aspect in an average field. Optimisation techniques are then used with this information to produce recommended fertiliser application rates for each area of the field according to the two goals of either maximising yields for the same amount of fertiliser applied, or reducing the amount of fertiliser used without reducing the overall yield.

METHODS

We have focused on the responses of yield to applied nitrogen, as N is a major input that the farmer has some control over. Moreover, it is the nutrient that will most often have an impact on crop production, and because of its soluble nature is prone to off-field movement which can impact negatively on ground and surface water if too much is applied. For the purposes of the study, we have considered only the effect that different levels of solar radiation, such as might be caused by variations in slope and aspect, have on determining the yield response to nitrogen applications. Slope has been shown previously to be an important factor in determining the response to nitrogen application (e.g., Kitchen & Hughes, 1995; Nolan et al., 1995). It is recognised that many other interacting factors are also likely to be influential, but it was decided to simplify the system initially in order to develop a methodology, and to consider additional complexity in future studies.

The model

The CERES-Wheat model (Godwin et al., 1989) was used to develop the response curves. CERES-Wheat is part of the DSSAT (Decision Support System for Agrotechnology Transfer) software package produced by the International Benchmark Sites Network for Agrotechnology Transfer (Harrison et al., 1990; IBSNAT, 1988), and has been widely tested in many parts of the world, providing some degree of confidence in its predictions.

The model operates on a daily time step and simulates the main processes of crop growth and development, including the timing of phenological events, the development of the canopy to intercept light, and its use to fix carbon which is converted into dry matter. Potential growth rates may be modified by shortages of water or nitrogen, but the effects of other nutrients, pests or diseases are not currently considered. The model is capable of simulating the growth of both winter wheat (requiring vernalisation) and spring wheat, and can also take into account photoperiodic sensitivity. Six variety-specific parameters are used; three of these control rates of phenological development, the fourth determines the number of grains set per plant, the fifth the maximum rate of grain-filling, and the sixth is used to calculate the number of ears per plant. Daily solar radiation receipts, minimum and maximum temperatures, rainfall, humidity and wind speed are required as inputs; information must also be provided on soil water release characteristics, soil depth and bulk density. Crop management details, such as time of planting, population density, and timing and amount of fertiliser applications, must also be input. The date of harvest is determined automatically by the model.

Developing response curves

The model was used to simulate the yield response of the winter-wheat cultivar Maris Funden to applied nitrogen, using 1974-75 weather data and soils data collected at Rothamstead, England. In each simulation, the crop was sown on 6 November 1974, and harvested on 30 July 1975, and eight levels of nitrogen fertiliser as ammonium nitrate were applied in a single application on 18 April 1975 at rates of 0, 30, 60, 90, 120, 150, 180, and 210 kg N ha^{-1}. Different sets of simulations were made, in which the observed daily totals of solar radiation were adjusted by a factor ranging from 50% to 140% using the Environmental Modifications option in the model's Management Details input file. This range of levels represents that shown by Jones (1983, p.23) for north- and south-facing slopes with a slope angle of up to 45°. Measured solar radiation values (on a level surface) ranged from less than 0.5 MJ m^{-2} d^{-1} in the winter to more than 25 MJ m^{-2} d^{-1} in the summer.

Optimising fertiliser application for specific goals

A hypothetical 100 ha field was used with its total area being classified into homogeneous regions receiving various levels of solar radiation ranging from 50% to 140% of that received by a level surface. For the purposes of this analysis, the areas in each class were assumed to be approximately normally distributed about the 100% class (see Table 1), but it is recognised that the distribution will vary widely from field to field.

A computer program was developed to optimise the application of fertiliser over the whole field to achieve two separate goals; (1) that of maximising yield for the same amount of fertiliser applied, and (2) that of minimising the amount of fertiliser applied but still achieving the same yield. The baseline for comparison in each case was the yield obtained by applying a set amount of fertiliser at a uniform rate across the whole field.

Optimisation was achieved for Goal 1 by taking a unit of fertiliser from the total amount available and applying it to the part of the field where the marginal response was the greatest. Thus, fertiliser would be applied first to the region which gave the greatest initial yield response, but as more and more fertiliser was applied to the region, the marginal response would decline until such time as it was lower than the initial marginal response for another region, to which further units would be applied. This was done reiteratively until the total amount of fertiliser was accounted for. The size of the unit of fertiliser used could be varied, but 1/1000th of the total amount to be applied to the field was found to work well in most cases.

Optimisation for Goal 2 was achieved by a similar procedure, except the reiterations were halted when the specified total production from the field was reached.

RESULTS

Yield response curves

The response curves of grain yield to applied nitrogen from 0-210 kg N ha[-1] predicted by the CERES-Wheat model for different solar radiation intensities are shown in Figure 1. At 40% light intensity, there is no response to nitrogen at all, indicating that light is the only factor limiting growth. As light intensity increases, nitrogen rather than light progressively becomes the limiting factor, so that a greater and greater response to applied N is predicted, with the plateau being reached at 150 kg N ha[-1] at the higher light intensities. At the higher light intensities, grain yields of 3000-4000 kg ha[-1] are predicted with no addition of fertiliser at all, indicating that a significant proportion of the N requirements of the crop are being met from soil reserves.

Maximising yields

The effect of optimising fertiliser application across the field is compared with that of a uniform application in Table 1. In both cases, the total amount of fertiliser applied was the same, and was equivalent to 150 kg N ha[-1] at the uniform rate. When N fertiliser was applied uniformly, an average grain yield of 6593 kg ha[-1] was achieved. However, when the N fertiliser was optimally distributed across the field, there was a barely perceptible rise to 6600 kg ha[-1]. It can be seen that the optimal distribution resulted in less N being applied to the lower light intensity regions of the field, and more to the 100% and 140% regions where the response was greatest. However, because the average application rate of 150 kg N ha[-1] is on the plateau of the response curve for most of the regions, there is little scope for improvement in yield due to optimisation.

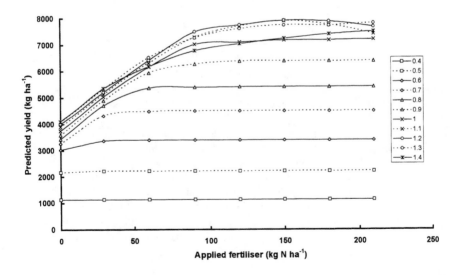

FIGURE 1. Response curves of wheat yields (kg ha[-1]) to applied nitrogen (kg N ha[-1]) predicted by the CERES-Wheat model for different fractions of observed solar radiation at Rothamstead, England. Genotype coefficients for Maris Funden were used.

TABLE 1. Effect of optimising N fertiliser applications to achieve higher production using the same amount of fertiliser.

fraction of area solar radiation (%)	(ha)	Uniform rate application				Variable rate application			
		total N applied (kg)	rate of N applied (kg ha⁻¹)	total grain produced (kg)	yield (kg ha⁻¹)	total N applied (kg)	rate of N applied (kg ha⁻¹)	total grain produced (kg)	yield (kg ha⁻¹)
50	1	150	150	2209	2209	30	30	2209	2209
60	3	450	150	10191	3397	180	60	10191	3397
70	6	900	150	27048	4508	540	90	27048	4508
80	12	1800	150	64920	5410	1440	120	64920	5410
90	25	3750	150	159625	6385	3750	150	159625	6385
100	30	4500	150	215340	7178	5490	183	215469	7182
110	12	1800	150	94800	7900	1800	150	94800	7900
120	6	900	150	47514	7919	900	150	47514	7919
130	3	450	150	23220	7740	450	150	23220	7740
140	2	300	150	14468	7234	420	210	15008	7504
TOTALS 100		15000		659335		15000		660004	
MEAN					6593				6600

Minimising fertiliser applications

The effect of optimising fertiliser application across the field to minimise the amount of fertiliser needed to achieve the same yield as that with a uniform application, is shown in Table 2. Again, the comparison is with that obtained from 150 kg N ha⁻¹ applied at a uniform rate. In both cases, an average grain yield of 6593 kg ha⁻¹ was obtained, but when the N fertiliser was applied optimally to each region of the field, this yield could be obtained with 15% less fertiliser. Again, the optimal distribution resulted in less N being applied to the lower light intensity regions due to their lower yield potential, and proportionately more being applied to the higher potential, higher light intensity regions.

DISCUSSION

The results presented in this paper indicate that, with some knowledge of the variability present in a field and of the factors that may be causing it, optimal management strategies can be developed to enhance the efficiency of use of input resources. Where the average level of the input is close to producing the maximum yield (e.g. Table 1), variable rate applications of the input are unlikely to be able to further increase productivity, although reductions in the amount of input applied are possible without decreasing the overall yield (Table 2). This represents a reduction in the cost of production to the farmer, and therefore an increase in profitability. The environment is also likely to benefit, as there should be less loss of N through runoff or leaching.

TABLE 2. Effect of optimising N fertiliser applications to achieve the same production using less fertiliser.

fraction of area solar radiation		Uniform rate application				Variable rate application			
		total N applied	rate of N applied	total grain produced	yield	total N applied	rate of N applied	total grain produced	yield
(%)	(ha)	(kg)	(kg ha^{-1})	(kg)	(kg ha^{-1})	(kg)	(kg ha^{-1})	(kg)	(kg ha^{-1})
50	1	150	150	2209	2209	30	30	2209	2209
60	3	450	150	10191	3397	180	60	10191	3397
70	6	900	150	27048	4508	540	90	27048	4508
80	12	1800	150	64920	5410	800	67	64537	5378
90	25	3750	150	159625	6385	3000	120	159474	6379
100	30	4500	150	215340	7178	4500	150	215340	7178
110	12	1800	150	94800	7900	1800	150	94800	7900
120	6	900	150	47514	7919	900	150	47512	7919
130	3	450	150	23220	7740	450	150	23218	7740
140	2	300	150	14468	7234	420	210	15006	7504
TOTALS 100		15000		659335		12620		659335	
MEAN					6593				6593

A central requirement to the process of optimisation and the development of spatially variable management strategies are the response curves describing the relationship between yield and the level of a particular input. Ideally, the range of the input should start at zero application and extend to a maximum beyond which it is known there is no yield response. As an example, Vetsch et al. (1995) experimentally determined the yield response curves for a range of N treatments from 0 to 235 kg N ha^{-1} across fields on four soil types in Minnesota, USA, and fitted a linear-plateau model to this data. They found that there was little spatial variation in maximum plateau level, but that there was considerable variation in the y-intercept, slope, and the point at which the plateau started. Experimental determination of response curves, however, is time-consuming, labour-intensive and expensive, and moreover, produces results that are site-specific and not easily transferable to other locations. Kitchen et al. (1995) approached the problem in a different way, by firstly determining the yield potential of different parts of a field when N was known not to limit growth. The optimal amount of N applied to each part of the field was then related to its yield potential, the lower potential areas receiving less fertiliser.

The results shown in this paper suggest that crop simulation models can provide a further way of determining yield response curves which is both cost-effective and potentially complementary to the methods described above. Traditionally, research into the response of crops to various inputs such as fertiliser has generally been conducted on small uniform plots, and recommendations then developed by combining results from a number of state- or region-wide studies into simplified 'universal' recommendations. Consequently, these recommendations in general cannot take into account response differences due to soil variability. However, crop simulation models represent the synthesis of much of the information accumulated from these small-plot experiments, and provide a way of extrapolating this information to fields which are spatially variable.

We recognise that the example we have used in this paper is simplistic in that it only considers the effect on variability of yield potential of one variable - that of incident solar radiation as influenced by slope and aspect. Indeed, slope and aspect may not only influence incident solar radiation; factors such as the temperature, water holding characteristics, depth, and fertility of the soil are also likely to be affected. Across the whole field, climate, soil properties, topography, hydrology and management history may all affect crop growth potentials. Response curves may even change with time over the life of the crop, and also between years, due to fluctuations in weather. Equally important may be how managing one variable affects another; adjusting field pH, for example, can affect a host of other variables such as nutrient uptake and herbicide application. Simple empirical correlations between these variables and yields are inadequate, as they are usually not extrapolatable to other fields, do not account for the dynamic nature of the relationships between many of the variables, and are not able to describe year-to-year variation or new management practices.

However, it is important not to be overwhelmed by the apparent complexity. Being able to understand a system requires it to be broken down into manageable components. We would therefore propose a framework for the systematic analysis of spatial variability based on that suggested for uniform monocultures by Penning de Vries et al. (1989) in which hierarchical levels of crop production are identified. At the first level, the potential production of a crop is assumed to be limited only by solar radiation, temperature, daylength, and varietal characteristics. Factors such as the topography of a field will influence the degree of spatial variability in the first two of these. The second production level is when water limits production; factors that will influence spatial variability of this level will be soil water release characteristics, soil and rooting depth, and possibly variations in evaporative demand due to differences in temperature across a field. The third production level is where water and nutrients (particularly nitrogen) are limiting; additional factors that need to be taken into account here are those causing variation in native soil nitrogen content, rates of N transformations in the soil, and crop N uptake. Nitrogen may not often be limiting yields since it is applied by the farmer, but responses to N will depend on the spatial variation in these factors. The fourth production level is where pests, weeds and diseases are taken into account; there may be considerable spatial variation in these, although the causes of this variability are not well understood.

Starting at the potential production level, and taking the other levels in turn, the first step is to use crop simulation models to determine the sensitivity of crop yields to each of the factors that cause spatial variability across fields. By identifying those factors that can explain the largest part of the yield variation, research effort can be directed to where it is most likely to have the greatest effect. However, we do not advocate that crop models replace field experimentation; rather that they should complement it by providing a way of applying to precision agriculture the considerable body of existing knowledge of crop responses to the environment.

CONCLUSION

Identifying yield response curves to different levels of a particular input are crucial to developing variable rate treatment maps for that input in a precision farming context. While response curves can be obtained by field experimentation, this can be time-consuming and expensive. Moreover, the results obtained are not always able to be extrapolated to other areas. However, by drawing on the considerable body of existing knowledge with which they were developed, process-based crop simulation models, combined with fewer well-designed traditional agronomic trials, offer a cheaper and quicker solution to the problem. The next step is to use such models to help identify and prioritise the factors responsible for variability in crop yields to focus future research effort more efficiently.

REFERENCES

Godwin, D.C., Ritchie, J.T., Singh, U., Hunt, L., (1989) *A User's Guide to CERES-Wheat v2.10*. Muscle Shoals, Alabama, USA, International Fertiliser Development Center.

Harrison, S.R., Thornton, P.K., Dent, J.B., (1990) The IBSNAT project and agricultural experimentation in developing countries. *Expl. Agric.*, **26**, 369-380.

IBSNAT, (1988) Technical Report 1: Experimental design and data collection procedures for IBSNAT. Department of Agronomy and Soil Science, College of Tropical Agriculture and Human Resources, University of Hawaii.

Jones, H.G., (1983) *Plants and Microclimate*. Cambridge, UK, Cambridge University Press, 323 pp.

Kitchen, N.R., Hughes, D.F., (1995) Comparison of variable rate to single rate nitrogen fertiliser application: corn production and residual soil NO_3-N. *Site-Specific Management for Agricultural Systems*, P.C. Robert, R.H. Rust and W.E. Larson (Eds). Minneapolis, Minnesota, American Society of Agronomy, Inc., pp 426-441.

Nolan, S.C., Goddard, T.W., Heaney, D.J., Penney, D.C., MacKenzie, R.C., (1995) Effects of fertiliser on yield at different soil landscape positions. *Site-Specific Management for Agricultural Systems*, P.C. Robert, R.H. Rust and W.E. Larson (Eds), Minneapolis, Minnesota, American Society of Agronomy, Inc., pp 553-558.

Penning de Vries, F.W.T., Jansen, D.M., Berge, H.F.M.T., Bakema, A., (1989) Simulation of ecophysiological processes of growth in several annual crops. *Simulation Monographs*, Wageningen, Netherlands, Pudoc/International Rice Research Institute, 271 pp.

Vetsch, J.A., Malzer, G.L., Robert, P.C., Huggins, D.R., (1995) Nitrogen specific management by soil condition: managing fertiliser nitrogen in corn. *Site-Specific Management for Agricultural Systems*, P.C. Robert, R.H. Rust and W.E. Larson (Eds). Minneapolis, Minnesota, American Society of Agronomy, Inc., pp 465-473.

USING COMPUTER SIMULATION TO COMPARE PATCH SPRAYING STRATEGIES.

M.E.R. PAICE, W. DAY

Silsoe Research Institute, Wrest Park, Silsoe, Beds, MK45 4HS.

ABSTRACT

A stochastic computer simulation model has been used to compare the long term effects of patch spraying. Two methods of spatially selective herbicide application have been investigated. On/off patch spraying delivers a high dose of herbicide to areas of the treatment map where weed density is above a decision threshold and no herbicide to areas below the threshold. Dual dose patch spraying allows high dose to be applied in areas which are above the threshold and a lower dose to be applied below the threshold.

These two strategies are considered in relation to initial weed distribution and spatial resolution of the sprayer. The economic effects in terms of herbicide usage and weed-induced yield loss are examined alongside changes in the extent and pattern of weed infestation.

The results of this analysis and implications for the design of practical patch spraying systems are discussed. At the estimated efficacies used in the analysis, dual-dose patch spraying is greatly preferable, as it both restricts the spread of patches and is more profitable than on/off spraying, particularly when spatial resolution is coarser.

INTRODUCTION.

In addition to being a major cost in modern arable farming, herbicides are potentially damaging to biodiversity in farming ecosystems, and possibly to human health through accumulation in water supplies. Prophylactic whole-field application of herbicide is also likely to encourage herbicide resistance (Maxwell, 1992).

Various methodologies for reducing whole field herbicide inputs have been proposed, and operated with some success. An economic optimum threshold can be calculated taking account of herbicide and yield loss costs and the long term changes in the weed seedbank (Cussans *et al.*, 1986). The field is only treated with herbicide if the mean weed density is above this threshold in the current year. Alternatively, by determining mean weed density and quantifying the main parameters that are controlling herbicide efficacy, an economic optimum dose can be set (Pandey and Medd, 1992). In practice, weed demography and herbicide efficacy factors are usually highly variable and may not have been determined at the time of spraying. Finney (1993) argues that the application of reduced doses will always lead to an increased risk of weed-induced yield loss and a rapid increase in weed population.

Patch spraying or spatially selective herbicide application has been proposed as a technique for minimising the risk of reducing herbicide inputs (Cussans, 1992). Herbicide application can either be switched on and off locally (by analogy to the whole field threshold approach) or alternatively optimum doses can be set in accordance with local conditions.

Miller and Stafford (1991) have discussed the technology required to implement patch spraying techniques and Paice *et al.* (1995) have described a practical patch spraying system. It is clear however (Paice *et al.* 1996), that there are a number of alternative approaches to patch spraying and many variable parameters. Rew *et al.* (1996) have quantified the short term herbicide savings from patch spraying (as against an even whole field application) of the perennial weed, *Elymus repens.* However for annual weed species with a seedbank, it is likely to be more important to take into account the longer term effects on the weed population and its spatial distribution. To do this by experiment would require many long term experiments replicated at several sites. The use of the simulation model allows a wide range of patch spraying scenarios to be evaluated, whilst making full use of limited experimental data to ensure that the model accurately describes the component processes.

SIMULATION MODELLING

The model

As a complementary strategy, to define generalised patch spraying system requirements, Day *et al.* (1996) have proposed a simulation modelling approach. The model is based on an operational research tool described by Audsley (1993) but in order to introduce a spatial component, it has been developed into a cellular model similar to that described by Perry and Gonzalez-Andujar (1993). Events in the weed life cycle, including germination, herbicide effect, seed production and seed mortality (Moss, 1990), are modelled as stochastic processes. The dispersal of seeds between spatially mapped cells by natural shedding, combine harvesting and cultivation are simulated using models and parameters derived from Howard *et al.* (1991) and Rew and Cussans (1997).

A number of patch spraying scenarios can be simulated over several years and compared with whole field herbicide application. For the purposes of this paper and for simplicity of comparison, on/off patch spraying is compared with dual dose application. In both cases, a local weed density decision threshold is used to decide which level of dose to apply. Weed-induced yield loss and present value are calculated according to Audsley (1993).

Paice and Day (1997) have discussed the factors that are likely to affect weed patchiness, spatial scale of density variation and frequency distribution shape, and are therefore likely to influence the long term profitability and environmental effects of patch spraying.

Comparison of on/off and dual dose patch spraying.

For the purposes of this analysis, we simulate the autumn, post-emergence control of *Alopecurus myosuroides* by the application of isoproturon (IPU). *A. myosuroides* is an important weed in winter cereals in the U.K. and IPU is most frequently used for its control. Christensen (1997) has produced empirically derived parameters (a, b) for the dose response relationship:-

$$R = \frac{1}{1 + \exp\left[-2\left(a + b.\log(X)\right)\right]}$$

where X is the dose in litres per hectare and R is the fractional dry weight of weed measured after application compared to a control. For IPU acting on A.myosuroides at an Autumn post-emergence stage these are a = -0.1 ; b = -3.0. For modelling purposes efficacy was converted from fractional dry weight to a notional expected proportional of surviving plants. IPU cost was taken as £25/ha for label recommended dose giving an efficacy of 98% and £8.33/ha for one-third dose giving an efficacy of 80%.

Patch spraying costs are likely to be sensitive to the initial weed distribution (in particular the proportion of the field infested) and the spatial resolution of the sprayer. Rew *et al.* (1997) have shown that the proportion of the field infested by *A. myosuroides* is very variable, but for the 13 fields that they surveyed, it had an upper quartile of around 70% and a lower quartile of around 40%. In addition, it was clear from their data and that described by Wilson and Brain (1991) that whilst *A. myosuroides* tends to form large contiguous patches of high density, weed seeds are often present at much lower density over a wider area of the field. Accordingly, we chose four initial weed distribution patterns for simulation. A: "small discrete"; a circular patch covering 40% of the modeled area at a density of 500 seeds/m². B: "large discrete"; a circular patch covering 70% of the modeled area (1 ha) at a mean density of 500 seeds/m². C: "small bimodal" as A but simulating a light infestation of 20 seeds/m² (mean) in the rest of the field. D: "large bimodal" as B but with 20 seeds/m² (mean) outside the patch.

Simulations were performed over ten years for on/off (full/zero dose) and dual dose (full/one third dose) patch spraying and repeated for sprayer resolution of 4 x 4 and 6 x 6 m. The full dose decision threshold was held at 1 plant/m² for all simulations.

Results of simulations

Tables 1 to 4 show the combined herbicide and yield loss costs of controlling each of the initial *A.myosuroides* infestations over ten years, calculated as described by Audsley (1993). In each case the cost of conventional blanket herbicide application was around 220 £/ha.

TABLE 1. Discounted costs (£/ha) of 10 years control of *A.myosuroides*. Initial pattern A.

	Patch spraying methodology	
	On/off	Dual dose
4 x 4 m Resolution	150	94
6 x 6 m Resolution	218	113

TABLE 2. Discounted costs (£/ha) of 10 years control of *A.myosuroides*. Initial pattern B.

	Patch spraying methodology	
	On/off	Dual dose
4 x 4 m Resolution	210	129
6 x 6 m Resolution	287	150

TABLE 3. Discounted costs (£/ha) of 10 years control of *A.myosuroides*. Initial pattern C.

	Patch spraying methodology	
	On/off	Dual dose
4 x 4 m Resolution	202	110
6 x 6 m Resolution	265	121

TABLE 4. Discounted costs (£/ha) of 10 years control of *A.myosuroides*. Initial pattern D.

	Patch spraying methodology	
	On/off	Dual dose
4 x 4 m Resolution	219	130
6 x 6 m Resolution	288	152

The lower costs of control with dual dose patch spraying can be partly explained by looking at the annual costs. Figures 1 and 2 show the annual herbicide costs, yield loss costs and combined costs for treating initial condition A at a resolution of 4 x 4m. On/off control leads to considerable instability of the weed population, which leads to a fluctuation of yield loss and herbicide costs. Under dual dose control, the annual fluctuation of costs is much less apparent. It is interesting to compare this result with Pandey and Medd's (1992) analysis of whole field control of *Avena fatua*. They showed more stable control and higher present value with a multi-period optimum dose treatment than with a single period threshold approach.

Dual dose patch spraying appears to be more effective at controlling the spread of weed infestation than the on/off approach. After ten years of dual dose patch spraying for initial distribution A (at a resolution of 4 x 4 m), 43.2% of the modelled area still had no seeds in the seedbank. On/off patch spraying allowed the weed infestation to spread until only 8.1% of the area was weed free.

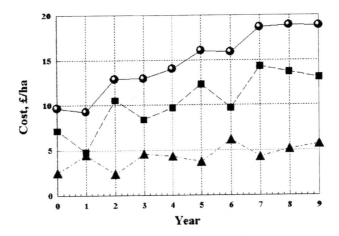

FIGURE 1. Annual herbicide ■, yield loss ▲ , and total costs for controlling *Alopecurus myosuroides* starting with initial seedbank distribution A and simulating on/off patch spraying at 4 x 4 m resolution.

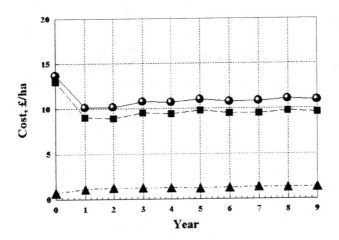

FIGURE 2. Annual herbicide ■, yield loss ▲ , and total costs for controlling *Alopecurus myosuroides* starting with initial seedbank distribution A and simulating dual dose patch spraying at 4 x 4 m resolution.

DISCUSSION AND CONCLUSIONS

In some circumstances, on/off patch spraying could be less profitable than whole field application of herbicide (Tables 1 to 4). At boundaries between lightly and heavily infested regions of the weed seedbank, the mean density within a quadrat or decision area may be below the spraying threshold. However, range of weed density within such an area will be high (relative to that in the rest of the field). Failure to treat such regions will allow the weed infestation to spread into uncolonised or lightly infested parts of the field. Savings in herbicide made in the first few years will be offset against increased yield loss in later years. Rew *et al.* (1997) have shown how this effect can be avoided by applying buffer regions around identified weed patches. This dilation process will reduce not only the proportion of the field to be left untreated but also the mean size of contiguous untreated regions. Thus, it is likely that on/off patch spraying systems will need to operate at relatively high spatial resolution if they are to generate substantial long term savings.

Dual dose control tends to inhibit the spread of weed patches without the need to introduce buffer regions. In addition, examination of Tables 1 to 4 shows that costs are less sensitive to spatial resolution under a dual dose regime. This implies that for equivalent projected long term savings, the specification of the spatial resolution of a patch spraying system can be relaxed if dose resolution is increased. It might also be deduced from this that dual and multi dose control would have a higher immunity to uncertainties in the weed identification process (Audsley and Beaulah, 1996) and the positioning of mapping and spraying equipment (Stafford and Ambler, 1994).

As might be expected dual dose patch spraying is more cost effective than on/off control where the weed infestation is not confined to discrete patches but is present at low density over much of the field . Comparison of Tables 1 and 3 shows that increases in costs for treating the bimodal initial seedbank distribution C compared with those for the discrete distribution A, are considerably less for the dual dose methodology. Both weed species and crop husbandry technique will influence the degree of discreteness or definition of the weed patches. Species which exhibit relatively low fecundity such as *Avena fatua* will normally be distributed more discretely than *A. myosuroides* and may cover smaller proportional areas of the field (Rew *et al.*, 1997). Perennial species such as *Elymus repens,* which is dispersed mainly by vegetative means, will also tend to form discrete patches especially in a zero or minimal tillage environment. On/off patch spraying might be more suited to these situations.

For simplicity of comparison we have considered only on/off and dual/dose control. It seems likely that application systems which are capable of applying multiple or continuously variable doses would have the potential to generate even greater herbicide savings, particularly when operating at lower spatial resolution. The move towards higher dose resolution would require not only more sophisticated application equipment (Paice et al., 1996) but also higher resolution weed density maps. A number of researchers (Dessaint and Caussanel, 1994, Audsley and Beaulah(1996)) have discussed techniques for making more effective use of weed survey data, so as to improve weed seedbank map accuracy and resolution without increasing surveying effort.

Although each event in the weed life cycle was treated as a stochastic process, this analysis has not examined the effect of gross temporal and spatial variation in the mean value of parameters (e.g. germination and herbicide efficacy). In practice reduced doses may generate greater annual variation in efficacy. This will lead to a greater risk of weed population increase in the current year but if reduced doses are confined to areas of the field which are only lightly infested, this does not imply an increased risk of yield loss. Balanced against this, Paice *et al.* (1996) concluded that annual fluctuations in herbicide efficacy will enhance the patchiness of the weed infestation and this effect may tend to maintain the long term advantages of the patch spraying approach.

ACKNOWLEDGEMENTS

The authors would like to thank Mr. G. W. Cussans and Dr. L. J. Rew of Rothamsted Experimental Station (U.K.) and Dr. S. Christensen and Dr. A. M.Walter of the Danish Institute of Plant and Soil Science for their support and agronomic advice. Development of the model was funded by the Ministry of Agriculture Fisheries and Food (MAFF) under project PA1707 and by the European Commission under project PATCHWORK (AIR3-CT93-1299). Application of the model to considerations of patch spraying strategy have also been supported by the LINK project funded by MAFF and Home Grown Cereals Authority.

REFERENCES.

Audsley, E. (1993) An operational research analysis of patch spraying. *Crop Protection,* **12**, 111 - 119.

Audsley, E., Beaulah, S.A. (1996) Combining weed maps to produce a treatment map for patch spraying. *Aspects of Applied Biology,* **46**, 1-8.

Christensen, S. (1997) Personal communication. *Danish Institute of Plant and Soil Science, Flakkebjerg, Denmark.*

Cussans, G.W., Cousens, R.D., Wilson, B.J. (1986) Thresholds for weed control - the concepts and their interpretation. *Proceedings of the European Weed Research Society Symposium on Economic weed control,* Pub:EWRS. 256-260.

Cussans, G.W. (1992) Identifying ways of optimising herbicide use in crops. *Proceedings of the First International Weed Control Congress, Melbourne,* Pub: Weed Research Society of Victoria, 208-214.

Day, W., Paice, M.E.R., Audsley, E. (1996) Modelling weed control under spatially selective spraying. *Acta Horticulturae,* **406**, 281-288.

Dessaint F, J.P., Caussanel, J.P. (1994) Trend surface analysis: A simple tool for modelling spatial pattern of weeds. *Crop Protection.* **13:6,** 433-438.

Finney, J. Risks and rewards from lower chemical inputs into agriculture. (1993) *Proceedings of the Home Grown Cereals Authority conference on cereals R & D, Maximising Profits with Lower Inputs,* Pub: HGCA. 144 - 158.

Howard, C.L., Mortimer AM, Gould P, Putwain PD; (1991) The dispersal of weeds: Seed movement in arable agriculture. *British Crop Protection Conference - Weeds,* Pub: British Crop Protection Council, 821 - 828.

Maxwell, B.D. (1992) Weed thresholds: the space component and considerations for herbicide resistance. *Weed Technology*, **6**, 205-212.

Miller, P.C.H., Stafford, J.V. (1991) Herbicide application to targeted patches. *British Crop Protection Conference - Weeds 1991*, Pub: British Crop Protection Council, **9A-9**, 1249-1256.

Moss, S.R. (1990) The seed cycle of Alopecurus Myosuroides in winter cereals: A quantitative analysis. *Proceedings of the European Weed Research Society symposium, Integrated Management in Cereals*, Pub:EWRS, 27-35.

Paice, M.E.R., Miller, P.C.H., Bodle, J.D. (1995) An experimental sprayer for the spatially selective application of herbicides. *Journal of Agricultural Engineering Research*, **60**, 107-116.

Paice, M.E.R., Miller, P.C.H., Day, W. (1996) Control requirements for spatially selective herbicide sprayers. *Computers and Electronics in Agriculture*, **14**, 163-177.

Paice, M.E.R., Day, W.(1997) A computer simulation analysis of the factors influencing weed patchiness and implications for patch spraying. *Proceedings of the Computer Modelling in Weed Science Workshop, Copenhagen 1996*, in press.

Pandey, S., Medd, R.W. (1992) A multi period economic model for the control of wild oats. *Proceedings of the 1st International Weed Control Congress, Melbourne 1992*, Pub: Weed Research Society of Victoria, **2**, 375-377.

Perry, J.N., Gonzalez-Andujar, J.L. (1993). Dispersal in a metapopulation neighbourhood model of an annual plant with a seedbank. Journal of Applied Ecology, **81:3**, 453-463.

Rew, L.J., Cussans, G.W., Mugglestone, M.A., Miller, P.C.H. (1996) A technique for mapping the spatial-distribution of *Elymus repens*, with estimates of the potential reduction in herbicide usage from Patch spraying. *Weed Research*, **36:4**, 283-292.

Rew, L.J., Cussans, G.W. (1997). Horizontal movement of seeds following tine and plough cultivation: implications for spatial dynamics of weed infestation. *Weed Research*, in press.

Rew, L.J., Miller, P.C.H., Paice, M.E.R. (1997). The importance of patch mapping resolution for sprayer control. *Aspects of Applied Biology*, **48**, 49-55.

Stafford, J.V.; Ambler, B. (1994). In-field location using GPS for spatially variable field operations. Computers and Electronics in Agriculture **11**, 23-36.

Wilson, B.J., Brain, P. (1991) Long-term stability of distribution of Alopecurus myosuroides Huds. within cereal fields. *Weed Research*. **31**, 367-373.

AUTOMATIC RECOGNITION OF WEEDS AND CROPS

M.R. Scarr, C.C. Taylor, I.L. Dryden

Department of Statistics, University of Leeds, LS2 9JT, UK.

ABSTRACT

Various discrimination techniques for the automatic classification and precision treatment of weeds in row crops are investigated. We discuss both parametric and non-parametric classifiers, and also the problem of predictor variable selection. Classification results are then presented for a variety of row crops, using several different classifiers and combinations of predictor variables.

INTRODUCTION

The aim of this research[1] is to develop image capture techniques for the detection of weeds in arable and horticultural row crops so that precision chemical or mechanical weed control may be applied. For the successful production of arable and horticultural crops, weed control is essential and is generally carried out by blanket spraying a field with liquid chemical herbicides. The development of systems to reduce and optimise the use of agro-chemicals will not only lead to a reduction in production costs, but also a reduction in pesticide residues, both on the crops and in the environment. The methods investigated here would allow areas of weeds in crops to be identified, and hence treated in isolation.

There are numerous approaches that can be used in the automatic classification of weeds. In this paper we examine several parametric and non-parametric classifiers including a neural network, using shape texture and shape information. We also address the non-trivial problem of selecting the classification variables. Results are then presented for these various methods.

THE DATA

The image data sets consist of four crops: onion, sugarbeet, cauliflower and carrot, taken in the field, using a camera at various dates between 10/6/96 and 17/7/96. Each data set contains up to three classes: crop, weed, or mixed crop and weed (not always present). Rather than use the raw images, we use the ratio of two images taken at different spectral wavelengths, in the red and near-infrared regions of the spectrum. This produces a good plant/soil contrast (Bull and Zwiggelaar, 1995). Each image is typically a $2\ m \times 2\ m$ square, containing 3-4 rows of a particular crop; Figure 1 shows an example of this type of image. From these large images sub-images are cut out by hand (Figure 2) and classified.

Depending on the classifier, the data are either partitioned into independent training and testing sets, or cross validation (Stone, 1974) is used on the complete data set to assess the performance. Cross validation omits each observation in turn, generates a classification

[1]This project is a joint collaboration with Silsoe Agricultural Research Institute. It is funded by the Ministry of Agriculture Fisheries and Food (MAFF).

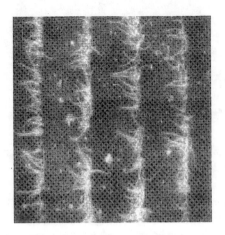

CARROT   ONION

ONION + WEED   WEED

FIGURE 1: Onions   FIGURE 2: The image data

rule, and then classifies the omitted observation according to that rule. The next section describes the various classification methods.

CLASSIFICATION METHODS

There exists a wealth of different classification techniques, one is only limited by the availability of the relevant software. Here we investigate several parametric (distribution based) and non-parametric (distribution free) classifiers.

Discriminant analysis (Mardia *et al*, 1979) is a way of classifying observations into groups based on specified predictors. The classification can be performed using parametric or non-parametric methods. When the within group distributions of the data are assumed to be multivariate Normal, the classification is based on the following generalised squared distance:

$$d_i^2(x) = (x - \bar{x}_i)^{\mathrm{T}} S_i^{-1} (x - \bar{x}_i) + \ln |S_i| - 2\ln \pi_i \tag{1}$$

i.e. an observation x is classified to group i if the squared distance of x to group i is smallest, x is a p-vector containing the values of the predictors for that observation, $\bar{x}_i$ is a p-vector of means of predictors from the data in group i, S_i is the sample covariance matrix for group i, and π_i is the prior probability that an observation belongs to group i. In our case the prior probabilities are proportional to the data. Equation 1 is called the Quadratic Discriminant Function (QDF). It is often appropriate to make the simplifying assumption that all the groups have the same covariance matrix, if this is the case we get:

$$d_i^2(x) = (x - \bar{x}_i)^{\mathrm{T}} S_p^{-1} (x - \bar{x}_i) - 2\ln \pi_i \tag{2}$$

where S_p is the pooled sample covariance matrix. Equation 2 is called the Linear Discriminant Function (LDF). If all the prior probabilities are assumed to be equal then the quadratic form in equation (1) is known as the Mahalanobis distance.

Non-parametric discriminant methods make no distributional assumptions about the data. One such non-parametric method is nearest neighbour (NN) classification (Dasarathy,

430

1991), which classifies an observation x to a particular group based on the information from its k nearest neighbours. We investigate 7 and 3 nearest neighbour classification rules. The QDF, LDF and NN classifiers are all implemented using SAS (SAS Institute Inc, 1990) on a SUN Sparc workstation.

Logistic discriminant analysis is a semi-parametric method[2] (Anderson, 1982). Essentially logistic discrimination models the group membership probabilities as linear-logistic functions of x. Another semi-parametric approach is SMART Smooth Additive Regression Technique (Friedman, 1984), which is based on projection pursuit regression (Friedman and Stuetzle, 1981).

With the rapid development of computing power in recent years, neural networks have become very popular as a classification method (Krzanowski and Marriott, 1996). We use a general purpose back propagation neural network here (Goodman, 1993) with one hidden layer consisting of 3-10 hidden units. The software for the logistic discrimination, SMART and the neural network is run on a SUN Sparc workstation.

CLASSIFICATION VARIABLES

We require predictor variables that will best summarise or characterise the structure of our particular images, so as to minimise the misclassification rates. Two different methods are adopted, the first is to compute a variety of shape and texture measures from each image, and use these variables in the classification routines. The second method uses the raw grey-level values of the pixels directly in the classification routines. The following is a list of the image measures computed for each 64×64 pixel image:

P1-P3 The perimeter of an image thresholded at $\bar{x} - s$ (P1), $\bar{x}$ (P2) and $\bar{x} + s$ (P3) where $\bar{x}$ is the sample mean and s is the sample standard deviation. The perimeter is obtained by counting the number of pixels on the boundary (defined by the threshold), and includes diagonal neighbourhoods.

A1-A3 Using the same three threshold levels as above the area is obtained by counting the number of pixels above the threshold.

PA1-PA3 The ratio of area to perimeter squared.

V1-V4096 Correlogram estimates (the variogram divided by the sample variance) for every possible direction. The variogram (Cressie, 1993), looks at the spatial correlation between the grey levels in an image for a given spatial lag vector h. For example low variogram values (neighbouring pixel grey-level values are similar) indicate a smooth texture whereas high values indicate a rough texture (neighbouring values tend to be different).

s^2 The sample variance of the grey levels.

σ^2_{local} The average variance of a local 4×4 pixel non-overlapping neighbourhood.

$\mu, \tau^2, \theta_1, \theta_2, \theta_3, \theta_4$ The second order Gaussian Markov Random Field (GMRF) parameter

---

[2]Logistic discrimination is sometimes referred to as a semi-parametric method, because although the logistic model is valid for various parametric distributions, it may be applied empirically avoiding any distributional assumptions about the data.

estimates. μ is the sample mean, τ^2 the conditional variance, and $\theta_1, \theta_2, \theta_3, \theta_4$ the spatial dependency parameters. The GMRF (Dubes and Jain, 1989) allows textures to be modelled mathematically. The spatial dependency parameters control the spatial correlations in the vertical, horizontal and diagonal directions.

We now apply a variable selection criterion to reduce the image measures to a suitable number for the classification routines. However, we do not want to exclude any "important" discriminatory variables. This procedure is carried out in two ways, the first is to use stepwise discriminant analysis. Forward selection is used to select a subset of the predictor variables to produce a good discrimination model. At each step a variable is entered based on the significance level of an F-test from an analysis of covariance. The within group distributions of the data are assumed to be multivariate Normal with a common covariance matrix. Approximately 70 variables are typically selected, of which we use the top 10.

The F-statistics from the first step of the stepwise discriminant analysis are also used to rank the correlogram variables. The ranks are used rather than the actual F-values as they produce a much smoother surface. Figure 3 shows plots of the ranked correlograms (white = rank 1) *i.e.* the light areas indicate those correlograms with large F-values. The pixel in the top left hand corner corresponds to a lag of $h = (0,0)$ and the bottom right to a lag of $h = (63,63)$. From these plots we identify four modes and use the corresponding correlograms along with the other shape and texture measures as our predictor variables.

The second method uses the raw grey-level pixel values themselves in the classification routines. In fact due to computing limitations rather than use all 4096 pixel values in a 64×64 image we generate a 32×32 "averaged" image by taking a 2×2 pixel block average across the image. This gives us 1024 new "averaged" pixel values. Principal components analysis (Mardia *et al*, 1979) is then performed on the 1024 "averaged" variables, to reduce the dimensionality. Figure 4 shows the first eigenvector (factor loadings) for the sugarbeet data 17/6/96 and 24/6/96 displayed as an image. The intensity at each pixel represents the magnitude of that particular factor loading, normalised onto a 0-255 grey scale. The dark area in the centre of the images indicates that these factor loadings are less "important" *i.e.*

FIGURE 3: Ranked correlograms for carrot 17/6/96 (left) and cauliflower 17/6/96 (right)

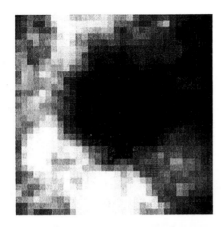

FIGURE 4: First eigenvector, sugarbeet 17/6/96 (left) and sugarbeet 24/6/96 (right)

there is a lower correlation between that variable and the associated Principal Component. This dark central area is a characteristic of the sugarbeet data, and is in fact a consequence of the sub-images being centred on a plant. As a result most of the variability is around the periphery of the centre.

We also take the scalar product of the first 10 eigenvectors with the "averaged" grey-level values to give 10 new variables which are then used in the various classification routines.

ERROR RATES

For a given data set we can define the default rule as classifying everything to the largest class (in our case crop), this will give us a default error rate e_1 say. We can then compare this default value with the error rates of our different classifiers.

The error rates for the different classifiers are computed as follows: Let the total number of misclassified observations for a particular test data set be a binomial random variable $\mathbf{X}$ *i.e.* $\mathbf{X} \sim \text{Bin}(n, p)$, where p is the probability of misclassification and n the number of observations in the data set. Now the Normal approximation to the Binomial distribution is, $\mathbf{X} \sim N(np, np(1-p))$, for $\min\{np, n(1-p)\} > 5$, say. An unbiased estimator of p is: $\hat{e}_2 = X/n$. Hence $\hat{e}_2$ can be modelled by a Normal distribution with mean p and $var(\hat{e}_2) = p(1-p)/n$. We want to test if $\hat{e}_2$ is statistically better than e_1, *i.e.* our null and alternative hypotheses are : $H_0 : \hat{e}_2 = e_1$ versus $H_1 : \hat{e}_2 < e_1$. The test statistic is:

$$\mathbf{Z} = (\hat{e}_2 - e_1)/\sqrt{var(e_1)} \tag{3}$$

where $\mathbf{Z} \sim N(0,1)$, from which we can compute a p-value, that can be used to decide if the particular classifier in question performs significantly better than the default rule.

RESULTS

The results are presented in tabular form based on the type of predictor variables and classification method used. The tables have the following column headings; date and crop type (those dates marked with † only have two classes), the number of observations (n), and the type of classifier used. The stars indicate significant p-values, labelled as follows: * p-val < 0.05, ** p-val < 0.01, and *** p-val < 0.001.

Table 1 shows the error rates ($\hat{e}_2$) for various classification methods using stepwise discriminant analysis to select 10 predictor variables. In this case half of the data are used for training, and the other half for testing. The training and testing data sets are swapped over and an average error rate computed. The error rates from Table 1 are graphed as time series plots in Figure 5. The plots indicate that in general the error rates are lower at the earlier dates, this appears to be true for all of the crops. Using this method of variable selection the best results are obtained for the sugarbeet, and cauliflower data with virtually all of the p-values highly significant. The carrot and onion data do particularly badly, in particular onions (17/7/96) and carrots (2/7/96). This is due to the fact that images from different classes for these dates are visually identical.

The error rates in Table 2 are computed using the shape and texture variables and the top four correlograms when ranked by their F-statistics. Here cross validation on the complete data set is used. This method performs very well with all the sugarbeet, and cauliflower data, all of the p-values are significant. The error rates for the onions are better than with the previous method, with the majority of the p-values significant. However there are still classification problems with the carrot data (2/7/96).

The ten image measures derived from the Principal Components Analysis are used in various classifiers with cross validation, the results are shown in Table 3. The classification results for the onion and carrot data using this method are very poor, with very few significant p-values, whereas all of the p-values for the sugarbeet and cauliflower data are highly significant. These last two approaches were tested on the LDF, QDF and NN classifiers only.

TABLE 1: Error rates using variables selected by stepwise discriminant analysis

Crop/date	n	e_1	LDF	QDF	7 NN	3 NN	LOG	NET	SMART
onions						$\hat{e}_2$			
17-Jun	145	.152	.125	.143	.119	.116	.166	.117	.152
24-Jun	253	.229	.133***	.145***	.146***	.147**	.162**	.150**	.178*
09-Jul	229	.323	.198***	.212***	.258*	.288	.240*∗	.258*	.271*
17-Jul	230	.317	.306	.307	.321	.330	.304	.296	.320
sugarbeet						$\hat{e}_2$			
10-Jun	183	.339	.159***	.108***	.100***	.096***	.147***	.126***	.197***
17-Jun	176	.501	.083***	.083***	.099***	.095***	.125***	.097***	.085***
24-Jun	199	.522	.154***	.142***	.153***	.154***	.200***	.136***	.161***
02-Jul	207	.367	.166***	.183***	.177***	.184***	.183***	.164***	.188***
09-Jul	149	.275	.143***	.195*	.175**	.196**	.121***	.134***	.134***
carrots						$\hat{e}_2$			
10-Jun	180	.105	.091	.106	.085	.066*	.089	.089	.083
17-Jun†	201	.164	.026***	.063***	.033***	.029***	.164	.040***	.015***
02-Jul†	191	.063	.054	.063	.073	.065	.063	.063	.079
cauliflower						$\hat{e}_2$			
17-Jun†	169	.550	.120***	.093***	.117***	.113***	.159***	.118***	.148***

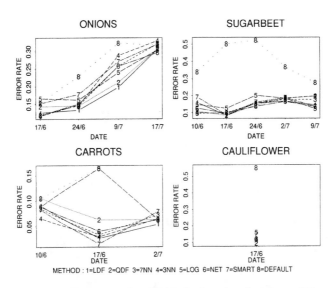

FIGURE 5: Error rates from Table 1 plotted as functions of time

We also investigated using one date as a training data set to generate an LDF and classify data for subsequent dates of the same crop. This approach led to disastrous results, but this was to be expected as the crops and weeds were growing rapidly and changing structure between the dates.

TABLE 2: Error rates using shape and texture variables, and top four ranked correlograms

Crop/date	n	e_1	LDF	QDF	7 NN	3 NN
onions			$\hat{e}_2$			
17-Jun	145	.152	.117	.152	.124	.090*
24-Jun	253	.229	.130***	.115***	.130***	.138***
09-Jul	229	.323	.236**	.275	.231**	.231**
17-Jul	230	.317	.270	.261*	.270	.257*
sugarbeet			$\hat{e}_2$			
10-Jun	183	.339	.104***	.093***	.087***	.082***
17-Jun	176	.501	.085***	.091***	.074***	.063***
24-Jun	199	.522	.131***	.146***	.126***	.147***
02-Jul	207	.367	.159***	.179***	.150***	.150***
09-Jul	149	.275	.154***	.269	.161***	.141***
carrots			$\hat{e}_2$			
10-Jun	180	.105	.068	.106	.050**	.044**
17-Jun†	201	.164	.035***	.025***	.035***	.020***
02-Jul†	191	.063	.079	.063	.058	.073
cauliflower			$\hat{e}_2$			
17-Jun†	169	.550	.053***	.036***	.036***	.030***

TABLE 3: Error rates using ten Principal Components based image measures

Crop/date	n	e_1	LDF	QDF	7 NN	3 NN
onions			$\hat{e}_2$			
17-Jun	145	.152	.140	.186	.166	.124
24-Jun	253	.229	.210	.198	.186	.182*
09-Jul	229	.323	.245**	.258*	.223***	.231**
17-Jul	230	.317	.313	.304	.278	.270
sugarbeet			$\hat{e}_2$			
10-Jun	183	.339	.137***	.120***	.148***	.120***
17-Jun	176	.501	.108***	.085***	.080***	.085***
24-Jun	199	.522	.191***	.176***	.186***	.161***
02-Jul	207	.367	.217***	.213***	.217***	.198***
09-Jul	149	.275	.175**	.235	.201*	.195*
carrots			$\hat{e}_2$			
10-Jun	180	.105	.133	.106	.100	.089
17-Jun†	201	.164	.040***	.030***	.030***	.040***
02-Jul†	191	.063	.052	.058	.063	.052
cauliflower			$\hat{e}_2$			
17-Jun†	169	.550	.107***	.124***	.095***	.154***

CONCLUSIONS

Texture and shape show good potential for discrimination of certain crops, so that precision weed control may be applied. The methods applied here have up to a 98% correct discrimination rate for weeds in crops, depending on the type of crop and approach used. There also appears to be an optimal discrimination window for each crop as is clear from Figure 5. This suggests that the precision treatment of weeds would be best on or around those dates. As far as the different classification methods are concerned, there is very little difference between the various approaches. There is however insufficient data to be really sure that a particular approach is best suited to a particular crop. These algorithms could be incorporated into an imaging device to automatically detect areas of weeds, and hence allow their precision treatment.

REFERENCES

Anderson, J . (1982), Logistic discrimination. *Handbook of Statistics*, Volume 2, Amsterdam, The Netherlands North-Holland pp. 169-191

Bull, C. Zwiggelaar, R. (1995), The potential of discriminating between crops and weeds on the basis of their spectral reflectance characteristics. Technical report, Silsoe Research Institute, Wrest Park, Silsoe, Bedfordshire, MK45 4HS, UK.

Cressie, N. (1993), *Statistics for spatial data*. New York Wiley series.

Dasarathy, B. (1991), *Nearest Neighbor (NN) Norms: NN Pattern Classification Techniques*. Los Alamos, California, USA IEEE Computer Society Press.

Dubes, R. Jain, A. (1989), Random field models in image analysis. *Journal of applied statistics*, **16**, pp. 131-164.

Friedman, J. (1984), Smart user's guide. Technical Report **1**, Department of Statistics, Stanford University.

Friedman, J. Stuetzle, W. (1981), Projection pursuit regression. *Journal of the American Statistical Association*, **76**, pp. 817-823.

Goodman, P. (1993), *NevProp version 1.xx*. University of Nevada Center for Biomedical Modelling Research.

Krzanowski, W. Marriott, F. (1996), *Multivariate Analysis: Classification, covariance structures and repeated measurements* (second ed.), Volume 2. 338 Euston Road, London NW1 3BH Arnold (Holder Headline Group).

Mardia, K. Kent, J. Bibby, J. (1979), *Multivariate analysis*. London Academic Press.

SAS Institute, Inc. (1990), *SAS User's Guide: Statistics* (6.04 ed.). Cary, North Carolina SAS Institute, Inc.

Stone, M. (1974), Cross-validatory choice and assessment of statistical predictions. *JRSS Series B*, **36**, pp. 111-113.

SPATIAL MODELING OF CROP YIELD USING SOIL AND TOPOGRAPHIC DATA

K.A. SUDDUTH

US Department of Agriculture, Agricultural Research Service, Cropping Systems and Water Quality Research Unit (USDA-ARS-CSWQRU), Agricultural Engineering Building, University of Missouri, Columbia, Missouri 65211, USA

S.T. DRUMMOND

University of Missouri, Department of Biological and Agricultural Engineering, Columbia, Missouri 65211, USA

S.J. BIRRELL

University of Illinois, Department of Agricultural Engineering, 1304 W. Pennsylvania Avenue, Urbana, Illinois 61801, USA

N.R. KITCHEN

USDA-ARS-CSWQRU, Agricultural Engineering Building, University of Missouri, Columbia, Missouri 65211, USA

ABSTRACT

The spatial relationship between crop yield and soil and topographic parameters was modeled using several methods. The goal was not only to estimate yield based on the other spatial data, but also to understand the shape of the yield response to changes in measured limiting factors. The approach involved empirical analysis of multivariate spatial datasets which included soil parameters, topography, and multi-year yield data obtained on research fields in Missouri. Projection pursuit regression, a multivariate nonparametric method, was able to produce estimated yield maps which were similar to actual yield maps. The estimated yield response curves generated by this method appeared to successfully model major yield-limiting factors. For the fields studied, yield responses to soil and topographic parameters were highly variable between years, due to differences in climatic conditions.

INTRODUCTION

Knowledge of the response of crop yield to changes in soil nutrient levels and other spatial factors is important for site-specific nutrient management. Established crop response and nutrient recommendation algorithms are generally averages based on data obtained at multiple locations over a large geographic area, such as an entire state. Although this approach has generated algorithms which can be used to model yield response over a range of locations and soils, the accuracy of the response relationship for any one location is reduced because of the generalized nature of the algorithm. To take full advantage of the precise data collection and application control methods available in precision agriculture, it is necessary to use the most precise nutrient recommendation algorithms applicable. If data and analysis methods are available, it may be possible and desirable to develop individual response algorithms tailored to particular soils or soil associations, or perhaps even to a particular field, or similar areas within a field.

Current nutrient recommendation procedures necessarily assume that all factors limiting yield are controlled or are included as inputs to the recommendation algorithm. This is a conservative approach, appropriate for field-level management of nutrient applications based on limited data. However, when the procedures are applied on a point-by-point basis within a field, there may be areas in which crop growth and yield are limited by other, uncontrolled factors, such as water availability. In these portions of the field, the current recommendation procedures will not be accurate, and may over-predict nutrient requirements. Considering the wealth of data available through precision agriculture, it is now becoming possible to develop more accurate recommendation algorithms, through the inclusion of multiple measured factors in the development of those algorithms.

Traditionally, nutrient response algorithms and recommendations have been developed based on field plot research. The confounding effects of controllable factors (such as other nutrients) have been minimized by controlling those factors to levels which are not yield-limiting. The effects of uncontrollable factors (such as water availability) have generally been regarded as part of the experimental error, and have been minimized by using small, homogeneous experimental areas. If sub-field response algorithms were to be developed using this approach, the number of experimental plots required would be highly impractical.

Another approach that can be applied to this problem is statistical analysis of the multivariate spatial data collected at the tens, or even hundreds, of soil sampling and yield data points contained within fields managed under precision agriculture. In a sense, each sampling point becomes a 'natural' test plot, where the relationship between yield and other factors can be examined. If the yield response relationships are stable over a number of site-years representing a range of growing conditions, these relationships could then be used to develop site-sensitive recommendations that permit more precise fertilizer application on sub-field areas.

This paper describes a study which investigated the use of this statistical approach for modeling crop grain yield and for developing yield response functions based upon spatial soil and topographic data. Earlier studies which modeled yield using other statistical methods and neural network analysis techniques have been reported by Drummond *et al.* (1995) and Sudduth *et al.* (1997).

MATERIALS AND METHODS

Data were collected on two fields, 36 ha and 28 ha in size, located in central Missouri. The soils of the area are characterized as claypan soils, primarily of the Mexico-Putnam association (fine, montmorillonitic, mesic Udollic Ochraqualfs). These soils are poorly drained and have a restrictive, high-clay layer (a claypan) occurring below the topsoil. The two fields were managed in a high yield goal, high input, minimum-till corn-soybean rotation. Fertilizer and chemical inputs were applied at a single rate.

Data were obtained for one field (Field 1) in 1993 (corn), 1994 (soybean), 1995 (grain sorghum), and 1996 (soybean). Grain sorghum was planted in this field in 1995, rather than corn, because planting was delayed by an excessively wet spring. Yield data for the other field (Field 2) were obtained only in 1995 (soybean). Conditions for crop production were quite different over the four years. The 1993 growing season was characterized by heavy and frequent rains, with an annual precipitation of 157 cm. Yield reductions were observed in lower portions of the landscape, due to excess water. The 1994 precipitation of 82 cm was only slightly below average, but less than 5 cm of rainfall was received in July and August and crops experienced drought stress during much of the growing season. In 1995, precipitation was 115 cm, with an excessively wet planting season which caused reduced crop stands and some yield reductions in the lower portion of the landscape. In 1996, conditions were nearly ideal for crop growth, with frequent and timely precipitation events during the growing season.

Data acquisition

Data obtained on the study fields included grain yield, elevation and slope, and a number of soil properties. Grain yield measurements were obtained using a full-size combine equipped with a commercial yield sensing system and global positioning system (GPS) receiver, using data collection and processing techniques described by Birrell et al. (1996).

Based on previous findings (Sudduth et al., 1995), topsoil depth above the claypan was estimated from soil conductivity. A mobile measurement system as described by Kitchen et al. (1996) was used to obtain this root-zone soil conductivity data with an electromagnetic induction sensor. The actual depth of topsoil was measured at a set of calibration points and a regression between topsoil depth and the inverse of soil conductivity was developed (Field 1: $r^2 = 0.90$, std. err. = 6.7 cm; Field 2: $r^2 = 0.89$, std. err. = 9.4 cm). These regressions were then applied to convert the sensor data to estimated topsoil depth.

Field 1 was soil sampled on a 30 m grid in March of 1995, prior to spring tillage. A hand soil probe was used to collect soil cores to a 20 cm depth. Three soil cores obtained within a 1 m radius of each sample position were combined, oven dried and analyzed for phosphorus, potassium, pH, organic matter, calcium, magnesium, and cation exchange capacity (CEC). Field 2 was soil sampled on a 25 m grid in March of 1996. Procedures were identical to those for Field 1, except that eight cores were combined at each sample position.

Data analysis

Yield and topsoil depth data were analyzed using geostatistics, and appropriate semi-variogram models and parameters were used to krige the data to a grid with a 10 m cell size. Yield and topsoil depth data from the grid cell centered closest to each soil sampling point were extracted and combined with the soil sample data for analysis. If any data were missing for a grid cell, that cell was eliminated from the analysis. The whole-field datasets ranged in size from 301 to 436 cells or observations.

Additional datasets were created by dividing each field into five sub-field areas based on elevation and topsoil depth. It was thought that the relationship of yield to soil and

441

topographic parameters might be more predictable within these sub-field areas than across whole fields. The static parameters of elevation and topsoil depth were chosen because in previous analyses these had the most consistent impact on yield of all the measured parameters in the dataset. To create the sub-field areas, each field was first divided into areas of low (<25 cm), medium, and high (>50 cm) topsoil depth. The medium and high topsoil depth areas were then subdivided into the lower one-third of the landscape and the higher two-thirds of the landscape (Figure 1). The sub-field datasets ranged in size from 14 to 232 observations.

Previous work (Drummond *et al.*, 1995; Sudduth *et al.*, 1997) had shown that linear correlation and regression techniques did not adequately model yield for these data. In this study, projection pursuit regression (PPR) analyses were completed on each whole-field and sub-field dataset. In this nonparametric regression method (Friedman and Stuetzle, 1981), the response (yield in this case) is modeled as the sum of a set of general (nonlinear) smooth functions of linear combinations of the independent (soil and topography) variables. In the PPR analyses, yield data for each field-year were regressed against seven soil and topographic variables – phosphorus, potassium, pH, organic matter, topsoil depth, CEC, and elevation. The other originally measured variables were not used in this analysis to reduce problems associated with colinearity. On these fields, we found that calcium and magnesium were highly correlated with CEC, and slope was correlated with topsoil depth.

RESULTS AND DISCUSSION

Yield patterns for Field 1 varied considerably from year to year. Visual comparison of yield maps and soil maps from this field allowed us to associate some, but not all, yield patterns with soil variations. For example, the 1994 yield pattern showed some similarity to the pattern of topsoil depth variation across the field (Drummond *et al.*, 1995).

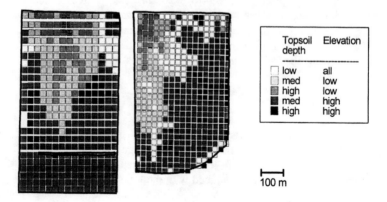

FIGURE 1. Sub-field areas classified by topsoil depth and elevation for Field 1 (left) and Field 2 (right).

Nonparametric regression analysis by PPR provided good estimates of yield (Table 1). Estimated yield maps based on PPR compared well with actual yield maps (Figure 2). The best PPR estimations were obtained for field-years with well-defined, relatively large-scale yield patterns (Table 1, Figure 2). For example, the areas of highest yield for Field 1 in 1994 were reproduced well, as were the areas of lowest yield for Field 2 in 1995. The correspondence of PPR estimates and actual yields was weaker when the spatial structure of actual yield was less well-defined, as was the case for Field 1 in 1993, 1995, and 1996.

Yield response curves

The response of the PPR yield estimates to variations in the input parameters was investigated on a point-by-point basis. Sensitivity analyses were conducted by holding all but one of the model input parameters constant and varying the other parameter from a minimum to a maximum value. All response curves for each sub-field area were then combined into a mean response curve for that area. For generation of the response curves, all variables were normalized to a field-year mean of zero and unity standard deviation. This facilitated comparison of yield responses between the different soil and topographic parameters measured for multiple field-years.

The response curves appeared to successfully model major yield-limiting factors. Data from Field 2 in 1995 are used for illustration in Figure 3, since the PPR analysis for that field-year provided the best fit to measured data. Higher soybean yields were related to increases in elevation within the field, with the strongest response found in the sub-field areas of lower elevation. This trend was caused by the excess rainfall in the spring of 1995, which caused significant problems with crop stands in the low-lying areas of the field. Yield decreases indicated at the highest elevations were likely caused by the presence of a tree line which reduced yield in that area. Yield response to topsoil depth was large and positive in the low and medium topsoil areas of the field. The response was negative in the high topsoil and high elevation areas because the locations of greater topsoil accumulation were also the locations where standing water reduced crop stand early in the season. Sub-field responses to higher levels of phosphorus and potassium were generally flat or negative for this field-year. Most areas of Field 2 were sufficient in these nutrients, and the negative relationship may have been due to mining of nutrients in the more productive areas of the field.

TABLE 1. Model and dataset statistics for projection pursuit regression (PPR) estimation of yield data as a function of seven soil and topographic parameters.

	Field 1 1993	Field 1 1994	Field 1 1995	Field 1 1996	Field 2 1995
PPR regression statistics					
r^2	0.54	0.66	0.49	0.58	0.80
standard error, t/ha	0.46	0.17	0.46	0.14	0.24
Dataset statistics					
number of data points	318	344	301	349	436
mean yield, t/ha	7.30	1.63	5.25	3.06	2.21
yield standard deviation, t/ha	0.68	0.29	0.64	0.22	0.54

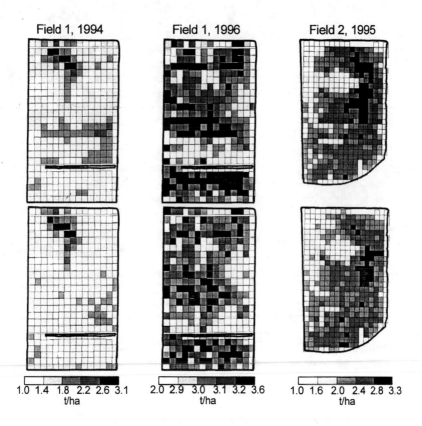

FIGURE 2. Measured (top) and PPR predicted (bottom) soybean yield for Field 1 in 1994 and 1996 and for Field 2 in 1995.

Differences were observed when comparing the 1995 soybean response curves for Field 2 (Figure 3) and the 1994 and 1996 soybean response curves for Field 1 (Figure 4). For example, in low-elevation areas the response of soybean yield to an increase in elevation was strongly negative for 1994, less strongly negative for 1996, and positive for 1995. Since crop growth was water-limited in 1994, run-on areas at low elevation benefited from additional water (Figure 4). In 1996, yield increases associated with water redistribution over the field landscape were smaller (Figure 4), due to timely, but not excessive, precipitation. In 1995 similar run-on areas at low elevation had crop growth limited by excess water (Figure 3). Yield response to soil test potassium (K) levels was different for Field 1 than for Field 2, with a positive response to higher K levels observed for Field 1 in both 1994 and 1996 (Figure 4). Mean soil test K levels were lower in Field 1, and the generalized fertility recommendations indicated that yields in significant portions of the field would be expected to respond to additions of K fertilizer.

Several cautions must be stated relative to these methods for the generation of yield response relationships. As with a more direct experimental approach, multiple years of data covering a variety of growing conditions are necessary for the development of

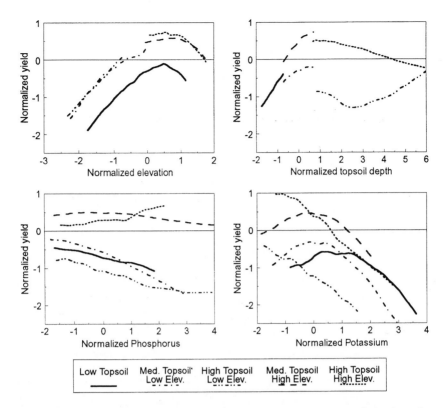

FIGURE 3. PPR-estimated soybean yield responses to elevation, topsoil depth, soil test phosphorous, and soil test potassium for Field 2 in 1995.

suitable predictive response functions which can be used for the estimation of crop response in future seasons. Care must be taken if strong correlations are observed among the independent variables used for developing the yield prediction equations. Yield response functions for those independent variables which are highly correlated may be suspect, particularly if they are contrary to expected results.

In addition, Mosteller and Tukey (1977) state the general caution that regression analysis by itself only indicates association between variables measured in uncontrolled experiments such as the one described here. They further state that for an association to reflect causation it must be consistent, responsive, and have a mechanistic basis. In this research, consistency was demonstrated across within-field areas having similar characteristics and across multiple field-years when climatic conditions were similar. Responsiveness was inferred from the shape of the estimated response curves, and could be confirmed through plot-scale response experiments at selected sites within the study fields. The extensive body of agronomic knowledge relating crop yield to differences in soils and topography can be thought of as a general mechanistic basis to support the idea of causation. If necessary, a more specific mechanistic basis could be developed and supported through component research studies initiated on a case-by-case basis, or through the application of well-documented and proven crop growth models.

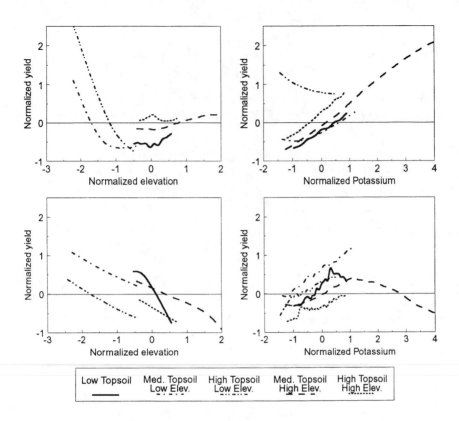

FIGURE 4. PPR-estimated soybean yield responses to elevation and soil test potassium for Field 1 in 1994 (top) and in 1996 (bottom).

CONCLUSIONS

Statistical models describing the spatial relationship between crop yield and soil and topographic parameters were developed. Independent variables included in the analysis included phosphorus, potassium, pH, organic matter, topsoil depth, cation exchange capacity (CEC) and elevation. Spatial yield estimates obtained with projection pursuit regression, a multivariate nonparametric method, agreed well with actual yields.

Response curves were generated to investigate the sensitivity of yield estimates to changes in the independent variables. The relationships identified by the response curves agreed with field observations and general crop response expectations. Variations in physical factors such as topsoil depth and elevation generally had a greater effect on yield than did soil nutrient differences. The specific response of yield to these physical factors varied between years, due to variable climatic conditions.

The methods presented in this paper may provide an alternative to the more conventional test-plot approach for the generation of response relationships. However, additional work is needed to further investigate these techniques and validate their applicability.

Particular areas of concern include the handling of collinear data, methods for assimilating responses from multiple field-years of data into a predictive function, and insuring that the relationships determined are causative as opposed to merely associative.

REFERENCES

Birrell, S.J., Sudduth, K.A., Borgelt, S.C. (1996) Crop yield mapping: comparison of yield monitors and mapping techniques. *Computers and Electronics in Agriculture*, **14**, 215-233.

Drummond, S.T., Sudduth, K.A., Birrell, S.J. (1995) Analysis and correlation methods for spatial data. Paper no. 951335, St. Joseph, Michigan, American Society of Agricultural Engineers.

Friedman, J.H., Stuetzle, W. (1981) Projection pursuit regression. *Journal of the American Statistical Association*, **76**, 817-823.

Kitchen, N.R., Sudduth, K.A., Drummond, S.T. (1996) Mapping sand deposition from 1993 midwest floods with electromagnetic induction measurements. *Journal of Soil and Water Conservation*, **51(4)**, 336-340.

Mosteller, F., Tukey, J.W. (1977) *Data Analysis and Regression: A Second Course in Statistics*. Reading, Massachusetts, Addison-Wesley.

Sudduth, K.A., Kitchen, N.R., Hughes, D.F., Drummond, S.T. (1995) Electromagnetic induction sensing as an indicator of productivity on claypan soils. *Site-Specific Management for Agricultural Systems*, P.C. Robert, R.C. Rust and W.E. Larson (Eds), Madison, Wisconsin, American Society of Agronomy, pp. 671-681.

Sudduth, K.A., Drummond, S.T., Birrell, S.J., Kitchen, N.R. (1997) Analysis of spatial factors influencing crop yield. *Proceedings of the 3rd International Conference on Precision Agriculture*, P.C. Robert, R.C. Rust and W.E. Larson (Eds), Madison, Wisconsin, American Society of Agronomy, pp. 129-140.

Author Index